A002747758

CW00548753

Mastering

Electronic and Electrical Calculations

Macmillan Master Series

Accounting
Advanced English Language
Advanced Pure Mathematics
Arabic
Banking
Basic Management
Biology
British Politics
Business Administration
Business Communication
Business Law
C Programming
Catering Science
Catering Theory
Chemistry
COBOL Programming
Commerce
Computers
Databases
Economic and Social History
Economics
Electrical Engineering
Electronic and Electrical Calculations
Electronics
English as a Foreign Language
English Grammar
English Language
English Literature
English Spelling
French
French 2
German
German 2
Human Biology
Italian
Italian 2
Japanese
Manufacturing
Markcting
Mathematics
Mathematics for Electrical and Electronic Engineering
Modern British History
Modern European History
Modern World History
Pascal Programming
Philosophy
Photography
Physics
Psychology
Restaurant Service
Science
Social Welfare
Sociology
Spanish
Spanish 2
Spreadsheets
Statistics
Study Skills
Word Processing

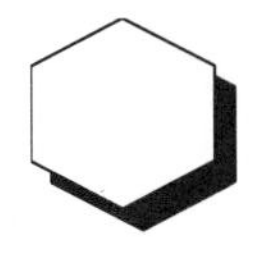

Mastering

Electronic and Electrical Calculations

Noel M. Morris

First published 1996 by
MACMILLAN PRESS LTD
Houndmills, Basingstoke, Hampshire RG21 6XS
and London
Companies and representatives
throughout the world

ISBN 0–333–63345–8

A catalogue record for this book is available from the British Library.

10 9 8 7 6 5 4 3 2 1
05 04 03 02 01 00 99 98 97 96

Copy-edited and typeset by Povey–Edmondson
Okehampton and Rochdale, England

Printed in Hong Kong

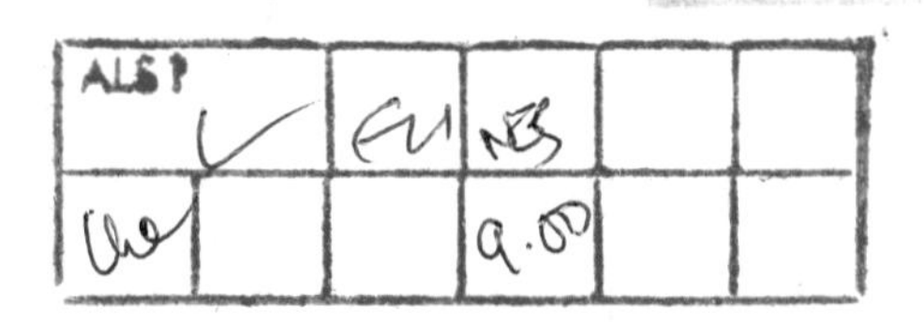

This book is dedicated to Barbara, Helen, Ian, Kathy and Anthony

Contents

List of figures

List of tables

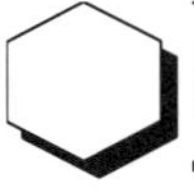

BASIC computer listings

Preface

This book deals with the methods and techniques used in problem solving in Electronic and Electrical circuits, ranging from manipulating equations to op-amp circuits, logic circuits and power engineering

The material is presented in the form of 'what you need to know first, is dealt with first'. The author has made every effort to cover all the topic area in a large subject matter, and the text is supported by many worked examples and informative diagrams.

Since each section is self-contained, the more experienced reader can turn to the section they are interested in and deal with that topic.

The book covers all the Principles for BTEC National Certificate in Electronic and Electrical Engineering courses, and a very considerable part of the Principles for Higher National Certificates. It will be invaluable for students requiring further explanation of the solution of problems in electronic and electrical engineering, since methods of carrying out computations are both fully and clearly explained.

As colleges move into the GNVQ philosophy, more emphasis is placed on students researching and learning for themselves, and this book will be an invaluable resource.

Each chapter commences with an introduction which describes the basic theory involved, followed by a range of worked examples. The concluding section of each chapter describes computer listings in the BASIC language, which are particularly suitable for the type of problems in that chapter.

This book will provide invaluable support for the following:

- NVQ and GNVQ Levels II and III
- BTEC National and Higher National Electronics/Electrical courses
- C & G LI Parts II and III
- 'A' level Electronics courses
- 1st Year Electronics/Electrical/Physics degrees
- Home constructors who need to understand the 'why' as well as 'how'.

Further information about the software is contained in 'A note about the computer programs' on p. xxii.

Chapter 20 describes the use of SPICE software, which is widely available, and can be used to solve almost any type of electronic or electrical problem, details of which are given on pp. xxii and xxiii.

Each chapter also contains a large number of exercises on which the reader can test their skill; the solution to the exercises is provided at the end of the book (p. 437). The exercises have been presented in the order in which they appear within the chapter and, within this grouping, they have been graded in order of difficulty. Finally, each chapter is concluded by a summary of important facts, which provides an easy to read summary of the vital data in the chapter.

A bibliography of other related titles in the Masters series is given at the end of the book.

The book is designed to suit the busy reader, so that you can quickly obtain the information you need about solving a particular type of problem.

The reviewers of the book deserve many thanks for their very helpful and constructive comments.

I would like to thank Mr P. Goss, Technical Manager of ARS Microsystems, for his advice and assistance in connection with Chapter 20, where the use of PSpice software is described. I am indebted to Mr A. Lewis, Head of Management Services, Longdean, Hemel Hempstead, for advice received. Finally, I would like to thank my wife for her support, without which the writing of this book would not have been possible.

NOEL M. MORRIS

A note about the computer programs

The computer programs in this book are written in QUICK BASIC, and some minor amendments may be required to the programs for use in other BASIC dialects. The programs have been written using line numbers (which, strictly speaking, is not necessary for QUICK BASIC) so that simple instructions such as GOTO can be used. On the whole, simple programming techniques have been adopted so that the programs are easy to understand.

Many of the programs introduce the reader to the use of menus and other topics. The programs enable the reader to gain expertise in program writing, and they may easily be modified by adding extra lines to them. If syntax errors occur during the loading of the programs, it may mean that some characters such as parenthesis, commas or semicolons have either been overlooked or changed by the reader. Any errors can be corrected by studying the manual for your computer.

To save the reader the effort of entering the programs into their computer, all of the programs in the book are available from:

AVP Educational Software
School Hill Centre
Chepstow
Gwent NP6 5PH
Tel: 01291 625439
Fax 01291 629671

The AVP catalogue number is COM 476. All the BASIC files on the disc have the extension '.BAS', and are stored in a text format so that they may be transferred to any word processor. The manual for the disc is also available on the disc in the file MANUAL.TXT, which is in a text format which may be read by any word processor.

Also included on the disc are a number of PSpice files associated with Chapter 20, including 17 '.OUT' files, together with other files relating to PSpice (a total of 74 files in all on the disc). The files which describe the circuits have the extension '.CIR', and other files have the extension '.CFG' or '.DAT'. Files which contain the results have the extension '.OUT'; these files also contain the PSpice circuit description which is identical to that in the corresponding '.CIR' file. An advantage of having the '.OUT' files is that they may be read by a word processor, and it is not necessary to have PSpice software installed in your own computer in order to study them. Readers need PSpice installed in their computer in order to run '.CIR' files.

All the programs in Chapter 20 (and for many problems in other chapters) were developed using PSpice software. Versions of this software for PCs usually have on-disc documentation, and is available from the address below. The software is also available through the pages of magazines dealing with computer shareware.

ARS Microsystems
Herriard Business Centre
Alton Road
Basingstoke RG25 2PN

The majority of the Worked Examples and Exercises in this book have been checked by the use of either the BASIC language programs or by PSpice.

Introduction to electronic and electrical equations and resistance

1.1 Introduction

This chapter provides an introduction to a broad range of topics essential for anyone studying electronic and electrical engineering, ranging from the manipulation of equations, through units in common usage, to calculations involving resistivity and resistance-temperature coefficient. By the end of this chapter, the reader will be able to

- Manipulate equations.
- Understand the units used in electrical and electronic engineering.
- Manipulate and change unit sizes associated with an equation.
- Solve equations relating to resistivity and resistance.
- Understand the meaning of resistance-temperature coefficient, and use it in the calculation of resistance over a range of temperature.
- Use the resistance colour code.
- Write and modify BASIC equations relating to the contents of this chapter.

1.2 Transposition of equations

When dealing with equations in electronic and electrical engineering, such as the Ohm's law expression

$$E = IR$$

where E is an e.m.f., I is a current and R a resistance. We often need to transpose the equation so that, for example, the resistance, R, becomes the ***subject of the equation***, i.e., R appears on the left-hand side of the equals sign. The golden rule when transposing an equation is to maintain the balance on both sides of the equal sign, that is

make the same changes to BOTH SIDES of the equation.

For example, if we wish to make R the subject of the above equation, we can proceed as follows:

1. If possible, re-write the equation so that the subject is contained in an expression on the left-hand side of the equation. That is

$$IR = E$$

2. Remove all other terms, one at a time, from the left-hand side of the equation. To remove I from the left-hand side, we **divide both sides by I** as follows:

$$\frac{IR}{I} = \frac{E}{I}$$

Since I can be cancelled in the numerators and in the denominator on the left-hand side, we are left with the final expression as follows:

$$R = \frac{E}{I}$$

Taking another example, if we wish to make R_2 the subject of the equation

$$R_E = R_1 + R_2 + R_3$$

the steps are as follows: Initially, as above, we re-write the equation so that the subject is contained within an expression on the left-hand side of the equals sign as follows:

$$R_1 + R_2 + R_3 = R_E$$

Next, we remove R_1 from the left-hand side of this equation by subtracting it from both sides of the equation as follows:

left-hand side: $R_1 + R_2 + R_3 - R_1 = R_2 + R_3$
right-hand side: $R_T - R_1$

That is $R_2 + R_3 = R_T - R_1$
Similarly, we remove R_3 from the left-hand side to give

$$R_2 = R_T - R_1 - R_3 = R_T - (R_1 + R_3)$$

Worked Example 1.1

Make the variables mentioned the subject of the following equations: (a) make α_0 the subject of $R = R_0(1 + \alpha_0 t)$, (b) make H the subject of $W = \mu H^2/2$, (c) make Z_B the subject of $Z_{AB} = Z_A + Z_B + Z_A Z_B / Z_C$ and (d) make R_1 the subject of $R_E = R_1 R_2/(R_1 + R_2)$.

Solution

(a) We commence by re-writing the equation in the form

$$R_0(1 + \alpha_0 t) = R$$

Next, to remove R_0 from the left-hand side of the equation, both sides are divided by R_0 as follows:

$$\frac{R_0}{R_0}(1+\alpha_0 t)=\frac{R}{R_0}$$

or

$$(1+\alpha_0 t)=\frac{R}{R_0}$$

Subtracting unity from both sides yields

$$(1+\alpha_0 t)-1=\alpha_0 t=\frac{R}{R_0}-1$$

and dividing both sides by t gives

$$\frac{\alpha_0 t}{t}=\alpha_0=\left[\frac{R}{R_0}-1\right]/t$$

It is left as an exercise for the reader to simplify the right-hand side of the equation to show that

$$\alpha_0=\frac{R-R_0}{R_0 t}$$

(b) In this case the equation is re-written in the form

$$\frac{\mu H^2}{2}=W$$

Multiplying both sides of the equation by 2 gives

$$\mu H^2=2W$$

Parameter μ is removed from the left-hand side of the equation by dividing both sides by μ as follows:

$$\frac{\mu H^2}{\mu}=H^2=\frac{2W}{\mu}$$

We are now left with an equation with H^2 on its left-hand side. To obtain H, we must take the square root of the left-hand side and, to balance the equation, we must also take the square root of the right-hand side, as follows

$$H=\sqrt{(2W/\mu)}$$

(c) The equation is re-written as follows:

$$Z_A+Z_B+\frac{Z_A Z_B}{Z_C}=Z_{AB}$$

Subtracting Z_A from both sides of the equation yields

$$Z_B + \frac{Z_A Z_B}{Z_C} = Z_{AB} - Z_A$$

Since Z_B is common to both terms on the left-hand side of the equation, we can take it outside of a bracketed expression as follows:

$$Z_B\left[1 + \frac{Z_A}{Z_C}\right] = Z_{AB} - Z_A$$

Dividing both sides of the equation by $(1 + Z_A/Z_C)$ leaves us with

$$Z_B = \frac{Z_{AB} - Z_A}{1 + \frac{Z_A}{Z_C}}$$

It is left as an exercise for the reader to show that this can be written in the form

$$Z_B = \frac{Z_C(Z_{AB} - Z_A)}{Z_A + Z_C}$$

(d) Although this equation seem straightforward, it needs a little more manipulation than the earlier equations before we can get the subject onto the left-hand side of the equals sign. The original equation is

$$R_E = \frac{R_1 R_2}{R_1 + R_2}$$

Initially, we multiply both sides of the equation by $(R_1 + R_2)$ as follows:

$$R_E(R_1 + R_2) = \frac{R_1 R_2}{R_1 + R_2} \times (R_1 + R_2) = R_1 R_2$$

Now, we multiply out the left-hand side of the equation to give the following equation

$$R_E R_1 + R_E R_2 = R_1 R_2$$

At this stage, we collect the terms containing R_1 on the left-hand side of the equation by subtracting $R_1 R_2$ from both sides of the equation

$$R_E R_1 + R_E R_2 - R_1 R_2 = R_1 R_2 - R_1 R_2 = 0$$

Next, we group the terms containing R_1 together as shown

$$R_E R_1 - R_1 R_2 + R_E R_2 = 0$$

or $$R_1(R_E - R_2) + R_E R_2 = 0$$

When we subtract $R_E R_2$ from both sides of the equation we are left with

$$R_1(R_E - R_2) = -R_E R_2$$

and dividing both sides of the equation by $(R_E - R_2)$ gives

$$R_1 = \frac{-R_E R_2}{R_E - R_2}$$

At first glance it may appear that R_1 has a negative value! However, in practice, R_E has a lower value than R_2, so that both the numerator and the denominator have negative values, and R_1 has a positive value.

In fact, we can alter the appearance of the equation simply by multiplying both the numerator and the denominator of the right-hand side by (-1) as follows:

$$R_1 = \frac{-R_E R_2 \times (-1)}{(R_E - R2) \times (-1)} = \frac{R_E R_2}{R_2 - R_E}$$

1.3 Units in electronics and electrical engineering

Many systems of units have been used in scientific systems, ranging from the Gaussian system (introduced by K. F. Gauss in 1832), though the CGS system (Centimetre–Gramme–Second, introduced in 1873) and the MKS system (Metre–Kilogramme–Second), to the SI system (Système International, introduced in 1960), which we now use.

In this chapter we take a broad view on units, and we mention units to be dealt with in greater depth in later chapters. A range of popular SI units used in electronics and electrical engineering are listed in Table 1.1.

Table 1.1 A list of typical SI units

Quantity	*Symbol*	*Unit*	*Unit Symbol*
length	ℓ, L	metre	m
mass	m, M	kilogramme	kg
time	t	second	s
current	i, I	ampere	A
voltage	V, E	volt	V
absolute temperature	T	kelvin	K
angle	ϕ, θ	radian	rad
		degree	°

Whilst some of the basic SI units, such as the volt, are in use for popular day-to-day applications, there are cases where we need to use multiples in order to deal with them. For example, a voltage impulse generator may produce many millions of volts, whereas the voltage generated by the human brain may only be a few microvolts. A list of SI multiples is given in Table 1.2.

Table 1.2 Multiples of 10 used with the SI system

Symbol	*Prefix*	*Multiple*
T	tera-	10^{12}
G	giga-	10^{9}
M	mega-	10^{6}
k	kilo-	10^{3}
c	centi-	10^{-2}
m	milli-	10^{-3}
μ	micro-	10^{-6}
n	nano-	10^{-9}
p	pico-	10^{-12}
f	femto-	10^{-15}
a	atto-	10^{-18}

1.4 Conversion of unit size

A problem which troubles many students throughout their academic career is how to convert one unit size to another. For example, the result of a calculation may give a numerical answer of, say, 5.6×10^{-5} A. If we need to convert the answer into mA, or into μA, how do we perform the conversion?

Suppose that we have a value expressed as an SI multiple in the form

$$x \times 10^m$$

where x is simply a number such as 378.5 and 10^m is an SI multiple; it could be, for example, 10^{-3} or 10^6. If we wish to convert the value into some other multiple such as 10^n (i.e., into a multiple such as 10^{-6} (micro-) or 10^9 (giga-), etc.), we simply say that

$$y \times 10^n = x \times 10^m$$

where y is the numerical value associated with the new multiple. Hence

$$y = \frac{x \times 10^m}{10^n} \tag{1.1}$$

Taking a relatively simple case, let us convert the dimensions of an area of x m^2 to one of y cm^2. Here we treat the number and its dimension as though they are 'multiplied' together so that

$$y\ \text{cm}^2 = x\,\text{m}^2$$

or

$$y = x\ \frac{\text{m}^2}{\text{cm}^2}$$

but 1 m = 100 cm, hence

$$y = x \times \frac{(100)^2}{1^2} = x \times 10^4 \text{ cm}^2$$

That is $1.8 \text{ m}^2 = 1.8 \times 10^4 \text{ cm}^2 = 18\,000 \text{ cm}^2$.

On other occasions we need to convert from one complex unit into another; we can do so using a similar technique to that described above. For example, if the resistivity of a material (see also section 1.6) is $x\ \Omega$ m, and we wish to alter its value to $y\ \Omega$ cm, then we merely say that

$$y\ \Omega \text{ cm} = x\ \Omega \text{ m}$$

or

$$y = x\ \frac{\Omega \text{ m}}{\Omega \text{ cm}} = x\ \frac{\text{m}}{\text{cm}}$$

and, since 1 m = 100 cm, then

$$y = x\ \frac{100}{1} = 100x$$

That is

$$\begin{aligned} 1.75 \times 10^{-8} \Omega \text{ m} &= (1.75 \times 10^{-8}) \times 100\ \Omega \text{ cm} \\ &= 1.75 \times 10^{-6}\ \Omega \text{ cm} \end{aligned}$$

Worked Example 1.2

Convert (a) 0.36 mV into (i) V, (ii) μV, (b) 8.5 kW into (i) MW, (ii) μW.

Solution

(a) $0.36 \text{ mV} = 0.36 \times 10^{-3}$ V. This is expressed in the form $x \times 10^m$ in the section above, where $x = 0.36$ and $m = -3$. We use (1.1) as follows:

(i) 0.36 mV voltage is converted into volts as shown

$$\begin{aligned} y &= \frac{0.36 \times 10^{-3}}{10^0} = \frac{0.36 \times 10^{-3}}{1} \\ &= 0.36 \times 10^{-3} \text{ V} \quad \text{or} \quad 0.00036 \text{ V} \end{aligned}$$

(ii) The voltage is converted into microvolts as follows:

$$\begin{aligned} y &= \frac{0.36 \times 10^{-3}}{10^{-6}} = 0.36 \times 10^{(-3+6)} = 0.36 \times 10^3 \\ &= 0.36 \times 10^3 \mu\text{V or } 360 \mu\text{V} \end{aligned}$$

(b) $8.5 \text{ kW} = 8.5 \times 10^3$ W, hence

(i) To convert 8.5 kW it into MW, we proceed as follows:

$$\begin{aligned} y &= \frac{8.5 \times 10^3}{10^6} = 8.5 \times 10^{(3-6)} = 8.5 \times 10^{-3} \text{ MW} \\ &= 0.0085 \text{ MW} \end{aligned}$$

(ii) Converting it into μW gives

$$y = \frac{8.5 \times 10^3}{10^{-6}} = 8.5 \times 10^{(3+6)} = 8.5 \times 10^9 \ \mu\text{W}.$$

(**Note**: The reader will find a BASIC language program in Computer listing 1.1 at the end of this chapter (p. 16) very useful for this type of calculation. To make life easier for the reader, this and other BASIC programs in the book are available on disc.)

1.5 Ohm's law

Ohm's law is one of the most important laws in electronic and electrical engineering, and states that the relationship between the applied e.m.f. E, the current I, and the circuit resistance R is as follows:

$$E = IR \qquad (1.2)$$

Sometimes we use V for voltage instead of E, but this can lead to confusion because the SI symbol for the unit of voltage is V. It is therefore usual to give the applied voltage a subscript, such as V_1, when we use it in an equation, so that we may write

$$V_1 = IR \ V$$

In this case there is no confusion for the applied voltage V_1, and the unit of voltage V.

Equation (1.2) can, alternatively, be written as

$$R = E/I$$

or as

$$I = E/R$$

(a goulish reminder of the latter is

***I*nterment = *E*arth over *R*emains**)

Worked Example 1.3

In an electronic circuit, a voltage of 10.6 mV is applied to a 16 kΩ resistor. Calculate the current in the circuit in (a) mA, (b) μA.

Solution

In this case $V_1 = 10.6$ mV$= 10.6 \times 10^{-3}$ V, and $R = 16$ kΩ $= 16 \times 10^3$ Ω, hence

$$I = V_1/R = 10.6 \times 10^{-3}/16 \times 10^3 = 6.625 \times 10^{-7} \text{ A}$$

(a) To convert this current to mA, we merely say that

$$I = \frac{6.625 \times 10^{-7}}{10^{-3}} \text{ mA} = 6.625 \times 10^{-4} \text{ mA}$$

(b) To convert the value to μA, we proceed as follows:

$$I = \frac{6.625 \times 10^{-7}}{10^{-6}} \text{ μA} = 6.625 \times 10^{-1} \text{ μA}$$
$$= 0.6625 \ \mu A.$$

1.6 Resistivity and resistance

The resistance of a conductor is given by

$$R = \rho \frac{l}{a}$$

where ρ is the resistivity of the conductor in Ω m
l is the length of the conductor in m
a is the area of the conductor in m^2

The resistivity of the material depends on many factors including the way in which it is produced, its heat treatment, etc. The resistivity of annealed copper is often quoted as 1.72×10^{-8} Ω m, whereas 'pure' copper has a slightly different value.

Worked Example 1.4

(a) Calculate the resistance of a 100 m length of copper wire of 3 mm diameter, the resistivity of the material being 1.72×10^{-8} Ω m. (b) Another conductor having the same length and diameter has a resistance of 0.3 Ω; determine the resistivity of the material in Ω m.

Solution

(a) The radius of the conductor is 1.5 mm = 1.5×10^{-3} m. The resistance of the conductor is

$$R = \rho \frac{l}{a} = 1.72 \times 10^{-8} \times \frac{100}{\pi (1.5 \times 10^{-3})^2}$$
$$= 0.243 \ \Omega$$

(b) Transposing the equation to make ρ the subject of the equation gives

$$\rho = R \frac{a}{l} = \frac{0.3 \times \pi (1.5 \times 10^{-3})^2}{100} = 2.12 \times 10^{-8} \ \Omega \text{ m}$$

From section 1.3, we know that

$$2.12 \times 10^{-8}\ \Omega\ \text{m} = (2.12 \times 10^{-8}) \times 100\ \Omega\ \text{cm}$$
$$= 2.12 \times 10^{-6}\ \Omega\ \text{cm}.$$

1.7 Temperature coefficient of resistance

The resistance of all pure metals increases with increase in temperature, whereas the resistance of electrolytes and semiconductors decreases with increase in temperature. The resistance of some metal alloys such as manganin do not alter significantly over quite a wide range of temperature.

In the case of conductors, the resistance changes linearly over the 'normal' range of temperature associated with electronic and electrical circuits.

A graph showing the way in which the resistance of a conductor changes with temperature is shown in Figure 1.1. The temperature, θ_x, at which the straight-line graph cuts the zero-resistance axis is known as the **inferred absolute zero of temperature** which, for copper, is −234.5°C.

Knowledge of the value of the gradient of the graph in Figure 1.1 (i.e., tan ϕ) enables us to calculate the resistance of the conductor at any temperature. The gradient of the graph between temperature θ_1 and θ_2 can be calculated as follows:

$$\tan\phi = \frac{R_2 - R_1}{\theta_2 - \theta_1}\ \Omega/°\text{C}$$

The **temperature coefficient of resistance referred to θ_1** is given the symbol α_1, where

$$\alpha_1 = \frac{\text{gradient of graph}}{R_1} = \frac{R_2 - R_1}{R_1(\theta_2 - \theta_1)} \tag{1.3}$$

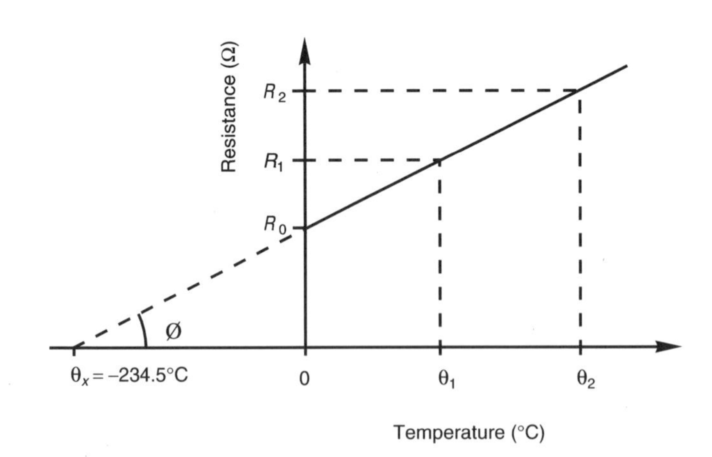

Figure 1.1 Variation of resistance of a conductor with temperature

and has units of 'per degree C' or $(°C)^{-1}$. Text books frequently refer to the **temperature coefficient of resistance referred to 0°C**, which is given by

$$\alpha_0 = \frac{\text{gradient of graph}}{R_0} = \frac{R_1 - R_0}{R_0(\theta_1 - 0)}$$
$$= \frac{R_1 - R_0}{R_0\theta_1}\,(°C)^{-1} \tag{1.4}$$

or, alternatively

$$\alpha_0 = \frac{\text{gradient of graph}}{R_0} = \frac{R_2 - R_0}{R_0(\theta_2 - 0)}$$
$$= \frac{R_2 - R_0}{R_0\theta_2} = (°C)^{-1}$$

Similarly

$$\alpha_0 = \frac{\text{gradient of graph}}{R_0} = \frac{R_0 - 0}{R_0(0 - \theta_x)}$$
$$= \frac{R_0}{-R_0\theta_x} = \frac{1}{-\theta_x}$$

Since the slope of the graph in (1.3) and (1.4) is divided by different values of resistance, it follows that α_1 and α_0 have different numerical values; the reader should carefully note this fact, since use of the wrong value of α can lead to an incorrect result. In fact, the BASIC program listed in Listing 1.2 offers one option which allows you to determine the value of α referred to any temperature.

The resistance, R_1, of a conductor at temperature θ_1 is given by

$$R_1 = R_0(1 + \alpha_0\theta_1) \tag{1.5}$$

where R_0 is its resistance at 0 °C, and α_0 is the temperature coefficient of resistance referred to 0 °C. Also

$$R_2 = R_0(1 + \alpha_0\theta_2)$$

hence

$$\frac{R_1}{R_2} = \frac{1 + \alpha_0\theta_1}{1 + \alpha_0\theta_2} \tag{1.6}$$

Equations (1.5) and (1.6) allow us to solve a wide range of problems relating to the effect of temperature change on the resistance of conductors.

Worked Example 1.5

The resistance of a length of aluminium wire at 20 °C is 5 Ω. If the resistance-temperature coefficient referred to 0 °C of aluminium is 0.004 $(°C)^{-1}$, determine the resistance of the conductor when it is (a) cooled to 0 °C, (b) heated to 50 °C.

Solution

(a) From (1.5)

$$R_1 = R_0(1 + \alpha_0\theta_1)$$

Transposing to make R_0 the subject of the equation gives

$$R_0 = R_1/(1 + \alpha_0\theta_1)$$
$$= 5/(1 + (0.004 \times 20)) = 4.63\ \Omega$$

(b) In this case we use (1.6), where

$$\frac{R_1}{R_2} = \frac{1 + \alpha_0\theta_1}{1 + \alpha_0\theta_2}$$

or

$$\frac{R_2}{R_1} = \frac{1 + \alpha_0\theta_2}{1 + \alpha_0\theta_1}$$

That is

$$R_2 = R_1(1 + \alpha_0\theta_2)/(1 + \alpha_0\theta_1)$$
$$= 5(1 + (0.004 \times 50))/(1 + (0.004 \times 20))$$
$$= 5.56\ \Omega.$$

Worked Example 1.6

The resistance of a coil of wire in an electrical machine increases from 45 Ω at 18 °C to 52.87 Ω at 60 °C. Determine the temperature coefficient of resistance referred to 0 °C of the conductor material.

Solution

Using (1.6) we have

$$\frac{R_1}{R_2} = \frac{1 + \alpha_0\theta_1}{1 + \alpha_0\theta_2}$$

or

$$\frac{45}{52.87} = 0.851 = \frac{1 + (\alpha_0 \times 18)}{1 + (\alpha_0 \times 60)}$$

hence

$$0.851(1 + 60\alpha_0) = 1 + 18\alpha_0$$

Multiplying out the left-hand side of the equation gives

$$0.851 + 51.06\alpha_0 = 1 + 18\alpha_0$$

That is

$$51.06\alpha_0 = 1 - 0.851 + 18\alpha_0 = 0.149 + 18\alpha_0$$

or

$$(51.06 - 18)\alpha_0 = 0.149$$

hence

$$\alpha_0 = 0.149/(51.06 - 18) = 0.0045(^\circ\text{C})^{-1}.$$

Worked Example 1.7

The resistance-temperature coefficient of phosphor bronze is 0.0039 $(^\circ\text{C})^{-1}$ referred to 0°C. Calculate the value of the coefficient of resistance referred to (a) 20 °C, (b) 100 °C.

Solution

It was shown earlier that the temperature coefficient of a conductor referred to θ_1 is

$$\frac{\text{slope of resistance-temperature graph}}{\text{resistance at } \theta_1}$$

The graph is simplified and re-drawn in Figure 1.2, where θ_x is the inferred absolute zero of temperature (**Note**: $\theta_x = -1/\alpha_0$), and R_1 is the resistance of the conductor at θ_1. Hence

$$\alpha_1 = \frac{R_1/(-\theta_x + \theta_1)}{R_1} = \frac{1}{-\theta_x + \theta_1}$$

Since we know that $\alpha_0 = 0.0039\ (^\circ\text{C})^{-1}$, then

$$\theta_x = -1/\alpha_0 = -256.4\ ^\circ\text{C}$$

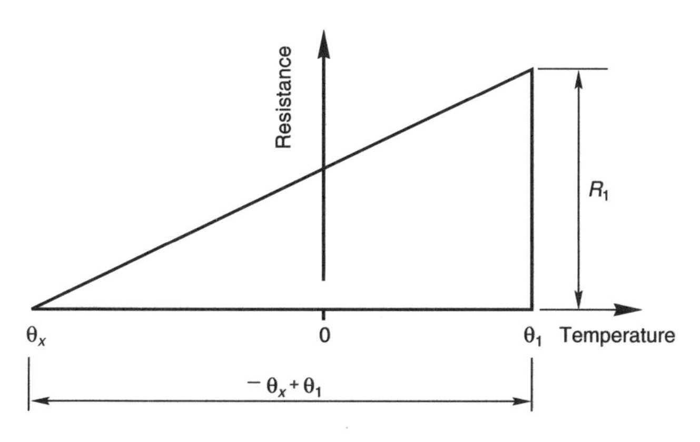

Figure 1.2 Graph for Worked Example 1.7

(a) From the above we see that, at 20 °C

$$\alpha_{20} = 1/(-\theta_x + 20) = 1/(256.4 + 20)$$
$$= 0.00362(°C)^{-1}$$

(b) At 100 °C

$$\alpha_{100} = 1/(-\theta_x + 100) = 1/(256.4 + 100)$$
$$= 0.0028(°C)^{-1}$$

1.8 Resistance colour code

An international colour code (see Table 1.3) is used to identify the value of resistors used in electronics. Resistance colour codes are defined in BS 1852. Table 1.3 also includes a simple mnemonic which allows you to remember the colour sequence.

Table 1.3 International colour code for resistor values

Colour	*Significant figure*	*Decimal multiplier*	*Mnemonic*
Black	0	1	Bye
Brown	1	10	Bye
Red	2	10^2	Rosie
Orange	3	10^3	Off
Yellow	4	10^4	You
Green	5	10^5	Go
Blue	6	10^6	Bristol
Violet	7	10^7	Via
Grey	8	10^8	Great
White	9	10^9	Western

The mnemonic 'Bye, Bye, Rosie, Off You Go, Bristol Via Great Western' refers to the defunct Great Western Railway line. In addition to the colours in Table 1.3, the colours silver and gold refer to the decimal multipliers 0.01 and 0.1, respectively. There are also *tolerance colours* of brown, red, gold, silver and 'no band' representing tolerances of 1, 2, 5, 10 and 20 per cent, respectively.

The way in which the bands are positioned on resistors with axial leads is shown in Figure 1.3. The values used are multiples of those in Table 1.4.

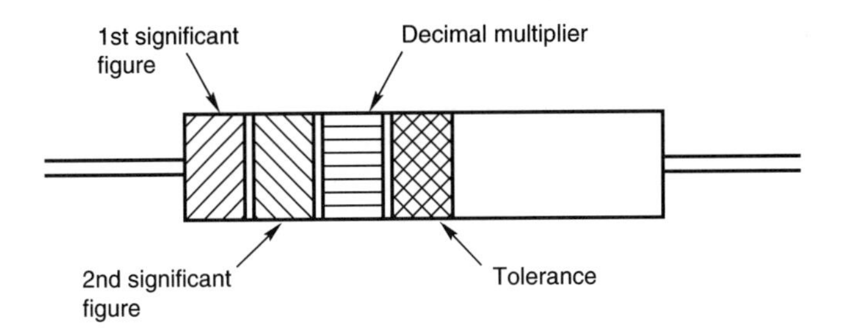

Figure 1.3 Resistance colour code

Table 1.4 Resistor values used with colour codes

Tolerance (%) 5	10	20
10	10	10
11		
12	12	
13		
15	15	15
16		
18	18	
20		
22	22	22
24		
27	27	
30		
33	33	33
36		
39	39	
43		
47	47	47
51		
56	56	
62		
68	68	68
75		
82	82	
91		

The values used in Table 1.4 are such that a resistor on its lower tolerance just overlaps with the next lower value on its upper tolerance. In this way Table 1.4 allows a continuous spectrum of resistor values.

Worked Example 1.8

What resistor values are associated with the following colour codes?

	Band 1	*Band 2*	*Band 3*	*Band 4*
(a)	yellow	violet	orange	none
(b)	brown	black	black	silver
(c)	red	red	red	red
(d)	violet	white	blue	gold

Solution

(a) Table 1.3 tells us that the value of the resistor is $47 \times 10^3\ \Omega \pm 20\%$ or $47\ \text{k}\Omega \pm 20\%$.
(b) In this case the resistor value is $(10 \times 10)\ \Omega \pm 10\%$ or $100\ \Omega \pm 10\%$.
(c) The value is $22 \times 10^2\ \Omega \pm 2\%$ or $2.2\ \text{k}\Omega \pm 2\%$.
(d) Here the value is $79 \times 10^6\ \Omega \pm 5\%$ or $79\ \text{M}\Omega \pm 5\%$.

1.9 Computer listings in the BASIC language

We include two computer listings in this chapter, both covering topics which are of interest to electronic and electrical engineers.

The programs are designed to solve specific problems and, as the reader will appreciate, not all problems are presented in a form which is soluble by these programs. In cases where the problem must be solved 'by hand' rather than by one of the programs, the program can be very helpful in verifying the solution simply by entering the calculated values to verify the original data supplied.

Listing 1.1 SI multiple converter

```
10 CLS : N = 0: P = 0 '**** LISTING CH1-1 ****
20 PRINT "SI multiple convertor": PRINT
30 PRINT "Symbol     Prefix    Power of 10"
40 PRINT "   T        Tera-          12"
50 PRINT "   G        Giga-           9"
60 PRINT "   M        Mega-           6"
70 PRINT "   k        Kilo-           3"
80 PRINT "   c        Centi-         -2"
90 PRINT "   m        Milli-         -3"
100 PRINT "            Micro-         -6"
110 PRINT " n          Nano-          -9"
120 PRINT " p          Pico-         -12"
130 PRINT " f          Femto-        -15"
140 PRINT " a          Atto-         -18"
```

```
150 PRINT
160 PRINT "Enter the number to be converted below, together"
170 PRINT "with its SI MULTIPLE, e.g., 62.3E-6 or 93.4E6."
180 INPUT "Number to be converted = ", N
190 INPUT "Power of 10 to be converted into = ", P
200 PRINT
210 PRINT "Required value = "; N / 10 ^ P; "x 10^("; P; ")"
220 END
```

Listing 1.1 is an SI multiple convertor which allows the reader to convert a value with a given SI multiple into another value with another SI multiple. Initially, the program presents you with a table of SI symbols, prefixes and powers of 10. You are prompted to supply the number to be converted, together with its SI multiple, and then you are asked to supply the SI multiple (power of 10) you want it converted to.

You can use the computer convention of 'E' to represent the exponent value. For example, the number 108.99×10^{-3} is entered into the computer as 108.99E-3. If the number is to be converted into 'micro-units', i.e., $\times 10^{-6}$, the computer gives the answer $108990 \times 10\hat{}(-6)$.

The reader will note that no 'symbol' is given for the prefix for 'micro-' in the table at the head of the listing. The reason is that not all computers use the same key codes for the Greek letter μ.

Listing 1.2 Temperature-dependent resistance calculations

```
10 CLS '**** CH1-2 ****
20 PRINT "VARIATION OF RESISTANCE WITH TEMPERATURE"
30 PRINT : Tn = 0: Tx = 0: R0 = 0: R1 = 0: T1 = 0: T2 = 0: A0 = 0
40 PRINT "Do you wish to:": PRINT
50 PRINT "1. Determine ALPHA at some temperature other than zero deg. C, or"
60 PRINT "2. Solve R1 = R0(1 + ALPHA 0 * THETA 1), or"
70 PRINT "                   R1(1 + ALPHA 0 * THETA 2)"
80 PRINT "3. Solve R2 = ---------------------"
90 PRINT "                     (1 + ALPHA 0 * THETA 1)"
100 PRINT
110 INPUT "Enter your selection here (1, 2 or 3): ", S: PRINT
120 IF S < 1 OR S > 3 THEN GOTO 10
130 IF S = 2 GOTO 240
140 IF S = 3 GOTO 290
150 PRINT "THETA n is the temperature (deg. C) at which"
160 PRINT " you wish to determine ALPHA n."
170 PRINT "ALPHA n is the value of ALPHA at temperature THETA n."
180 PRINT "THETA x is the inferred absolute zero of temperature."
190 PRINT "Note: for copper, THETA x = -234.5 deg C.": PRINT
200 INPUT "THETA n (deg. C) = ", Tn
210 INPUT "THETA x (deg. C) = ", Tx: PRINT
220 PRINT "ALPHA n = "; 1 / (Tn + ABS(Tx)); " per deg. C"
```

```
230 END
240 INPUT "RO (ohms) = ", RO
250 INPUT "ALPHA 0 (per deg. C) = ", A0
260 INPUT "THETA 1 (deg. C) = ", T1: PRINT
270 PRINT "R1 = "; RO * (1 + (A0 * T1)); " ohms"
280 END
290 INPUT "R1 (ohms) = ", R1
300 INPUT "THETA 1 (deg. C) = ", T1
310 INPUT "THETA 2 (deg. C) = ", T2
320 INPUT "ALPHA 0 (per deg. C) = ", A0: PRINT
330 PRINT "R2 = "; R1 * (1 + (A0 * T2)) / (1 + (A0 * T1)); " ohms."
340 END
```

Listing 1.2 is a simple menu-driven program for resistance-temperature calculations, in which you are offered a choice of three options from a simple menu in lines 50 to 90. Line 120 acts as a 'filter' to eliminate a random value outside the range 1, 2 or 3. The options offered cover the work in section 1.7 on the temperature coefficient of resistance.

Both Listings are written (as are the other Listings in the book) so that the data and the answers appear on the screen at the same time. This means that by pressing the 'PRINT SCREEN' key gives you a hard copy of the data together with the answer.

Exercises

1.1 How many volts are there in (a) 0.83 kV, (b) 79.8 kV, (c) 3.6 MV?

1.2 How many mA are there in (a) 9.7 kA, (b) 82 A, (c) 5.8 nA, (d) 81.7 fA?

1.3 Convert a resistivity of 1.8 Ω mm into (a) Ω m, (b) Ω cm.

1.4 Make the following the subject of the equations listed: (a) R_3 from $R_E = R_1 + R_2 + R_3$, (b) e from $W = D^2/2e$, (c) S_2 from $\Phi = F/(S_1 + S_2)$, (d) μ_r from $\mu = \mu_0\mu_r$.

1.5 Make the following the subject of the equations listed:

(a) r_2 from $C = 4\pi e/\left(\frac{1}{r_1} - \frac{1}{r_2}\right)$

(b) d from $E = Q/(4\pi e d^2)$

(c) h_r from $V_1 = h_i I_1 + h_r V_2$

(d) C_2 from $w_0 = \sqrt{\left(\frac{1}{L}\left[\frac{1}{C_2} + \frac{1}{C_2}\right]\right)}$.

1.6 Determine the resistance of a copper wire whose resistivity is 1.73×10^{-8} Ω m whose dimensions are (a) 1 mm^2 cross-sectional area and length 12 m, (b) radius 1 mm and length 150 m.

1.7 A resistor is wound with 50 m of 1.2 mm diameter wire of resistivity 12×10^{-6} Ω cm. Determine the diameter of a new resistance wire of resistivity $50 \times 10^{-6}\ \Omega$cm of length 29 m which is to replace the original.

1.8 If the resistance-temperature coefficient of a conductor is $45 \times 10^{-4}\ (°C)^{-1}$ referred to 0°C, calculate its resistance-temperature coefficient referred to (a) 30°C, and (b) 80°C.

1.9 A metallic strip has a resistance of 135.85 Ω at 15°C. If its resistance at 0°C is 130 Ω, determine its resistance-temperature coefficient referred to (a) 0°C, (b) 15°C.

1.10 The resistance of a coil of wire increases from 47.7 Ω at 15°C to 54 Ω at 50°C. Calculate the resistance-temperature coefficient of the material referred to 0°C.

1.11 Write a BASIC program to calculate the resistance of a conductor from the equation $R = \rho\ell/a$. The user must supply values of ρ, ℓ and a.

1.12 Modify the programs in Listings 1.1 and 1.2 to combine them into a single menu-driven program offering a selection of any one of the topics in the two original listings.

Summary of important facts

When **transposing equations**, it is important to **make the same changes to both sides of the equation**; this maintains the equation in balance.

Electronic and electrical engineering use the **SI (Système International) system** of units and dimensions; typical units and multiples are given in Tables 1.1 and 1.2, respectively.

Quite often we need to convert from one unit size to another. The method of doing this is explained in section 1.4 (see also the BASIC language Computer listing 1.1).

The resistance of a conductor is given by

$$R = \rho\frac{l}{a}$$

where ρ is the **resistivity** of the conductor material, l is its **length**, and a the **cross-sectional area**.

The **temperature coefficient of resistance** of a conductor is the **change in resistance per unit resistance per unit change in temperature**, and the ratio in the increase in resistance per °C rise in temperature at 0°C is the **temperature coefficient of resistance referred to 0°C**, or α_0.

If R_0 is the resistance of a conductor at 0°C, then its resistance at temperature θ_1 is

$$\boldsymbol{R_1 = R_0(1 + \alpha_0\theta_1)}$$

Similarly, at θ_2 its resistance is

$$\boldsymbol{R_2 = R_0(1 + \alpha_0\theta_2)}$$

hence

$$\frac{R_1}{R_2} = \frac{1 + \alpha_0\theta_1}{1 + \alpha_0\theta_2}$$

The resistance-temperature graph of a conductor is a straight line which cuts the zero-resistance axis at temperature axis at θ_x, the latter being known as the **inferred absolute zero of temperature**. It can be shown that

$$\theta_x = -1/\alpha_0$$

The **temperature coefficient of a conductor material referred to temperature** θ_n is

$$\alpha_n = 1/(\theta_n - \theta_x)$$

2 Resistor circuits

2.1 Introduction

In this chapter we look at important resistor circuits, including series circuits, parallel circuits and series-parallel circuits. A grasp of these is vital to a full understanding of electronic and electrical circuits. By the end of this chapter, the reader will be able to

- Determine the equivalent resistance of a series circuit.
- Evaluate the equivalent resistance of a parallel circuit.
- Calculate the resistance of series-parallel circuit.
- Determine the voltage distribution in a string of series-connected resistors.
- Compute the way in which current divides between parallel-connected resistors.
- Use the Wheatstone bridge to determine the value of an unknown resistor.
- Solve series- and parallel-circuits using BASIC language programs.

2.2 Resistors in series

A series circuit is one in which the **same current flows through each resistor in the circuit**. Series-connected resistors are sometimes described as a **chain** of resistors.

The total resistance or **equivalent resistance** of a series circuit is the *sum of the individual resistance values*. In the case of the circuit in Figure 2.1, the equivalent resistance is

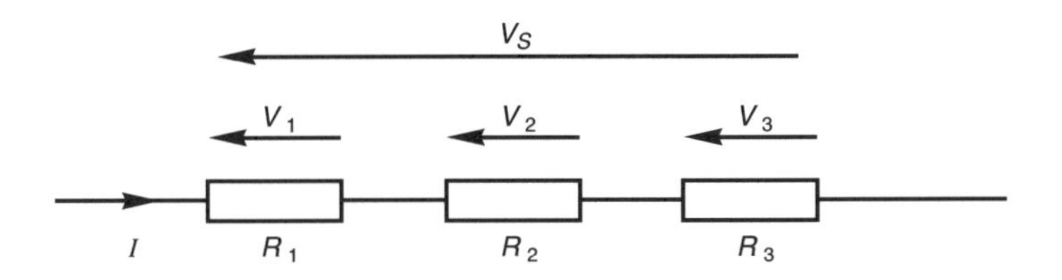

Figure 2.1 Three resistors in series

$$R_E = R_1 + R_2 + R_3 \qquad (2.1)$$

The equivalent resistance of a series-connected circuit is always greater than the largest value of resistance in the circuit.

If there are n resistors connected in series (where n may have any value), then

$$R_E = R_1 + R_2 + R_3 + \ldots + R_n \qquad (2.2)$$

From Ohm's law, the current flowing through the circuit is

$$I = V_S/R_E \qquad (2.3)$$

where V_S is the voltage applied to the circuit. The voltage across R_1 is $V_1 = IR_1$, across R_2 is $V_2 = IR_2$, across R_3 is $V_3 = IR_3$, etc. The voltage across the nth resistor is

$$V_n = IR_n \qquad (2.4)$$

That is

$$I = \frac{V_S}{R_E} = \frac{V_n}{R_n}$$

That is the voltage across the nth resistor can be calculated from

$$V_n = V_S\frac{R_n}{R_E} \qquad (2.5)$$

That is we can use (2.5) to calculate the p.d. across *any resistor* in a series-connected circuit.

Worked Example 2.1

Determine the resistance of a series circuit containing three resistors of 100 Ω, 330 Ω and 680 Ω. What current flows in the circuit and what is the p.d. across each resistor when 100 V is applied to the circuit?

Solution

From (2.1), the equivalent resistance is

$$R_E = R_1 + R_2 + R_3 = 100 + 330 + 680 = 1110\ \Omega$$

From Ohm's law, the current flowing in the circuit is

$$I = V_S/R_E = 100/1110 = 0.09009\ \text{A or } 90.09\ \text{mA}$$

and the p.d. across each resistor is

$$V_1 = IR_1 = 0.09009 \times 100 = 9.009 \approx 9.01\ \text{V}$$

$$V_2 = IR_2 = 0.09009 \times 330 = 29.73\ \text{V}$$

$$V_3 = IR_3 = 0.09009 \times 680 = 61.26\ \text{V}$$

We can check the calculation by adding the three p.d.s together as follows:

$$V_1 + V_2 + V_3 = 9.01 + 29.73 + 61.26 = 100V = V_S$$

Alternatively we can use (2.5) to calculate the voltage across each resistor. We do this for R_2 as follows:

$$V_2 = V_S R_2 / R_E = 100 \times 330/1110 = 29.73\,\text{V}$$

The reader should note that *the largest value of resistance supports the highest value of voltage.*

Worked Example 2.2

Three resistors are connected in series to a d.c. supply. One of the resistors has a resistance of 2 kΩ, and the p.d. across it is 4 V. If the p.d. across the other resistors are 2 V and 6 V, respectively, determine (a) the current in the circuit, (b) the supply voltage, (c) the resistance of the other two resistors, (d) the equivalent resistance of the circuit.

Solution

(a) From Ohm's law, the current in the circuit is

$$I = V_1/R_1 = 4/2000 = 2 \times 10^{-3}\text{A or } 2\text{ mA}$$

(b) The supply voltage is

$$V_S = V_1 + V_2 + V_3 = 4 + 2 + 6 = 12\text{ V}$$

(c) Ohm's law tells us that

$$R_2 = V_2/I = 2/2 \times 10^{-3} = 1000\ \Omega \text{ or } 1\text{k}\Omega$$

$$R_3 = V_3/I = 6/2 \times 10^{-3} = 3000\ \Omega \text{ or } 3\text{k}\Omega$$

(d) The equivalent resistance of the circuit is

$$R_E = R_1 + R_2 + R_3 = 2000 + 1000 + 3000$$
$$= 6000\ \Omega \text{ or } 6\text{ k}\Omega$$

2.3 Parallel circuits

A typical parallel circuit is shown in Figure 2.2. All the **branches** in the circuit meet at a common **node**. If there are more than two wires meeting at the node, it is called a **principal node**; nodes A and B in Figure 2.2 are principal nodes.

By Ohm's law, the current in R_1 is $I_1 = V_S/R_1$, in R_2 is $I_2 = V_S/R_2$, and in R_3 is $I_3 = V_S/R_3$. By observation, we see that the current flowing into node A (and from node B) is

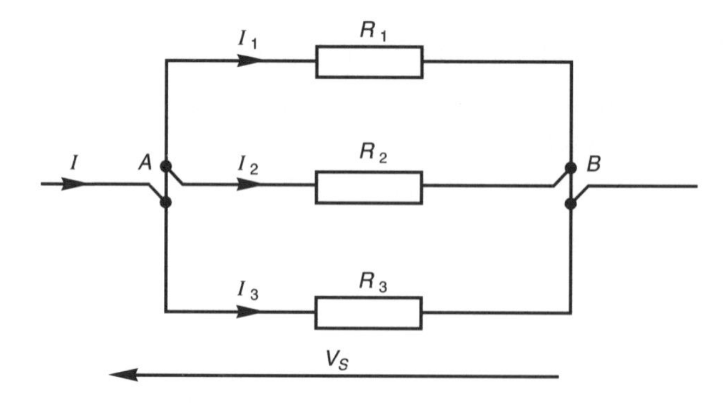

Figure 2.2 A three-branch parallel circuit

$$I = I_1 + I_2 + I_3$$

and the **reciprocal of the equivalent resistance** of the parallel circuit in Figure 2.2 is

$$\frac{\mathbf{1}}{\boldsymbol{R_E}} = \frac{\mathbf{1}}{\boldsymbol{R_1}} + \frac{\mathbf{1}}{\boldsymbol{R_2}} + \frac{\mathbf{1}}{\boldsymbol{R_3}} \tag{2.6}$$

In the *general case* of an n-branch parallel circuit, the reciprocal of the equivalent resistance is

$$\frac{\mathbf{1}}{\boldsymbol{R_E}} = \frac{\mathbf{1}}{\boldsymbol{R_1}} + \frac{\mathbf{1}}{\boldsymbol{R_2}} + \frac{1}{\boldsymbol{R_3}} + \ldots + \frac{\mathbf{1}}{\boldsymbol{R_n}} \tag{2.7}$$

where R_n is the resistance of the nth branch. *The equivalent resistance of a parallel circuit is always less than the smallest value of resistance in any branch.*

The reciprocal of resistance is **conductance**, G. The equivalent conductance of the complete circuit is therefore

$$G_E = G_1 + G_2 + G_3 \text{ siemens (S)} \tag{2.8}$$

where $G_E = 1/R_E, G_1 = 1/R_1, G_2 = 1/R_2$, etc.

In the *special case* of two resistors in parallel

$$\frac{1}{R_E} = \frac{1}{R_1} + \frac{1}{R_2} = \frac{R_1 + R_2}{R_1 R_2}$$

or

$$\boldsymbol{R_E} = \frac{\boldsymbol{R_1 R_2}}{\boldsymbol{R_1 + R_2}} \tag{2.9}$$

In a parallel circuit, **the same voltage appears across each branch of the circuit**. That is for the complete circuit

$$V_S = IR_E$$

where I is the total current drawn by the circuit. For R_1 we get $V_S = I_1 R_1$, for R_2 we get $V_S = I_2 R_2$, and for R_n we get $V_S = I_n R_n$. Since V_S is the same for each branch, it follows that

$$V_S = IR_E = I_n R_n$$

hence the current in the nth branch is

$$I_n = I\frac{R_E}{R_N} \tag{2.10}$$

Worked Example 2.3

Resistors of 20 Ω, 50 Ω and 80 Ω are connected in parallel to a 100 V supply. Determine (a) the equivalent resistance of the circuit, (b) the current in each branch and (c) the current drawn by the circuit.

Solution

(a) The reciprocal of the equivalent resistance is determined from

$$\frac{1}{R_E} = \frac{1}{20} + \frac{1}{50} + \frac{1}{80} = 0.05 + 0.02 + 0.0125$$
$$= 0.0825 \text{ S}$$

hence

$$R_E = 1/0.0825 = 12.12\ \Omega$$

(b) The current in each branch is

$$20\ \Omega \text{ branch: } I_1 = V_S/R_1 = 100/20 = 5 \text{ A}$$
$$50\ \Omega \text{ branch: } I_2 = V_S/R_2 = 100/50 = 2 \text{ A}$$
$$80\ \Omega \text{ branch: } I_3 = V_S/R_3 = 100/80 = 1.25 \text{ A}$$

(c) The current drawn by the complete circuit is

$$I = I_1 + I_2 + I_3 = 5 + 2 + 1.25 = 8.25 \text{ A}$$

Alternatively, we may say

$$I = V_S/R_E = 100/12.12 = 8.25 \text{ A}$$

Using the values calculated in the above we can also verify the answers in part (b) as follows:

$$I_1 = IR_E/R_1 = 8.25 \times 12.12/20 = 5 \text{ A}$$
$$I_2 = IR_E/R_2 = 8.25 \times 12.12/50 = 2 \text{ A}$$
$$I_3 = IR_E/R_3 = 8.25 \times 12.12/80 = 1.25 \text{ A.}$$

Worked Example 2.4

A two-branch parallel circuit has an effective resistance of 4.444 Ω, and is supplied by a 20 V d.c. source. If the resistance of one branch is 10 Ω, calculate the resistance in the other branch, the current in both branches, and the total current drawn by the circuit.

Solution

Since this is a two-branch circuit, its equivalent resistance is

$$R_E = R_1R_2/(R_1 + R_2)$$

That is

$$4.444 = 10R_2/(10 + R_2)$$

Cross-multiplying gives

$$4.444(10 + R_2) = 10R_2$$

or

$$44.44 + 4.444R_2 = 10R_2$$

Therefore

$$44.44 = (10 - 4.444)R_2 = 5.556R_2$$

That is

$$R_2 = 44.44/5.556 = 8\ \Omega$$

The current in the 10 Ω branch is

$$I_1 = V_S/R_1 = 20/10 = 2\ \text{A}$$

and the current in R_2 is

$$I_2 = V_S/R_2 = 20/8 = 2.5\ \text{A}$$

The current supplied to the complete circuit is

$$I = V_S/R_E = 20/4.444 = 4.5\ \text{A}$$

or

$$I = I_1 + I_2 = 2 + 2.5 = 4.5\ \text{A}$$

2.4 Series-parallel circuits

Many electrical circuits consist of a combination of series- and parallel-connected resistors. No two circuits are the same, and each circuit must be treated on its merits. We illustrate this by means of Worked Examples 2.5 and 2.6.

Worked Example 2.5

For the circuit in Figure 2.3, calculate the value of the current in, and the voltage across, each resistor. Also determine the effective resistance of the circuit, and the current drawn by the complete circuit.

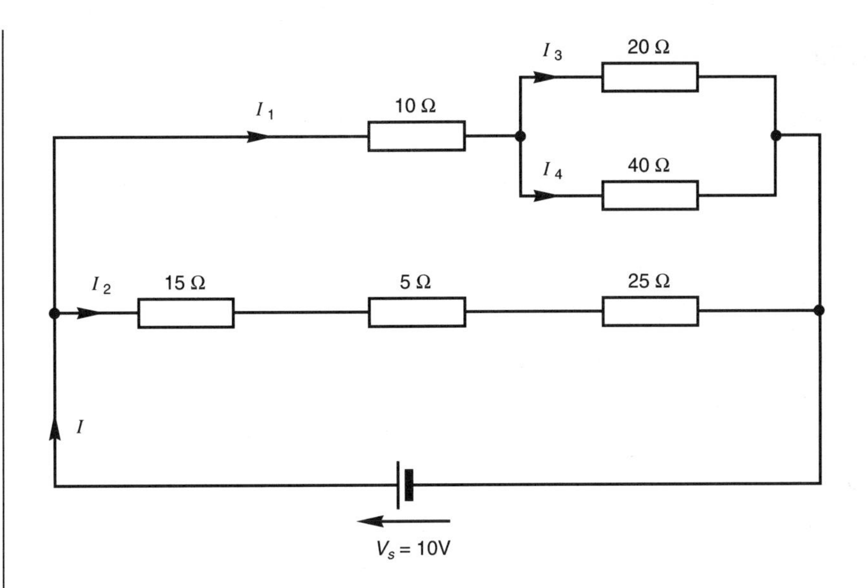

Figure 2.3 Circuit for Worked Example 2.5

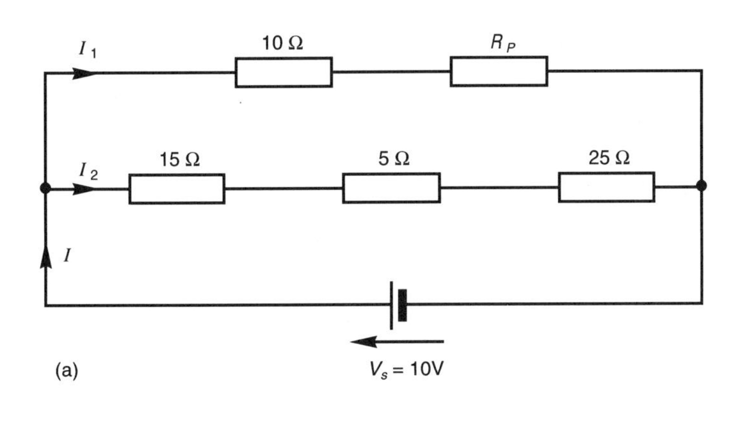

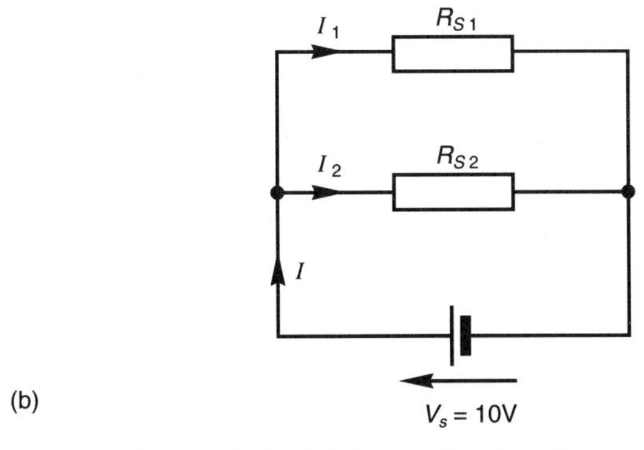

Figure 2.4 Solution of Worked Example 2.5

Solution

Initially, we must simplify the circuit in Figure 2.3 into its series and parallel groups. We do this in Figure 2.4(a); in the first case we group the 20 Ω and 40 Ω resistors together as a parallel pair, and we evaluate the equivalent resistance, R_P, of the pair as follows:

$$R_P = \frac{20 \times 40}{20 + 40} = 13.33\ \Omega$$

Next we group the 10 Ω resistor in series with R_P to give the equivalent resistance, R_{S1}, as shown in Figure 2.4(b)

$$R_{S1} = 10 + 13.33 = 23.33\ \Omega$$

The equivalent resistance, R_{S2}, of the lower branch in Figure 2.3 (see also Figure 2.4(b)) is

$$R_{S2} = 15 + 5 + 25 = 45\ \Omega$$

Since R_{S1} and R_{S2} are connected in parallel with one another, the equivalent resistance of the complete circuit is

$$\begin{aligned} R_E &= R_{S1}R_{S2}/(R_{S1} + R_{S2}) \\ &= 23.33 \times 45/(23.33 + 45) = 15.36\ \Omega \end{aligned}$$

From this it follows that the current drawn by the circuit is

$$I = V_S/R_E = 10/15.36 = 0.651\ \text{A}$$

The current I_1 in R_{S1} is

$$I_1 = V_S/R_{S1} = 10/23.33 = 0.429\ \text{A}$$

and the current I_2 in R_{S2} is

$$I_2 = V_S/R_{S2} = 10/45 = 0.222\ \text{A}$$

(**Note**: $I_1 + I_2 = 0.651 = I$)

We determine the value of I_3 and I_4 in Figure 2.3 using (2.10) as follows:

$$I_3 = I_1 \times R_P/20 = 0.429 \times 13.33/20 = 0.286\ \text{A}$$

and

$$I_4 = I_1 \times R_P/40 = 0.429 \times 13.33/40 = 0.143\ \text{A}$$

(**Note**: $I_3 + I_4 = 0.429\ \text{A} = I_1$)

Since we now know the current in every element in the circuit, we can determine the voltage across each resistor using Ohm's law. In the following, we write down the value of each resistor in Figure 2.3 as a subscript, so that V_{10} means the voltage across the 10 Ω resistor, etc.

$$\begin{aligned} V_{10} &= 10I_1 = 10 \times 0.429 = 4.29\ \text{V} \\ V_{20} &= 20I_3 = 20 \times 0.286 = 5.72\ \text{V} \\ V_{40} &= 40I_4 = 40 \times 0.143 = 5.72\ \text{V} \end{aligned}$$

(**Note**: $V_{10} + V_{20}$ (or V_{40}) $= 4.29 + 5.72 = 10.01 \approx V_S$)

The voltages across the resistors in the lower branch of Figure 2.3 are

$$V_{15} = 15I_2 = 15 \times 0.222 = 3.33 \text{ V}$$
$$V_5 = 5I_2 = 5 \times 0.222 = 1.11 \text{ V}$$
$$V_{25} = 25I_2 = 25 \times 0.222 = 5.55 \text{ V}$$

Worked Example 2.6

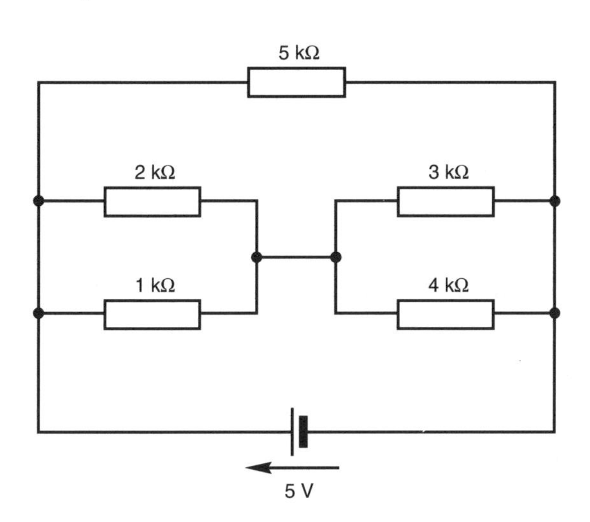

Figure 2.5 Circuit for Worked Example 2.6

Determine the equivalent resistance of the circuit in Figure 2.5, and the current in each resistor.

Solution

Since the lower part of the circuit comprises two series-connected parallel branches, we will initially evaluate the equivalent resistance of each parallel circuit.

For the 2 kΩ and 1 kΩ parallel section (see Figure 2.5 and 2.6(a))

$$R_{P1} = 2 \times 1/(2+1) = 0.667 \ \Omega$$

and for the 3 kΩ and 4 kΩ parallel section (see Figure 2.5 and 2.6(a))

$$R_{P2} = 3 \times 4/(3+4) = 1.71 \ \Omega$$

The equivalent resistance of the lower branch in Figure 2.6(b) is

$$R_S = R_{P1} + R_{P2} = 0.667 + 1.71 = 2.377 \ \Omega$$

R_S is in parallel with the 5 kΩ resistor in the upper branch of Figure 2.6(a), so that the equivalent resistance of the complete circuit is

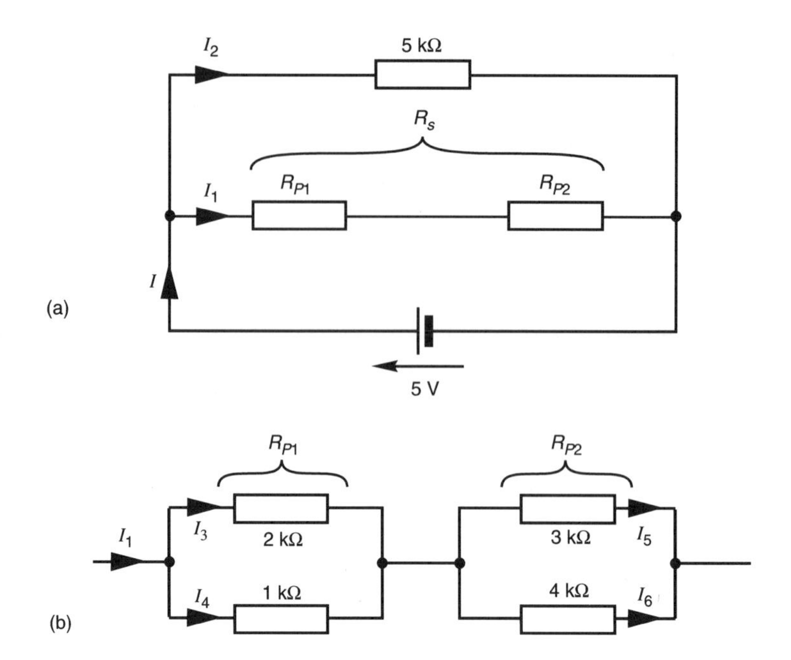

Figure 2.6 Solution of Worked Example 2.6: (a) resistance of upper branch, (b) resistance of lower branch

$$R_E = 5 \times 2.377/(5 + 2.377) = 1.611\ \Omega$$

and the current drawn by the complete circuit is

$$I = V_S/R_E = 5/1.611 \times 10^3$$
$$= 3.104 \times 10^{-3}\ \text{A or } 3.104\ \text{mA}$$

The current in the lower branch of Figure 2.6(a) is

$$I_1 = 5/R_S = 5/2.377 \times 10^3$$
$$= 2.104 \times 10^{-3}\ \text{A or } 2.104\ \text{mA}$$

and that in the upper branch is

$$I_2 = 5/5 \times 10^3 = 1 \times 10^{-3}\ \text{A or } 1\ \text{mA}$$

Referring to Figure 2.6(b), we calculate the current in each of the other resistors using the principle of current division as follows:

$$I_3 = I_1 \times R_{P1}/2\ \text{k}\Omega$$
$$= 2.104 \times 10^{-3} \times 0.667 \times 10^3/2 \times 10^3$$
$$= 0.701 \times 10^{-3}\ \text{A or } 0.701\ \text{mA}$$
$$I_4 = I_1 \times R_{P1}/1\ \text{k}\Omega = 1.403/1 \times 10^3$$
$$= 1.403 \times 10^{-3}\ \text{A or } 1.403\ \text{mA}$$
$$I_5 = I_1 \times R_{P2}/3\ \text{k}\Omega$$

$= 2.104 \times 10^{-3} \times 1.71 \times 10^3/3 \times 10^3$

$= 1.2 \times 10^{-3}$ A or 1.2 mA

$I_6 = I_1 \times R_{P2}/4\ \text{k}\Omega = 3.6/4 \times 10^3$

$= 0.9 \times 10^{-3}$ A or 0.9 mA

The above calculations tells us that the voltage across R_{P1} is

$$I_1 \times R_{P1} = 2.104 \times 10^{-3} \times 0.667 \times 10^3 = 1.403\ \text{V}$$

and the voltage across R_{P2} is

$$I_1 \times R_{P2} = 2.104 \times 10^{-3} \times 1.71 \times 10^3 = 3.6\ \text{V}$$

That is $1.403 + 3.6 = 5.003\ \text{V} \approx V_S$.

Where possible, the reader is advised to spend a little time checking the results of the calculation using a slightly different, and independent, method.

2.5 The Wheatstone bridge

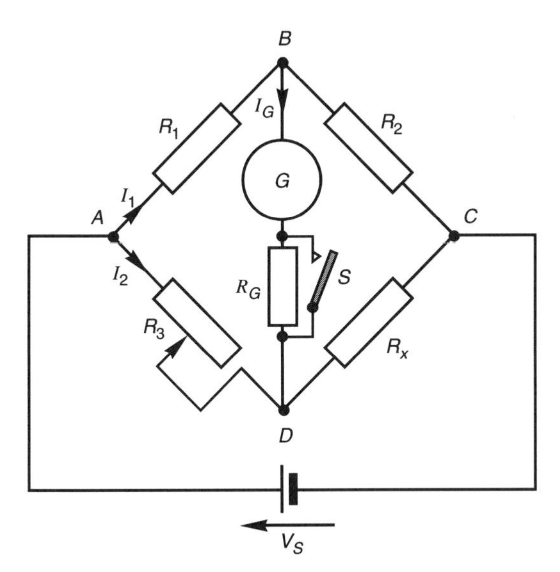

Figure 2.7 The Wheatstone bridge

The **Wheatstone bridge** (see Figure 2.7) is widely used for the determination of an unknown value of resistance (which is R_X in Figure 2.7).

The bridge comprises four resistors as follows: R_1 and R_2 are tapped resistors (usually a group of fixed resistors) whose value can be altered in a fixed ratio to one another (because of the fact that R_2/R_1 is a ratio, the name **ratio arm** is given to the upper arm of the bridge). Resistor R_3 is a continuously variable resistor, and R_X is the 'unknown' resistor.

A galvanometer (or a microammeter) is connected in series with resistor R_G between the nodes B and D of the bridge. The function of R_G is to limit the current which flows through the galvanometer during the initial set-up period, thereby protecting the galvanometer from damage. When the bridge is finally 'balanced', the contact of switch S is closed, so that R_G is short-circuited, giving the galvanometer its greatest sensitivity.

The bridge is said to be *balanced* when the galvanometer needle does not deflect from its centre-zero position, that is when

$$\text{p.d. across } R_1 = \text{ p.d. across } R_3$$

and

$$\text{p.d. across } R_2 = \text{ p.d. across } RX$$

It can be shown that this occurs when

$$R_X R_1 = R_2 R_3$$

That is when **the product of the resistance in opposite pairs of arms of the bridge are equal to one another**, or when

$$R_X = R_2 R_3 / R_1 \tag{2.11}$$

This shows that the bridge at balance is independent of the battery voltage.

Worked Example 2.7

In a Wheatstone bridge $ABCD$, the galvanometer is connected between B and D. The bridge is supplied by a 10 V battery of internal resistance 2.5 Ω, connected between A and C. A resistor of unknown value is connected between D and C. The resistance in the other arms when the bridge is at balance is

branch AB : 15 Ω

branch BC : 60 Ω

branch AD : 300 Ω

Determine (a) the value of the unknown resistance, (b) the current drawn from the battery when the bridge is at balance and (c) the p.d. across the unknown resistor.

Solution

The circuit is shown in Figure 2.8(a). The internal resistance, R, of the battery does not affect the operation of the bridge, and does not alter the balance sensitivity.

(a) Applying the rule for the condition at balance of the bridge gives

$$15R_X = 60 \times 300$$

or

$$R_X = 60 \times 300/15 = 1200\ \Omega$$

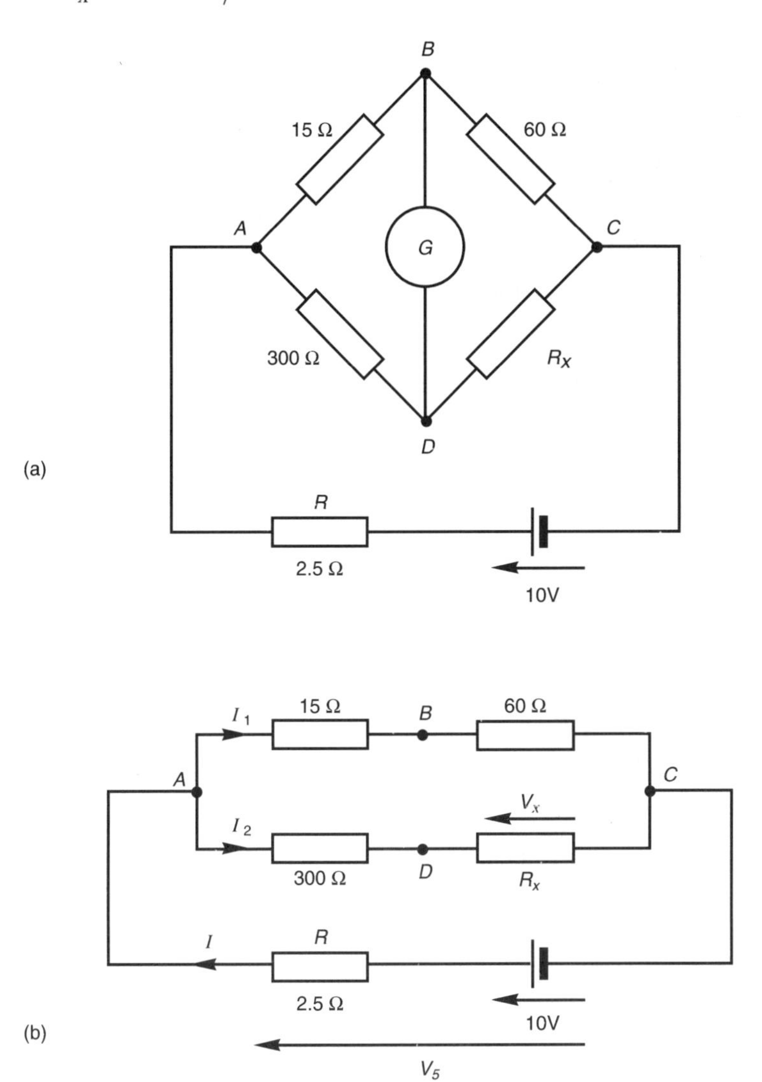

Figure 2.8 Circuit for Worked Example 2.7: (a) the circuit, (b) the bridge at balance

(b) At balance the galvanometer deflection is zero, so that the p.d. between B and D is zero, i.e. no current flows between the two nodes.

For all practical purposes we can represent the bridge at balance as shown in Figure 2.8(b). The resistance of branch ABC at balance is $15 + 60 = 75\ \Omega$, and the resistance of branch ADC is $300 + 1200 = 1500\ \Omega$. The effective resistance between A and C at balance is therefore

$$\frac{75 \times 1500}{75 + 1500} = 71.43\ \Omega$$

The total resistance of the circuit under these conditions is

$$R_E = R + 71.43 = 2.5 + 71.43 = 73.93\ \Omega$$

By Ohm's law, the current drawn from the battery is

$$I = V_S/R_E = 10/73.93 = 0.135\ \text{A}$$

(c) Using (2.10) for the current division between branches in parallel in Figure 2.8(b) gives

$$\begin{aligned} I_2 &= I \times 71.43/(300 + 1200) \\ &= 0.135 \times 0.0476 = 6.43 \times 10^{-3}\ \text{A or } 6.43\ \text{mA} \end{aligned}$$

The voltage across R_X at balance is

$$V_X = I_2 R_X = 6.43 \times 10^{-3} \times 1200 = 7.716\ \text{V}$$

An *alternative solution* is as follows: The p.d. across the internal resistance of the battery at balance is

$$IR = 0.135 \times 2.5 = 0.3375\ \text{V}$$

so that the p.d. across the parallel circuit is

$$10 - IR = 10 - 0.3375 = 9.6625\ \text{V}$$

Using (2.5) for the potential division between series-connected resistors, the p.d. across R_X is

$$\begin{aligned} V_X &= V_S R_X/(300 + R_X) = 9.6625 \times 1200/1500 \\ &= 7.73\ \text{V} \end{aligned}$$

where V_S is the voltage across $(300 + R_X)$. The difference between the two calculated values of V_X is about 0.18 per cent, which is small enough to be neglected.

2.6 Computer listings

Listing 2.1 Resistance of series- and parallel-connected resistors

```
10 CLS '**** CH2-1 ****
20 PRINT "RESISTANCE OF RESISTOR NETWORKS"
30 PRINT : S = 0: Re = 0: Ye = 0: Y = 0
40 DIM R(5)
50 FOR N = 0 TO 5
```

```
60   R(N) = 0
70 NEXT N
80 PRINT "Do you wish to:": PRINT
90 PRINT "1.  Calculate the equivalent resistance of"
100 PRINT "    a series-connected system, or"
110 PRINT "2. Calculate the equivalent resistance of"
120 PRINT "    a parallel-connected system?"
130 PRINT
140 INPUT "Enter your selection here (1 or 2): ", S
150 IF S < 1 OR S > 2 THEN GOTO 10
160 IF S = 2 THEN GOTO 290
170 CLS
180 INPUT "Number of resistors in series (Max. 5) = ", N
190 PRINT
200 IF N < 2 OR N > 5 THEN GOTO 170
210 FOR Y = 1 TO N
220   PRINT "For resistor"; Y
230   INPUT " Value in ohms = ", R(Y)
240   Re = Re + R(Y)
250 NEXT Y
260 PRINT
270 PRINT "Equivalent resistance = "; Re; " ohms."
280 GOTO 400
290 CLS
300 INPUT "Number of branches in parallel (Max. 5) = ", N
310 PRINT
320 IF N < 2 OR N > 5 THEN GOTO 290
330 FOR Y = 1 TO N
340   PRINT "For branch"; Y
350   INPUT " Value in ohms = ", R(Y)
360   Ye = Ye + 1 / R(Y)
370 NEXT Y
380 PRINT
390 PRINT "Equivalent resistance = "; 1 / Ye; " ohms."
400 PRINT
410 PRINT "Do you wish to:": PRINT
420 PRINT "1. return to the main menu, or"
430 PRINT "2. exit from the program?": PRINT
440 INPUT "Enter your selection here (1 or 2): ", S
450 IF S < 1 OR S > 2 THEN GOTO 440
460 IF S = 1 THEN GOTO 10
470 END
```

Listing 2.1 shows a menu-driven program which calculates the equivalent resistance of either a series circuit containing up to five resistors, or of a parallel circuit with up to five branches.

All you need to do is to supply the number of resistors (or branches), and the resistance of each resistor (or branch). Lines 200 and 320 prevent you from entering either less than two resistors (or branches) or more than five. You will find it an interesting exercise to extend the program to deal with more than five resistors (or branches).

The program uses an array called R (see line 40), which must be DIMensioned in order that the computer knows how much memory it must allocate. It is dimensioned as DIM D(5) which, strictly speaking, means we have allowed it to deal with *six items*, namely R(0), R(1), R(2), R(3), R(4) and R(5), of which we shall only use five. Lines 50 to 70 initially 'clear' the array by inserting zero into each location in the array.

If you need to deal with a series-parallel circuit, the subsidiary menu in lines 410 to 460 enable you to keep returning to the main menu until you have completed the calculation.

Listing 2.2 Voltage distribution between series resistors

```
10 CLS '**** CH2-2 ****
20 PRINT "VOLTAGE DISTRIBUTION BETWEEN SERIES RESISTORS"
30 PRINT : Re = 0
40 DIM R(5): FOR N = 0 TO 5: R(N) = 0: NEXT N
50 INPUT "Number of resistors (max. 5) = ", N: PRINT
60 IF N < 2 OR N > 5 THEN GOTO 10
70 FOR Y = 1 TO N
80   PRINT "For resistor"; Y
90   INPUT "          value in ohms = ", R(Y)
100   Re = Re + R(Y)
110 NEXT Y
120 PRINT
130 INPUT "Applied voltage (volts) = ", V
140 PRINT "Current in circuit = "; V / Re; "A"
150 PRINT
160 FOR Y = 1 TO N
170   PRINT "Voltage across R"; Y; " = "; V * R(Y) / Re; "V"
180 NEXT Y
190 END
```

The program in Listing 2.2 allows you to determine the voltage distribution between up to five resistors in a series circuit. You are asked to supply the number of resistors in series, the resistance of each resistor, and the supply voltage. The program calculates the current in the circuit and the voltage across each resistor.

Listing 2.3 Current distribution between parallel resistors

```
10 CLS '**** CH2-3 ****
20 PRINT "CURRENT DISTRIBUTION BETWEEN PARALLEL RESISTORS"
30 PRINT : Re = 0: Ye = 0
40 DIM R(5): FOR N = 0 TO 5: R(N) = 0: NEXT N
50 INPUT "Number of parallel branches (max. 5) = ", N: PRINT
60 IF N < 2 OR N > 5 THEN GOTO 10
70 FOR Y = 1 TO N
80   PRINT "For branch"; Y
90   INPUT "          value in ohms = ", R(Y)
100   Ye = Ye + 1 / R(Y)
110 NEXT Y
120 Re = 1 / Ye
130 PRINT
140 INPUT "Current flowing into parallel circuit (A) = ", I
150 PRINT "Voltage across circuit = "; I * Re; "V"
160 PRINT
170 FOR Y = 1 TO N
180 PRINT "Current in branch"; Y; " = "; I * Re / R(Y); "A"
190 NEXT Y
200 END
```

The program in Listing 2.3 enables you to determine the current distribution between up to five branches in a parallel circuit. You are asked to supply the number of branches, the resistance of each branch, and the current entering the circuit. The program calculates the voltage across the circuit and the current in each branch.

Exercises

2.1 The following resistors are connected in series: 0.04 Ω, 0.07 Ω, 2500 μΩ, 3500 μΩ, 25 mΩ, 50 mΩ, and 30 000 000 nΩ. Calculate the effective resistance of the circuit.

2.2 Three resistors are connected in series, their effective resistance being 0.1324 MΩ. If the resistance of the first resistor is 6800 Ω, and that of the second is 120 kΩ, determine the resistance of the third resistor.

2.3 The field circuit of a d.c. machine consists of four field coils, a rheostat, and connecting wires, all connected in series. Determine the resistance of the circuit if the resistance of each field coil is 85.5 Ω, that of the rheostat is 68.3 Ω, and that of the connecting wires is 0.2 Ω.

2.4 Resistors of 5 Ω, 4.5 Ω and 6.2 Ω are connected in series with a fourth resistor. If the equivalent resistance of the circuit is 19.8 Ω, determine the resistance of the fourth resistor.

2.5 Resistors of 10 kΩ and 15 kΩ are connected in parallel with one another. Calculate the equivalent resistance of the circuit.

2.6 A 10 Ω resistor is connected in parallel with a second resistor, the equivalent resistance of the circuit being 5.4545 Ω. Determine the resistance of the second resistor.

2.7 If resistors of 100 kΩ, 150 kΩ and 60 kΩ are connected in parallel with one another, what is the equivalent resistance?

2.8 A resistor of 8 Ω is connected in parallel with a 6 Ω resistor. The combination is in series with a resistor of unknown value. If the equivalent resistance of the circuit is 5 Ω, calculate the value of the unknown resistor.

2.9 A resistance of 10 Ω is in parallel with a conductance of 0.125 S, the circuit being supplied by a 10 V source. Calculate (a) the equivalent conductance of the circuit, (b) the equivalent resistance of the circuit, (c) the current drawn by the circuit, (d) the current in each branch.

2.10 Resistors of 10 Ω, 15 Ω and 20 Ω are connected in series with a fourth resistor. If the current drawn by the circuit is 2 A, and the supply voltage is 140 V, determine the resistance of the fourth resistor and the voltage across each resistor in the circuit.

2.11 Calculate the equivalent resistance of the circuit in Figure 2.9.

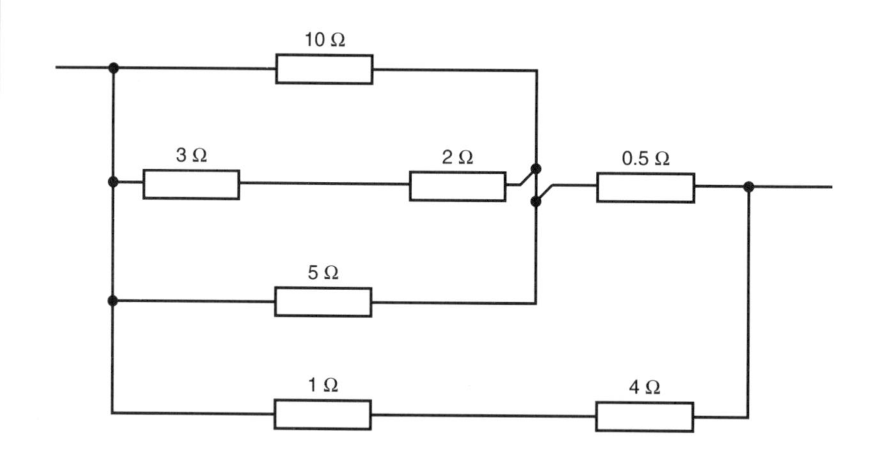

Figure 2.9 Circuit for Exercise 2.11

2.12 If the circuit in Figure 2.9 is supplied by a 10 V source, calculate (a) the current in each resistor, (b) the voltage across each resistor.

2.13 Calculate the equivalent resistance of the circuit in Figure 2.10, and the current in each resistor when it is supplied by a 100 V d.c. source.

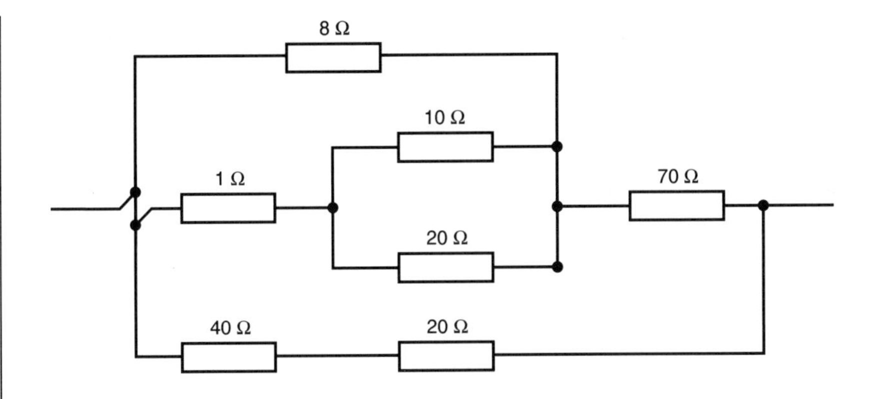

Figure 2.10 Circuit for Exercise 2.13

2.14 Figure 2.11 shows part of an electronic circuit. Calculate the resistance between terminals *A* and *B*. (**Note**: The circuit can be reduced to a single equivalent resistor by repeatedly using the parallel-resistor equation. The first step is to calculate the equivalent resistance of (2 + 4) kΩ in parallel with 1 kΩ. Next, this resistance should be considered to be in series with the 0.5 kΩ resistance, the combination being in parallel with the 1 kΩ resistance.)

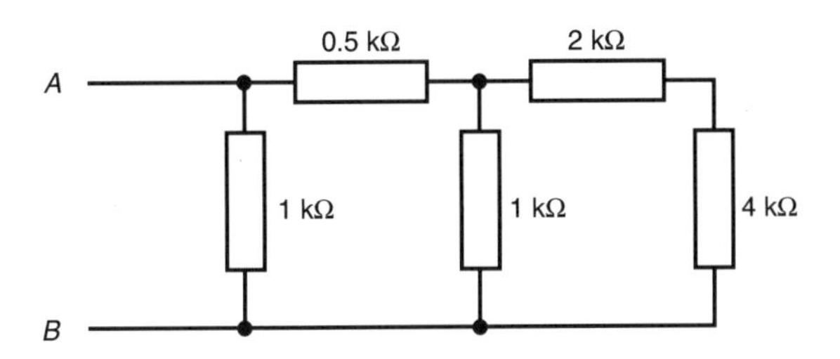

Figure 2.11 Circuit for Exercise 2.14

2.15 An unknown resistance is measured by means of a Wheatstone bridge. At balance, the ratio arms *AB* and *BC* have respective resistances of 1000 Ω and 10 Ω, and the variable resistance *AD* is 2.5 Ω at balance. Determine the unknown resistance between *C* and *D*.

2.16 If, in Exercise 2.15, the resistance between *AB* and *BC* is 10 Ω and 1000 Ω, respectively, and the variable resistance between *A* and *D* is 9500 Ω, what is the value of the unknown resistance between *C* and *D*?

2.17 Four resistors *AB*, *BC*, *CD* and *DA* are connected in the form of a Wheatstone bridge, the circuit being supplied between *A* and *C*. The supply is a cell of e.m.f. 1.5 V and internal resistance 0.9 Ω. A galvanometer is connected between B and D. When the bridge is at balance, the resistances in the circuit are as follows:

$AB = 2.1\ \Omega, BC = 2.0\ \Omega, CD = 4.0\ \Omega$

Determine

(a) the resistance between D and A,
(b) the current in sections ABC and ADC,
(c) the current drawn from the supply.

2.18 Modify Listing 2.1 to deal with up to ten resistors in series, and eight in parallel.

2.19 Combine Listings 2.2 and 2.3 to operate as a menu-driven program which allows the user the option of calculating either the voltage division between series-connected resistors, or current division between parallel branches.

Summary of important facts

A **series circuit** is one in which the **same current flows through all the elements in the circuit**.

The **equivalent resistance of a series circuit is equal to the sum of the individual resistances in the circuit**. That is

$$R_E = R_1 + R_2 + \ldots + R_n$$

The equivalent resistance of a series circuit is **always greater than the largest individual resistance in the circuit**.

When a voltage V_S is applied to a series circuit, the p.d. across resistor R_n is

$$V_n = V_S \frac{R_n}{R_E}$$

A **parallel circuit** is one in which **the same voltage appears across each branch of the circuit**. The reciprocal of the **equivalent resistance** of an n-branch parallel circuit is given by

$$\frac{1}{R_E} = \frac{1}{R_1} + \frac{1}{R_2} + \ldots + \frac{1}{R_n}$$

In the special case of a two-branch parallel circuit, the equivalent resistance is

$$R_E = \frac{R_1 R_2}{R_1 + R_2}$$

The value of the **equivalent resistance of a parallel circuit is always less than the lowest resistance of any one branch**.

If a current I enters a parallel circuit, the current in the nth branch is

$$I_n = IR_E/R_n$$

where R_n is the resistance of the nth branch.

Series-parallel circuits should be treated on their merits, and it is not possible to offer a specific method of solution.

The **Wheatstone bridge** is a circuit used for the measurement of resistance (see Figure 2.7). Two arms of the bridge are known as the **ratio arms**, another arm contains a variable resistor, and the third arm contains the unknown resistor. At balance, **the product of the resistance in diagonally opposite arms are equal to one another**.

3 Power, energy and efficiency

3.1 Introduction

In this chapter we look at power, energy and efficiency in electrical and electronic circuits, which are of great concern to all engineers. By the end of this chapter, the reader will be able to

- Calculate the power consumed in d.c. circuits.
- Determine the energy used by circuits.
- Calculate the efficiency of electrical and electronic systems.
- Determine power, energy and efficiency using computer software.

3.2 Power

The power consumed by a circuit can be calculated from one of the following equations.

$$P = VI$$
$$P = I^2R$$
$$P = V^2/R$$

where V is the voltage in volts applied to the circuit, I is the current in amperes flowing in the circuit, and R is the resistance of the circuit in ohms. The power consumed is in watts.

Worked Example 3.1

A transistor amplifier is energized by a 9 V battery, and it draws 4.5 mA from the battery. (a) What power is consumed by the amplifier, (b) what is the equivalent resistance of the amplifier as 'seen' by the battery?

Solution

(a) Since $P = VI$, then

$$P = 9 \times (4.5 \times 10^{-3}) = 0.0405 \text{ W or } 40.5 \text{ mW}$$

(b) By Ohm's law, the equivalent resistance of the amplifier is

$$R = V/I = 9/4.5 \times 10^{-3} = 2000\ \Omega \text{ or } 2\ \text{k}\Omega$$

Worked Example 3.2

A 60 W soldering iron is connected to a 240 V supply. Calculate the current drawn by the soldering iron, and determine its 'hot' resistance.

Solution

Since $P = VI$, then

$$I = P/V = 60/240 = 0.25 \text{ A}$$

and the 'hot' resistance is

$$R = V/I = 240/0.25 = 960\ \Omega$$

Also, since $P = V^2/R$, then

$$R = V^2/P = 240^2/60 = 960\ \Omega$$

and since $P = I^2R$ then

$$I = \sqrt{(P/R)} = \sqrt{(60/960)} = 0.25 \text{ A.}$$

Worked Example 3.3

If 100 mA flows through the human body, the effect is likely to be fatal. If, under certain circumstances, the hand-to-hand resistance of a person is 4 kΩ, what value of voltage is required to produce a current of 100 mA between the hands, and what is the power consumed?

Solution

From Ohm's law

$$V = IR = 100 \times 10^{-3} \times 4 \times 10^3 = 400 \text{ V}$$

and the power consumed is

$$P = VI = 400 \times 100 \times 10^{-3} = 40 \text{ W.}$$

Worked Example 3.4

A tungsten-filament lamp which has a 'hot' resistance of 576 Ω is connected to a 240 V supply. Determine (a) the power consumed by the lamp, (b) the current taken by the lamp, (c) the cost of running the lamp for 6 hours if the cost of electricity is 20 pence per kilowatt-hour.

Solution

(a) The power consumed by the lamp is

$$P = V^2/R = 240^2/576 = 100 \text{ W or } 0.1 \text{ kW}$$

(b) The current drawn by the lamp is

$$I = V/R = 240/576 = 0.417 \text{ A}$$

(c) When the lamp operates for 1 hour it consumes

$$0.1 \text{ kW} \times 1 \text{ h} = 0.1 \text{ kW h}$$

The cost of operating the lamp for 6 hours is

$$\text{cost} = 0.1 \text{ kW} \times 6 \text{ h} \times 20 \text{ p} = 12 \text{ p}.$$

Worked Example 3.5

Four heaters A, B, C, and D are rated at 500, 1000, 1500 and 2000 W, respectively, and are designed to operate from a 250 V supply. Calculate the resistance of each heater.

(a) If the heaters are connected in parallel with one another, determine the total power consumed, and the current drawn from the supply.
(b) Assuming that the resistance of each heater is unchanged when heaters C and D are connected in series with the parallel combination of A and B, calculate
  (i) the effective resistance of the combination
  (ii) the current drawn from a supply voltage of 500 V
  (iii) the total power consumed.

Solution

The resistance of each heater is

$$500 \text{ W heater: } R_A = V^2/P_A = 250^2/500 = 125 \ \Omega$$
$$1000 \text{ W heater: } R_B = V^2/P_B = 250^2/1000 = 62.5 \ \Omega$$
$$1500 \text{ W heater: } R_C = V^2/P_C = 250^2/1500 = 41.67 \ \Omega$$
$$2000 \text{ W heater: } R_D = V^2/P_D = 250^2/2000 = 31.25 \ \Omega$$

(a) The total power consumed by the parallel combination is

$$P_T = P_A + P_B + P_C + P_D$$
$$= 500 + 1000 + 1500 + 2000 = 5000 \text{ W}$$

and the total current consumed is

$$I = P_T/V_S = 5000/250 = 20 \text{ A}$$

(b) The circuit diagram for this part of the Worked Example is shown in Figure 3.1.

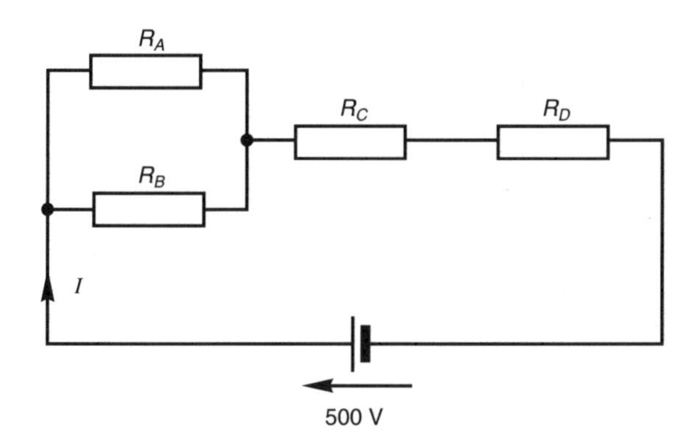

Figure 3.1 Circuit for Worked Example 3.5(b)

(i) The effective resistance of R_A and R_B in parallel is

$$R_P = R_A R_B/(R_A + R_B)$$
$$= 125 \times 62.5/(125 + 62.5) = 41.67 \ \Omega$$

This resistance is in series with R_C and R_D to give an effective circuit resistance of

$$R_E = R_P + R_C + R_D = 41.67 + 41.67 + 31.25 = 114.59 \ \Omega$$

(ii) The current drawn from a 500 V supply by this combination is

$$I = V_S/R_E = 500/114.59 = 4.36 \text{ A}$$

(iii) The total power consumed by the combination is

$$P = V_S I = 500 \times 4.36 = 2180 \text{ W.}$$

3.3 Energy consumption

The energy consumed by a circuit is

Energy = Power × Time

or

$$\mathbf{W = Pt}$$

Depending on the units of power and time, we use one of the following:

energy in joules or watt-seconds = power (W) × time (s)

energy in watt-hours = power (W) × time (h)

energy in kilowatt-hours = power (kW) × time (h)

The basic cost of electrical energy is given by

cost = energy (kWh) × cost per kWh

(**Note**: one kWh is said to be *one unit* of energy.)

Worked Example 3.6

(a) What current is consumed by a 250 V, 75 W tungsten filament lamp when it is illuminated, (b) what is its resistance at its working temperature, (c) what energy does it consume in 2 hours in (i) kWh, (ii) joules?

Solution

(a) Since $P = VI$, then

$$I = P/V = 75/250 = 0.3 \text{ A}$$

(b) From Ohm's law, the 'hot' resistance is

$$R = V/I = 250/0.3 = 833.3$$

(c) The energy consumed in two hours is

(i) $W = P(\text{kW}) \times t(\text{h}) = (75/1000) \times 2 = 0.15$ kWh

(ii) $W = P(W) \times t(\text{s}) = 75 \times (2 \times 60 \times 60) = 540\,000$ J.

Worked Example 3.7

A current of 6 A flows through a coil with an air core, the p.d. across the coil being 15 V. Determine

(a) the resistance of the coil

(b) the power consumed by the coil

(c) the energy consumed in a 12 hour period.

Solution

(a) From Ohm's law

$$R = V/I = 15/6 = 2.5\ \Omega$$

(b) The power consumed is

$$P = VI = 15 \times 6 = 90 \text{ W}$$

(c) The energy consumed in a 12 hour period is

$$W = Pt = 90 \times 12 = 1080 \text{ Wh or } 1.08 \text{ kWh.}$$

Worked Example 3.8

Two resistors R_1 and R_2 are connected in parallel with one another to a 500 V supply. If the effective resistance of the circuit is 25 Ω, and the power consumed by R_1 is 3.75 kW, determine

(a) the resistance of each resistor, and the power dissipated in R_2
(b) the current in each resistor
(c) the energy consumed by R_2 in a 2 hour period.

Solution

(a) The power dissipated in R_1 is $P_1 = V_S{}^2/R_1$, where V_S is the supply voltage. Hence

$$R_1 = V_S{}^2/P_1 = 500^2/(3.75 \times 1000) = 66.67\ \Omega$$

The power consumed by the circuit is $P_T = V_S{}^2/R_E$, where R_E is the equivalent resistance of the circuit, hence

$$P_T = 500^2/25 = 10\,000 \text{ W}$$

The power consumed by R_2 is, therefore

$$P_2 = P_T - P_1 = 10\,000 - 3\,750 = 6\,250 \text{ W}$$

Since $P_2 = V_S{}^2/R_2$, it follows that

$$R_2 = V_S{}^2/P_2 = 500^2/6250 = 40\ \Omega$$

(b) The power dissipated in R_1 is $P_1 = I_1{}^2R_1$, hence

$$I_1 = \sqrt{}(P_1/R_1) = \sqrt{}(3750/66.67) = 7.5 \text{ A}$$

and

$$I_2 = \sqrt{}(P_2/R_2) = \sqrt{}(6250/40) = 12.5 \text{ A}$$

(c) The energy consumed by R_2 in 2 hours is

$$W = I_2{}^2R_2t = 12.52 \times 40 \times 2$$
$$= 12\,500 \text{ Wh or } 12.5 \text{ kWh.}$$

Worked Example 3.9

A d.c. motor takes a steady current of 15 A at 400 V for 8 hours. What is the cost of running the motor if the cost of energy is 10p per kWh?

Solution

The energy consumed in an 8 hour period is

$$W = VIt = 400 \times 15 \times 8 = 48\,000 \text{ Wh or } 48 \text{ kWh}$$

and the cost of running the motor is

$$\text{cost} = \text{energy in kWh} \times \text{cost per kWh}$$
$$= 48 \times 10 = 480\,\text{p or £4.80.}$$

3.4 Efficiency

Efficiency is the ratio of the power output from a device or system to the power input. If the efficiency is expressed in terms of a power ratio, then it is known as the per-unit efficiency, that is

$$\textbf{per-unit efficiency} = \frac{\textbf{output power}}{\textbf{input power}} \text{ p.u.}$$

The per-cent efficiency is

$$\textbf{per-cent efficiency} = \frac{\textbf{output power}}{\textbf{input power}} \times \mathbf{100\,\%}$$

These ratios apply equally to an electronic system as they do to an electrical machine. If the circuit or system has a power loss, then

$$\text{output power} = \text{input power} - \text{power loss}$$

or

$$\text{input power} = \text{output power} + \text{power loss}$$

The per-unit efficiency can therefore be expressed in the form

$$\textbf{per-unit efficiency} = \frac{\textbf{input power} - \textbf{power loss}}{\textbf{input power}}$$
$$= \mathbf{1} - \frac{\textbf{power loss}}{\textbf{input power}}$$

or as

$$\textbf{per-unit efficiency} = \frac{\textbf{output power}}{\textbf{output power} + \textbf{power loss}}$$

The per-cent efficiency value is 100 times greater than the per-unit value.

Worked Example 3.10

If the input power to a motor is 5 kW, and the efficiency of the motor and its load is 75 per cent, what is the mechanical output power?

Solution

Since

$$\text{per-cent efficiency} = \frac{\text{output power} \times 100}{\text{input power}}$$

then

$$\text{output power} = \frac{\text{per-cent efficiency} \times \text{power}}{100}$$

$$= 75 \times 5000/100 = 3\,750 \text{ W or } 3.75 \text{ kW}$$

Worked Example 3.11

A d.c. motor consumes an electric current of 24 A at 400 V. What is the input power in kW? If the efficiency of the motor is 80 per cent, what is its output power?

Calculate the cost of running the motor continuously under these conditions for 8 hours at a cost of 9 p per kWh, and what is the cost of the losses?

Solution

From the data supplied

$$\text{input power} = VI = 400 \times 24 = 9\,600 \text{ W or } 9.6 \text{ kW}$$

Since p.u. efficiency = output power/input power, then

$$\text{output power} = \text{input power} \times \text{p.u. efficiency}$$

$$= 9.6 \times 0.8 = 7.68 \text{ kW}$$

The cost of running the motor for 8 hours under these conditions is

$$\text{cost} = \text{input power (kW)} \times \text{time (h)} \times \text{cost per kWh}$$

$$= 9.6 \times 8 \times 9 = 691.2 \text{ p or £}6.91$$

The power loss in the system is

$$\text{power loss} = \text{input power} - \text{output power}$$

$$= 9.6 - 7.68 = 1.92 \text{ kW}$$

and the cost of the power loss in an 8 hour period is

$$\text{cost} = \text{power loss (kW)} \times \text{time (h)} \times \text{cost per kWh}$$
$$= 1.92 \times 8 \times 9 = 138.2\,\text{p or £1.38.}$$

Worked Example 3.12

The power loss in a transformer can be subdivided into two types of loss, namely the I^2R loss or **copper loss**, and the **core loss** (that is, the power required to magnetize the iron circuit).

If a transformer supplies a power output of 8 kW, and its copper loss and core loss are, respectively, 250 W and 600 W, calculate (a) the input power of the transformer, (b) its overall efficiency.

Solution

(a) The total power loss in the transformer is

$$\text{losses} = \text{copper loss} + \text{core loss} = 250 + 600 = 850\ \text{W}$$

hence

$$\text{input power} = \text{output power} + \text{losses} = 8000 + 850 = 8850\ \text{W}$$

(b) Overall efficiency = output power/input power

$$= 8000/8850 = 0.904\ \text{p.u. or 90.4 per cent}$$

3.5 Computer listing

Listing 3.1 Calculation of power, energy and efficiency

```
10 CLS '**** CH3-1 ****
20 PRINT "CALCULATION OF POWER, ENERGY AND EFFICIENCY"
30 PRINT : P = 0: V = 0: I = 0: R = 0: W = 0: t = 0: PI = 0: PO = 0: PL = 0
40 PRINT "Which of the following do you wish to solve:": PRINT
50 PRINT "1. P = V * I"
60 PRINT "2. P = I^2 * R"
70 PRINT "3. P = V^2 / R"
80 PRINT "4. W = P (W) * t (s) Joules or W s"
90 PRINT "5. W = P (W) * t (hours) W h"
100 PRINT "6. W = P (kW) * t (hours) kW h"
110 PRINT "                    Output Power"
```

```
120 PRINT "7. Efficiency = ------------"
130 PRINT "                          Input Power"
140 PRINT "                          Input Power - Loss"
150 PRINT "8. Efficiency = -----------------"
160 PRINT "                                Input Power"
170 PRINT "                              Output Power"
180 PRINT "9. Efficiency = -----------------"
190 PRINT "                          Output Power + Loss"
200 PRINT
210 INPUT "Please enter your selection (1 - 9): ", S
220 IF S < 1 OR S > 9 THEN GOTO 10
230 IF S = 1 THEN GOTO 320
240 IF S = 2 THEN GOTO 360
250 IF S = 3 THEN GOTO 400
260 IF S = 4 THEN GOTO 440
270 IF S = 5 THEN GOTO 480
280 IF S = 6 THEN GOTO 520
290 IF S = 7 THEN GOTO 560
300 IF S = 8 THEN GOTO 640
310 IF S = 9 THEN GOTO 720
320 CLS : PRINT "The equation is P = V * I": PRINT
330 INPUT "V (volts) = ", V
340 INPUT "I (amperes) = ", I: PRINT
350 PRINT "P (watts) = "; V * I: GOTO 800
360 CLS : PRINT "The equation is P = I^2 * R": PRINT
370 INPUT "I (amperes) = ", I
380 INPUT "R (ohms) = ", R: PRINT
390 PRINT "P (watts) = "; I ^ 2 * R: GOTO 800
400 CLS : PRINT "The equation is P = V^2 / R": PRINT
410 INPUT "V (volts) = ", V
420 INPUT "R (ohms) = ", R: PRINT
430 PRINT "P (watts) = "; V ^ 2 / R: GOTO 800
440 CLS : PRINT "The equation is W = P (watts) * t (s)": PRINT
450 INPUT "P (watts) = ", P
460 INPUT "t (seconds) = ", t: PRINT
470 PRINT "W (joules or W s) = "; P * t: GOTO 800
480 CLS : PRINT "The equation is W = P (watts) * t (h)": PRINT
490 INPUT "P (watts) = ", P
500 INPUT "t (hours) = ", t: PRINT
510 PRINT "W (W h) = "; P * t: GOTO 800
520 CLS : PRINT "The equation is W = P (kW) * t (h)": PRINT
530 INPUT "P (kW) = ", P
540 INPUT "t (h) = ", t: PRINT
550 PRINT "W (kW h) = "; P * t: GOTO 800
560 CLS : PRINT "The equation is": PRINT
570 PRINT "                      Output Power"
580 PRINT "Efficiency = ------------"
```

```
590 PRINT "                Input Power": PRINT
600 INPUT "Power Output = ", PO
610 INPUT "Power input = ", PI: PRINT
620 PRINT "Efficiency = "; PO / PI; "per unit"
630 PRINT "           = "; PO * 100 / PI; "per cent": GOTO 800
640 CLS : PRINT "The equation is": PRINT
650 PRINT "                Input Power - Loss"
660 PRINT "Efficiency = ----- -----------"
670 PRINT "                    Input Power": PRINT
680 INPUT "Power Input = ", PI
690 INPUT "Power Loss = ", PL: PRINT
700 PRINT "Efficiency = "; (PI - PL) / PI; "per unit"
710 PRINT "           = "; (PI - PL) * 100 / PI; "per cent": GOTO 800
720 CLS : PRINT "The equation is": PRINT
730 PRINT "                   Output Power"
740 PRINT "Efficiency = ----------------"
750 PRINT "                Output Power + Loss": PRINT
760 INPUT "Power Output = ", PO
770 INPUT "Power Loss = ", PL: PRINT
780 PRINT "Efficiency = "; PO / (PO + PL); "per unit"
790 PRINT "           = "; PO * 100 / (PO + PL); "per cent"
800 PRINT
810 PRINT "Do you wish to:": PRINT
820 PRINT "1. return to the main menu, or"
830 PRINT "2. exit from the program?": PRINT
840 INPUT "Enter your selection here (1 or 2): ", S
850 IF S < 1 OR S > 2 THEN GOTO 840
860 IF S = 1 THEN GOTO 10
870 END
```

The calculations involved in this chapter are fairly straightforward, and Computer listing 3.1 contains a comprehensive menu-driven program which solves many of the important equations in the chapter.

On the conclusion of each calculation, you are given the option either of returning to the main menu, or leaving the program.

Exercises

3.1 A 50 Ω resistor is connected to a 15 V d.c. supply. Determine (a) the current in the resistor, (b) the power consumed, (c) the energy consumed in a 5 hour period of time, (d) the cost of the energy consumed in 6 hours if the cost of energy is 10 p per kWh.

3.2 A 2-core cable having a 'loop' resistance of 0.25 Ω consumes 18 W of power. Determine (a) the current in the cable, (b) the total voltage drop in the cable, (c) the energy consumed in the cable in an 8 hour working day, (d) the cost of the power loss in the cable if the cost of electricity is 10 p per kWh.

3.3 A current of 6 A flows for 2 hours through a resistance of 3.5 Ω. Determine (a) the p.d. across the resistance, (b) the power consumed, (c) the energy consumed.

3.4 If the cost of operating a 240 V heater for 10 hours is £3.00, and the cost of electricity is 10p per kWh, calculate (a) the power consumed, (b) the current in the load, (c) the resistance of the heater.

3.5 If a 10 kW electrical motor has an input power on full load of 11.75 kW, what is the efficiency of the motor?

3.6 A brake test on a d.c. motor indicates that its output power is 8 kW, the current drawn by the motor being 25 A when the supply voltage is 400 V. What is the efficiency of the motor?

3.7 A common-emitter amplifier has a collector supply voltage of 8 V, the quiescent collector current being 2.6 mA and the quiescent base current being 25 μA. Determine the total power dissipation of the circuit. If the quiescent collector voltage is 4.1 V, calculate the quiescent collector power dissipation.

3.8 Modify the BASIC program listing in Computer listing 3.1 so that each calculation deals with up to five calculations of the same kind, i.e. option 1 deals with up to five separate calculations of $P = V * I$, etc.

Summary of important facts

The power, P, consumed by a d.c. circuit can be determined from the following

$$P = VI$$

$$P = I^2R$$

$$P = V^2/R$$

where V is the voltage across the circuit, I is the current in the circuit and R is the resistance of the circuit.

The energy, W, consumed by a circuit is

energy = power × time

Since the units used in engineering for both power and time may vary, we use the following

energy in joules (J) = power (W) × time (s)

energy in watt hours (Wh) = power (W) × time (h)

energy in kWh = power (kW) × time (h)

The basic cost of electrical energy is

cost = energy in kWh × cost per kWh

The **efficiency** of an item of electrical plant is

$$\textbf{efficiency} = \frac{\textbf{output power}}{\textbf{input power}} \text{ per unit (p.u.)}$$

or

$$\textbf{efficiency} = \frac{\textbf{output power} \times \textbf{100}}{\textbf{input power}} \text{ per cent (\%)}$$

Alternatively, we may express the efficiency as

$$\begin{aligned}\textbf{efficiency} &= \textbf{1} - \frac{\textbf{power loss}}{\textbf{input power}} \text{ p.u.}\\ &= \frac{\textbf{output power}}{\textbf{output power} + \textbf{power loss}} \text{ p.u.}\\ &= \frac{\textbf{input power} - \textbf{power loss}}{\textbf{input power}} \text{ p.u.}\end{aligned}$$

4 Network theorems

4.1 Introduction

A **network theorem** is the combination of basic laws to solve a particular type of problem. For example, using Ohm's law on its own does not allow us to 'look' at a circuit and write down its equation directly. By combining basic laws in a common-sense way, we arrive at basic network theorems, and we look at several important theorems in this chapter. By the end of this chapter, the reader will be able to

- Understand and apply Kirchhoff's laws.
- Write down and solve circuit equations.
- Understand and apply Thévenin's theorem.
- Understand and apply Norton's theorem.
- Convert a Thévenin's theorem equivalent circuit to a Norton's equivalent circuit, and vice versa.
- Understand and apply the Maximum Power Transfer theorem.
- Use a computer program to solve two simultaneous equations, and to convert a Thévenin equivalent circuit into a Norton equivalent circuit, and vice versa.

4.2 Kirchhoff's laws

The laws laid down by Gustav Kirchhoff allow us to determine not only the current flowing in complex circuits, but also the voltage across each element in the circuit. Kirchhoff proposed two electric circuit laws, namely

1. Kirchhoff's current law (KCL)
2. Kirchhoff's voltage law (KVL).

Kirchhoff's current law (KCL) states that the current entering a node or point in a circuit is equal to the current flowing away from it. That is, *current cannot accumulate at a point in the circuit*. In the case of node N in Figure 4.1, the current flowing towards the node is $I_A + I_C$, and the current flowing away from it is $I_B + I_D$. KCL tells us that

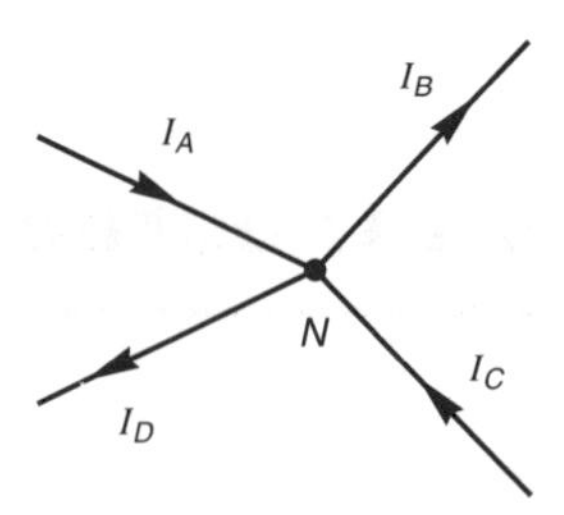

Figure 4.1 Illustration of Kirchhoff's current law (KCL)

$$I_A + I_C = I_B + I_D \tag{4.1}$$

Alternatively, we may say that **the sum of current flowing towards any node is zero.** In the case of Figure 4.1, $I_A + I_C$ *flows towards* node N, and $I_B + I_D$ flows away from it (or $(-I_B + -I_D)$ flows towards the node). Hence we may say that

$$I_A - I_B + I_C - I_D = 0 \tag{4.2}$$

Mathematically we see that (4.1) is equivalent to (4.2).

Kirchhoff's voltage law (KVL) states that **the algebraic sum of the e.m.fs and p.ds around any close path is zero.**

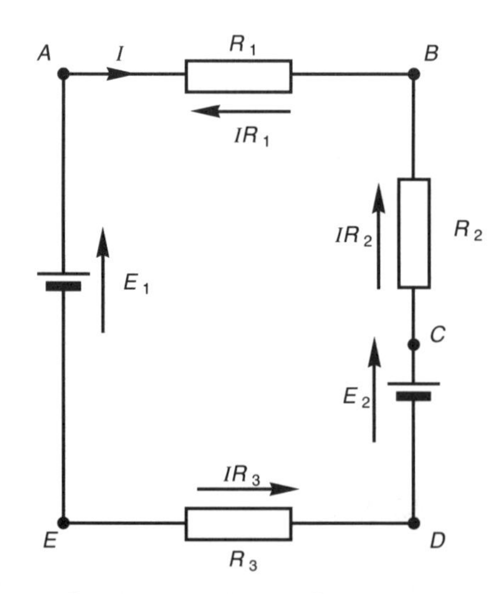

Figure 4.2 Illustrating Kirchhoff's Voltage Law (KVL)

When we apply KVL to the circuit in Figure 4.2 we get

$$E_1 - IR_1 - IR_2 - E_2 - IR_3 = 0$$

The equation may, alternatively, be written in the form

$$E_1 - E_2 = IR_1 + IR_2 + IR_3$$

The reader should note that since E_1 and E_2 act in opposite directions, the sum of the e.m.fs (acting in the direction of I) is $E_1 - E_2$. This leads us to an alternative form of KVL as follows:

In any closed path, the sum of the e.m.fs is equal to the sum of the p.ds.

The steps which should be followed when writing down the equation for a closed loop are:

1. Draw the assumed direction of current flow in each branch of the circuit.
2. Draw an e.m.f. arrow by the side of each source, pointing towards the positive terminal.
3. Draw a p.d. arrow by the side of each resistor in the circuit, the arrow pointing in the opposite direction to the current flow through the resistor.
4. Write down the equation for the p.d. across each resistor by the side of the p.d. arrow, i.e. IR_1, IR_2, IR_3, etc.
5. *Starting at any node in the loop and returning to the same node*, go around the loop and add together all the e.m.fs and p.ds, and equate the sum to zero.

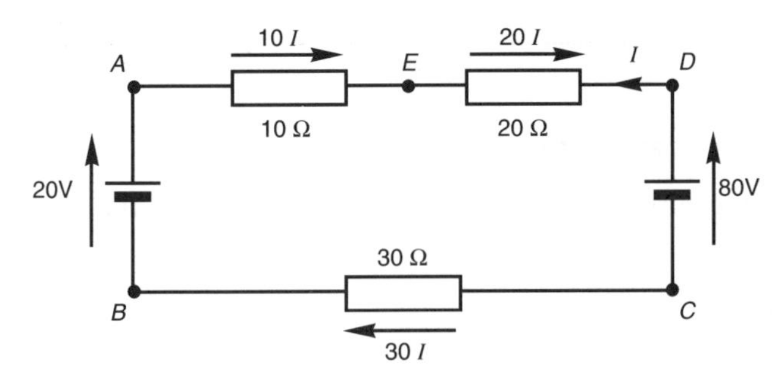

Figure 4.3 A simple application of KVL

Consider the circuit in Figure 4.3. Steps 1–3, inclusive, have already been carried out on the figure. In some cases, we do not know what way the current flows, in which case we simply assume a direction for the current (see Worked Example 4.3).

Commencing at node A (which is selected quite arbitrarily), we proceed around the loop in the direction $ABCDEA$, and *add together* the e.m.fs and p.ds which point in the direction in which we move around the loop, and *subtract* e.m.fs and p.ds which oppose our movement, as follows:

$$-20 - 30I + 80 - 20I - 10I = 0$$

that is $\quad 60 - 60I = 0$

or

$$I = 60/60 = 1 \text{ A}.$$

Similarly, if we start from node C and proceed in the direction $CBAEDC$ we get the equation

$$30I + 20 + 10I + 20I - 80 = 0$$

or $\quad 60I - 60 = 0$

hence

$$I = 60/60 = 1 \text{ A}$$

Worked Example 4.1

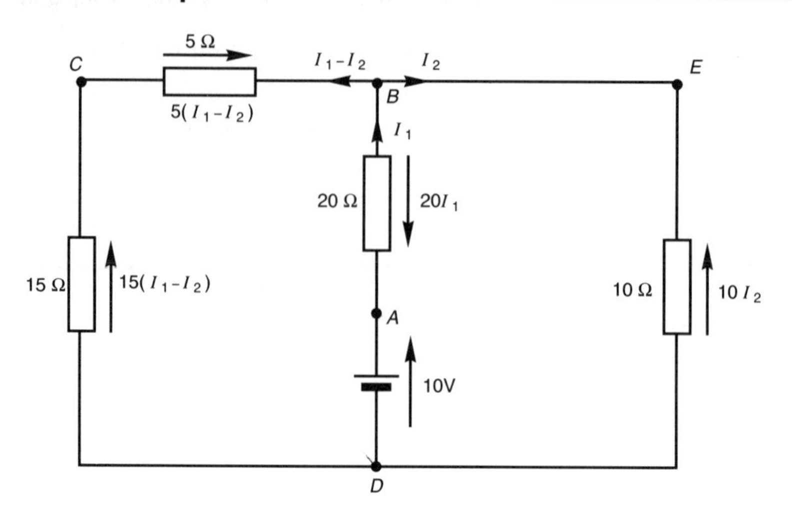

Figure 4.4 Circuit for Worked Example 4.1

Determine the current in each branch in Figure 4.4, and the p.d. across the 15 Ω resistor.

Solution

Since there is only one e.m.f. in the circuit, we can assume that current I_1 flows into node B. It is assumed that I_2 flows into the 10 Ω resistor, therefore KCL tells us that $(I_1 - I_2)$ flows in branch BCD. Since there are only two unknown currents, we need two equations (known as *simultaneous equations*) to solve for them. We obtain these by writing down the equations for two *independent loops*.

There are three loops in the circuit (loops $ABCDA$, $ABEDA$ and $BEDCB$), we select *any two* of these. We select the first two.

Loop ABCDA

Applying KVL to the loop we get

$$-20I_1 - 5(I_1 - I_2) - 15(I_1 - I_2) + 10 = 0$$

or $$-40I_1 + 20I_2 + 10 = 0$$

that is

$$10 = 40I_1 - 20I_2 \tag{4.3}$$

Loop ABEDA

Applying KVL to the loop we get

$$-20I_1 - 10I_2 + 10 = 0$$

or $$10 = 20I_1 + 10I_2 \tag{4.4}$$

We can eliminate one of the variables from the two equations by making the coefficient of one of the variables have the same value in the two equations, and then subtracting one equation from the other (this is known as *solving simultaneous equations by elimination*). We will eliminate I_1 by multiplying (4.4) by two, and subtracting it from (4.3) as follows:

$$10 = 40I_1 - 20I_2 \qquad \text{(4.3) repeated}$$
$$20 = 40I_1 + 20I_2 \qquad \text{(4.4)} \times 2$$

SUBTRACT $\quad -10 = \quad - 40I_2$

hence $\quad I_2 = -10/(-40) = 0.25\text{ A}$

Substituting this value into one of the original equations allows us to determine the value of I_1. We will use (4.3)

$$10 = 40I_1 - 20 \times 0.25 = 40I_1 - 5$$

that is

$$I_1 = (10 + 5)/40 = 0.375\text{ A}$$

The current in the 5 Ω and 15 Ω resistors is

$$I_1 - I_2 = 0.375 - 0.25 = 0.125\text{ A}$$

Finally, we *verify the results* by inserting the values in (4.4) as follows:

$$20I_1 + 10I_2 = (20 \times 0.375) + (10 \times 0.25) = 10$$

which agrees with (4.4). It is an important part of circuit analysis to verify the results.

Worked Example 4.2

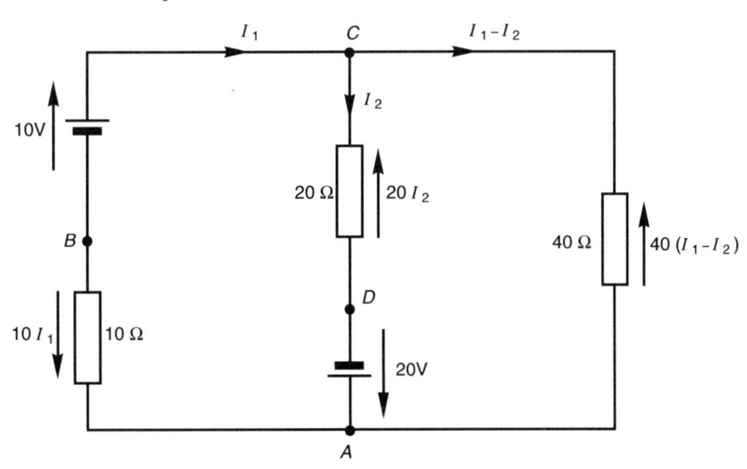

Figure 4.5 Circuit for Worked Example 4.2

Determine the current in each branch of Figure 4.5.

Solution

In this case it seems reasonable to assume that current flows out of the positive pole of each battery. Since I_1 flows towards node C, and I_2 flows away from it, KCL tells us that $(I_1 - I_2)$ flows away from node C and into the 40 Ω resistor.

Having decided on the current direction, we can write down the expression for the p.d. across each resistor (which has already been done in Figure 4.5), each p.d. arrow pointing in the opposite direction to the current flow through the resistor.

Once again, there are three possible loops, namely *ABCDA* (left-hand loop), *ADCA* (right-hand loop) and *ABCA* (outer loop). Selecting the latter two gives the following loop voltage equations.

Loop ADCA

$$-20 + 20I_2 - 40(I_1 - I_2) = 0$$

or $$-20 - 40I_1 + 60I_2 = 0$$

that is $$-20 = 40I_1 - 60I_2 \qquad (4.5)$$

Loop ABCA

$$-10I_1 + 10 - 40(I_1 - I_2) = 0$$

hence $$10 - 50I_1 + 40I_2 = 0$$

therefore $$10 = 50I_1 - 40I_2 \qquad (4.6)$$

We eliminate I_1 from the above equations by multiplying (4.5) by 1.25 and subtracting (4.6) from it as follows:

$$-25 = 50I_1 - 75I_2 \qquad (4.5) \times 1.25$$

$$10 = 50I_1 - 40I_2 \qquad (4.6) \text{ repeated}$$

SUBTRACT $$-35 = -35I_2$$

hence $$I_2 = -35/(-35) = 1 \text{ A}$$

Substituting this into (4.5) gives

$$-20 = 40I_1 - (60 \times 1)$$

or $$I_1 = (-20 + 60)/40 = 1 \text{ A}$$

The current in the 40 Ω resistor is

$$I_1 - I_2 = 1 - 1 = 0 \text{ A}$$

This apparently unexpected result can be explained by the fact that the 10 V battery makes node C positive with respect to node A, whilst the 20 V battery makes node C negative! The particular conditions in the circuit combine to result in zero p.d. between nodes A and C.

As a matter of interest, let us evaluate the p.d. of node C with respect to node A. We do this by following any path between node C and node A. Taking path *ABC* we get

$$V_{CA} = -10I_1 + 10 = -(10 \times 1) + 10) = 0 \text{ V}$$

In fact, the 40 Ω resistor can be replaced by *any value of resistance* (ranging from a short-circuit to an open-circuit), without altering the value of I_1 or I_2. The reader should try solving the circuit with (a) a 5 Ω resistor and (b) a 1000 Ω resistor in place of the 40 Ω resistor.

If either the 10 Ω resistor, or the 20 Ω resistor, or the polarity or magnitude of either battery was changed, then current would flow in the 40 Ω branch. For example, if the 20 V source were replaced by a 15 V source, the reader would find that $I_1 = 0.857$ A and $I_2 = 0.821$ A, so that $I_1 - I_2 = 0.036$ A.

Worked Example 4.3

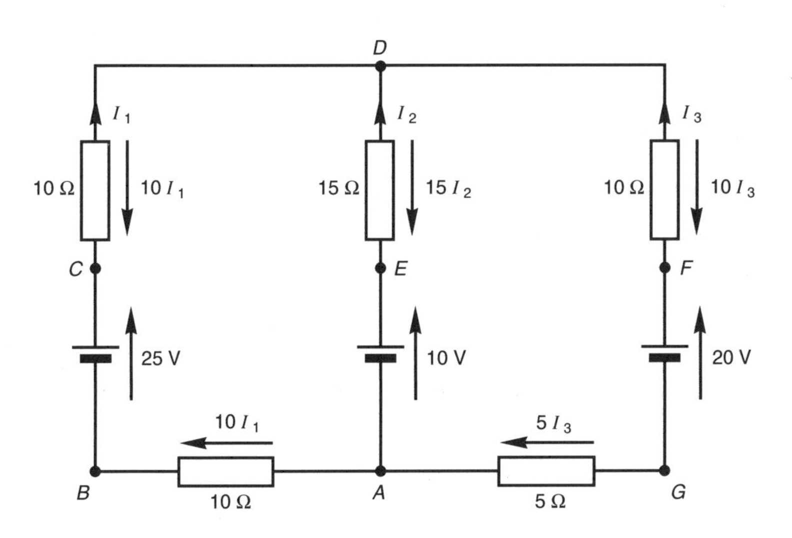

Figure 4.6 Circuit for Worked Example 4.3

Determine the value of the current in each branch of the circuit in Figure 4.6, the p.d. across and the power consumed by each resistor, and the potential of node D with respect to node B.

Solution

Initially it seems reasonable to assume that current flows out of the positive terminal of each battery, and this has been adopted in Figure 4.6. However, a moment's consideration suggests that current 'accumulates' at node D; since KCL states that this cannot occur, it follows that one (or two) of the currents in the figure flow in the opposite direction to those shown, i.e. one (or two) batteries are being 'charged'. Since we do not know the true direction of current flow, we will retain the direction of current in Figure 4.6.

Applying KCL to node D gives

$$I_1 + I_2 + I_3 = 0$$

or $$I_3 = -(I_1 + I_2)$$

We therefore need two simultaneous equations to determine the value of I_1 and I_2, from which we can calculate I_3. Once again, there are three loops in the circuit, namely $ABCDEA$, $AEDFGA$ and $ABCDFGA$; we will address the first two.

Loop ABCDEA

The loop equation is

$$-10I_1 + 25 - 10I_1 + 15I_2 - 10 = 0$$

or $$15 - 20I_1 + 15I_2 = 0$$

hence $$15 = 20I_1 - 15I_2 \quad (4.7)$$

Loop AEDFGA

The equation here is

$$10 - 15I_2 + 10I_3 - 20 + 5I_3 = 0$$

or $$10 - 15I_2 + 10(-I_1 - I_2) - 20 + 5(-I_1 - I_2) = 0$$

that is $$-10 - 15I_1 - 30I_2 = 0$$

hence $$-10 = 15I_1 + 30I_2 \quad (4.8)$$

Current I_2 can be eliminated from the above equations as follows:

$$30 = 40I_1 - 30I_2 \quad (4.7) \times 2$$

$$-10 = 15I_1 + 30I_2 \quad \text{(4.8) repeated}$$

ADD $$20 = 55I_1$$

hence $$I_1 = 20/55 = 0.364 \text{ A}$$

Substituting this value into (4.7) gives

$$15 = (20 \times 0.364) - 15I_2 = 7.28 - 15I_2$$

therefore

$$I_2 = (7.28 - 15)/15 = -0.515 \text{ A}$$

That is, I_2 flows in the opposite direction to that shown in Figure 4.6, i.e. the 10 V battery is being charged by the 25 V battery. The current, I_3, is given by

$$I_3 = -(I_1 + I_2) = -(0.364 + (-0.515)) = 0.151 \text{ A}$$

That is, the 10 V battery is being charged by both the 25 V battery and the 20 V battery!

Calculation of the potential across each resistor, and the power consumed

10 Ω resistor between A and B

The voltage V_{BA} (the potential of B relative to A) is

$$V_{BA} = -10I_1 = -10 \times 0.364 = -3.64 \text{ V}$$

and the power consumed is

$$V_{BA}{}^2/R_{BA} = (-3.64^2)/10 = 1.33 \text{ W}$$

The reader should note that the current flows from A to B, so that the potential of node A is greater than that of node B. That is the value of VBA (the potential of B with respect to A) is negative.

10 Ω resistor between C and D

The voltage V_{CD} (the potential of C relative to D) is

$$V_{CD} = -10I_1 = -10 \times 0.364 = -3.64 \text{ V}$$

and the power consumed is

$$V_{CD}{}^2/R_{CD} = (-3.64^2)/10 = 1.33 \text{ W}$$

15 Ω resistor between E and D

The voltage V_{DE} (the potential of D relative to E) is

$$V_{DE} = -15I_2 = -15 \times (-0.515) = 7.725 \text{ V}$$

and the power consumed is

$$V_{DE}{}^2/R_{DE} = 7.725^2/15 = 3.98 \text{ W}$$

The reader will observe that since I_2 flows into the positive terminal of the battery, the value of V_{DE} (the potential of D relative to E) has a positive value.

5 Ω resistor between A and G

The voltage V_{GA} (the potential of G relative to A) is

$$V_{GA} = -5I_3 = -5 \times (-0.151) = 0.755 \text{ V}$$

and the power consumed is

$$V_{GA}{}^2/R_{GA} = 0.755^2/5 = 0.114 \text{ W}$$

10 Ω resistor between D and F

The voltage V_{DF} (the potential of D relative to F) is

$$V_{DF} = -10I_3 = -10 \times (-0.151) = 1.51 \text{ V}$$

and the power consumed is

$$V_{DF}{}^2/R_{DF} = 1.51^2/10 = 0.228 \text{ W}$$

Check on the solution

We can check the solution obtained by calculating the potential between nodes D and A in each of the three branches by adding the voltage and p.d. between node A and D (commencing at node A) for any of the three nodes.

For the left-hand branch (ABCD)

$$V_{DA} = -10I_1 + 25 - 10I_1 = 25 - 20I_1$$
$$= 25 - (20 \times 0.364) = 17.72 \text{ V}$$

For the centre branch (AED)

$$V_{DA} = V_{EA} + V_{DE} = 10 + 15I_2 = 10 + (15 \times 0.515)$$
$$= 17.725 \text{ V}$$

For the right-hand branch (AGFD)

$$V_{DA} = V_{GA} + V_{FG} + V_{DF} = -5I_3 + 20 - 10I_3$$
$$= 20 - 15I_3 = 20 - (15 \times 0.151) = 17.735 \text{ V}$$

Since the difference between the three answers is less than 0.1 per cent, we can assume that the solution of the problem is correct.

Finally we have been asked to calculate the value of V_{DB}, i.e. the potential of node D with respect to node B. We do this by adding the p.d. and e.m.f. values *following any path* between node B and node D (commencing at node B). Let us take the path BCD; the equation is

$$V_{DB} = V_{CB} + V_{DC} = 25 - 10I_1 = 25 - (10 \times 0.364)$$
$$= 21.36 \text{ V}$$

Any other path may be chosen, such as $BAGFD$, and the answer should be within a fraction of 1 per cent of the above value.

4.3 Thévenin's theorem

Thévenin's theorem says that any active network, no matter how complex, **can be replaced by a simple source of e.m.f., E_T, which has a resistance, R_T, in series with it**.

By 'active network' we mean one containing one or more sources of e.m.f.; for example, the circuit in Figure 4.7(a) can be replaced by its equivalent Thévenin equivalent circuit in diagram (b).

Since the two circuits are equivalent to one another, once we know the value of E_T and R_T, we can quickly calculate the current flowing in any value of resistance connected between terminals A and B.

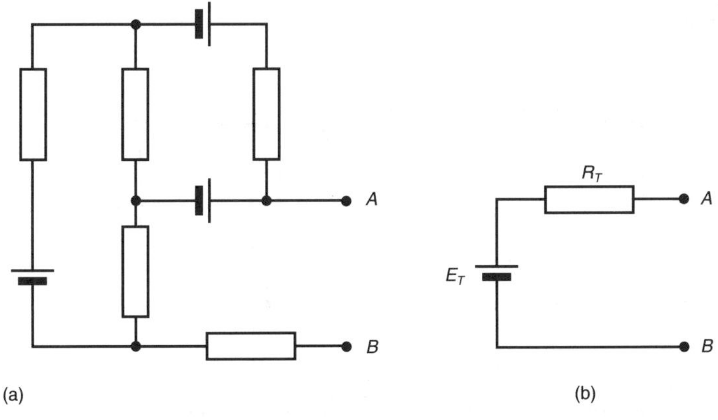

Figure 4.7 Thévenin's theorem: (a) original circuit, (b) Thévenin equivalent circuit

To calculate the value of E_T and R_T, we proceed as follows:

1. **When the load between *A* and *B* is disconnected, we calculate the voltage V_{AB} between the terminals *A* and *B*. This gives the Thévenin voltage E_T, where E_T has the same polarity as V_{AB}.**
2. **With *all sources of e.m.f.* within the network replaced by their internal resistance, and any external load between *A* and *B* disconnected, we determine the resistance between terminals *A* and *B*. This value is the Thévenin 'internal resistance' R_T.**

Having obtained E_T and R_T, we can calculate the current I_L flowing in the load from the equation

$$I_L = E_T/(R_T + R_L).$$

Worked Example 4.4

Determine the Thévenin equivalent circuit of the network between *A* and *B* in Figure 4.8, and calculate the current flowing in a resistor of (a) 5 Ω, (b) 10 Ω connected between *A* and *B*.

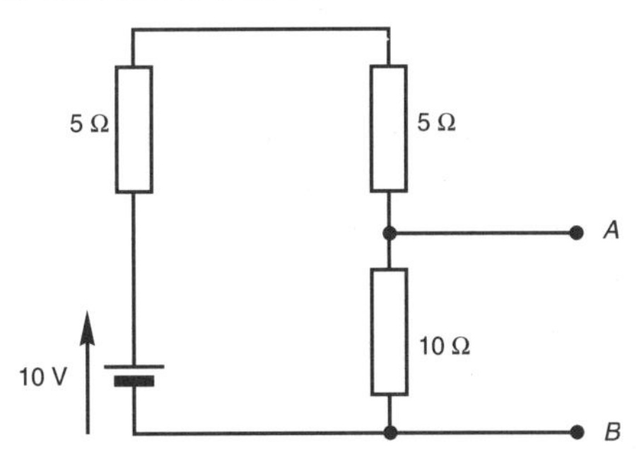

Figure 4.8 Circuit for Worked Example 4.4

Solution

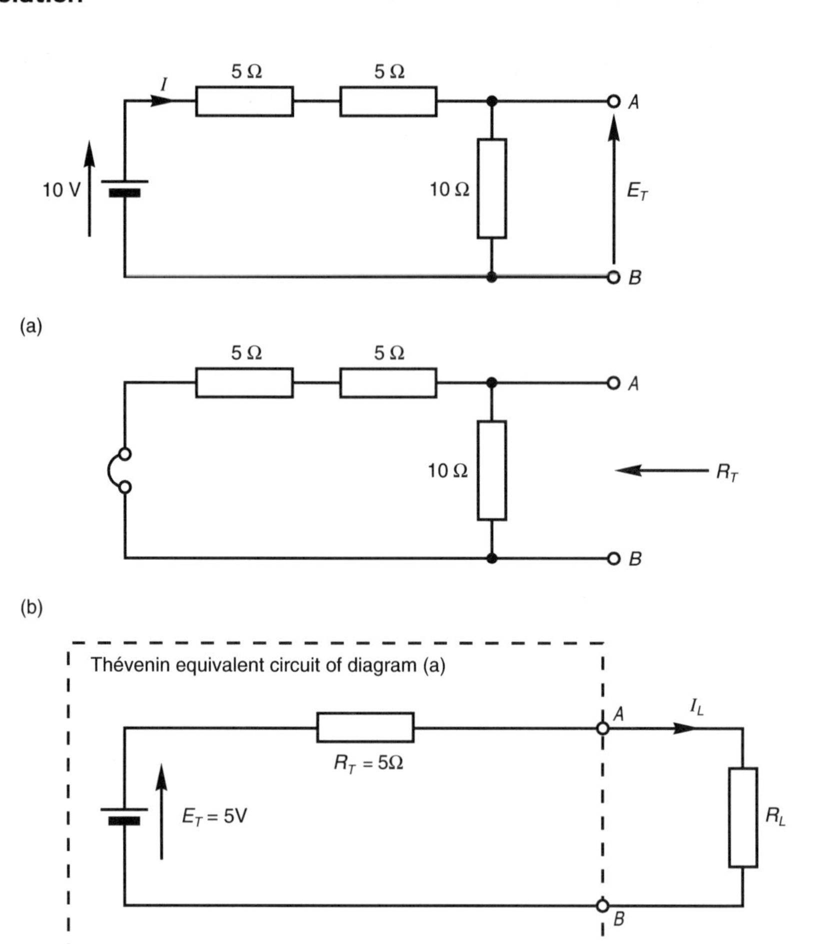

Figure 4.9 Solution of Worked Example 4.4: (a) re-drawn circuit; (b) calculation of values of R_T; (c) Thévenin equivalent circuit

The method used to determine the Thévenin equivalent circuit is shown in Figure 4.9. The circuit is re-drawn in Figure 4.9(a) with the load disconnected; the value of E_T is the voltage across the 10 Ω resistor. The current in the 10 Ω resistor is

$$I = 10/(5 + 5 + 10) = 0.5 \text{ A}$$

hence

$$E_T = E_{AB} = 10I = 5 \text{ V}$$

with terminal A being positive with respect to terminal B (this polarity *must be retained in the Thévenin equivalent circuit*).

The value of R_T is calculated as shown in Figure 4.9(b). With the external load being disconnected, and the battery replaced by its internal resistance (which is zero, since it is an ideal voltage source), the circuit in Figure 4.9(b) remains. The value of R_T is the resistance between the terminals A and B, which is equivalent to 10 Ω in parallel with (5 + 5) Ω, or

$$R_T = \frac{10 \times (5+5)}{10 + (5+5)} = 5\ \Omega$$

The Thévenin equivalent circuit of Figure 4.8 is shown enclosed in a broken line in Figure 4.9(c), together with the external load, R_L. The current in the load is given by

$$I_L = \frac{E_T}{R_T + R_L} + \frac{5}{5 + R_L}$$

(a) When $R_L = 5\ \Omega$ then

$$I_L = 5/(5+5) = 0.5\ \text{A}$$

(b) When $R_L = 10\ \Omega$

$$I_L = 5/(5+10) = 0.333\ \text{A}.$$

Worked Example 4.5

Determine the Thévenin equivalent circuit between terminals A and B for Figure 4.10. What current flows in a resistor of (a) 40 Ω, (b) 100 Ω connected between A and B?

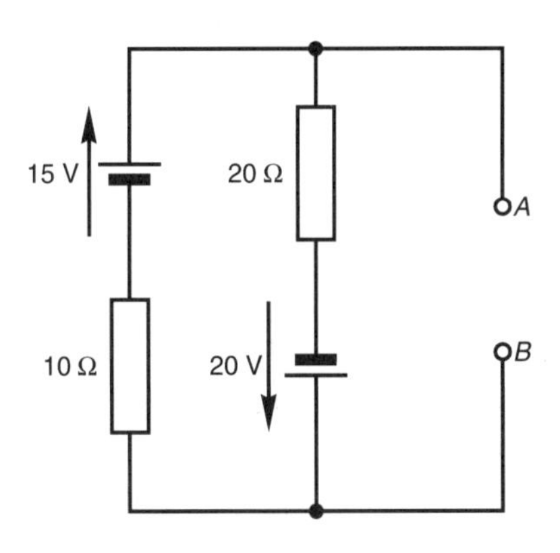

Figure 4.10 Circuit for Worked Example 4.5

Solution

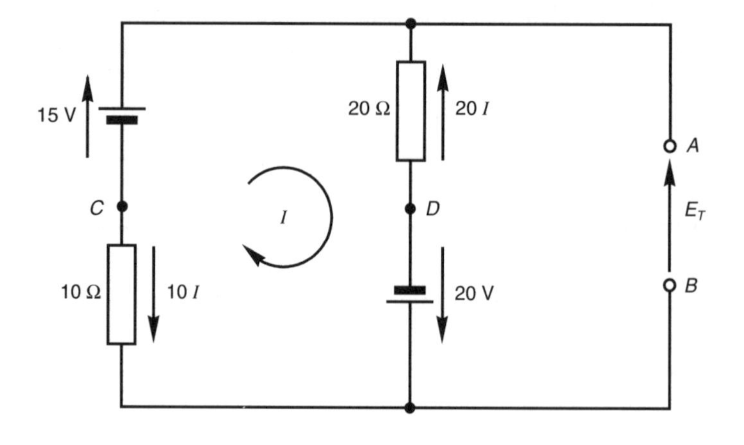

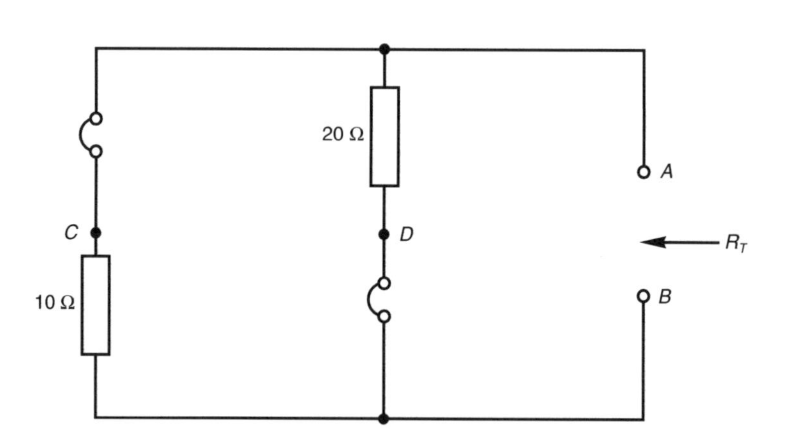

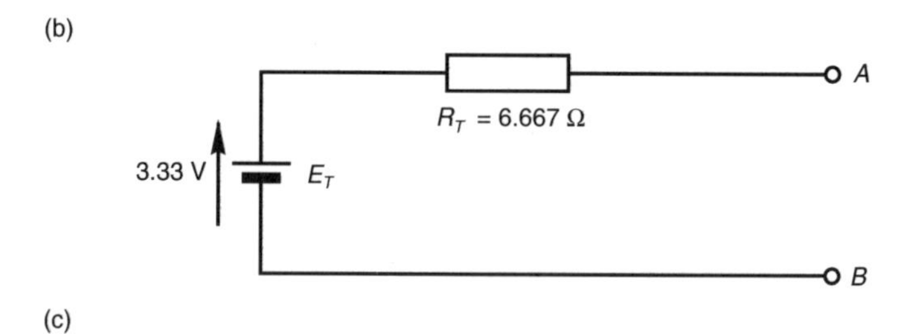

Figure 4.11 The Thévenin determination of (a) the Thévenin voltage, E_T (b) the Thévenin internal resistance, R_T (c) the Thévenin equivalent circuit of Figure 4.10

Initially we calculate the Thévenin voltage, E_T, as shown in Figure 4.11(a). Since the terminals A and B are initially open-circuited, the current, I, circulates between the two batteries, and is

$$I = \frac{(15 + 20)\ \text{V}}{(10 + 20)\ \Omega} = 1.167\ \text{A}$$

We can evaluate the Thévenin source voltage, $E_T = V_{AB}$, by determining the potential at node A relative to node B. To do this, we start at node B and move to node A *via any path* between the nodes. Nominating the path BCA we get

$$E_T = V_{AB} = -10I + 15 = -(10 \times 1.167) + 15 = 3.33\ \text{V}$$

The reader will find it an interesting exercise to obtain E_T using the path BDA.

Next we obtain the Thévenin source resistance as shown in Figure 4.11(b), where we replace each source by its internal resistance (which is zero, since we are dealing with ideal sources). The Thévenin source resistance is equal to the resistance between terminals A and B, which is 10 Ω in parallel with 20 Ω, so that

$$R_T = \frac{10 \times 20}{10 + 20} = 6.667\ \Omega$$

The resulting Thévenin equivalent circuit for Figure 4.10 is shown in Figure 4.11(c).

(a) When a 40 Ω resistor is connected between terminals A and B, the current in the resistor is

$$I = \frac{3.33}{6.667 + 40} = 0.0714\ \text{A or } 71.4\ \text{mA}$$

(b) When a 100 Ω resistor is connected between terminals A and B, the current in the resistor is

$$I = \frac{3.33}{6.667 + 100} = 0.0312\ \text{A or } 31.2\ \text{mA}$$

Whilst it is possible to solve for the value of the current in either one of the above cases using simultaneous equations obtained by KCL and KVL, it is certainly much quicker to determine the current for a number of load resistors using Thévenin's theorem.

Worked Example 4.6

Determine Thévenin's equivalent circuit between terminals B and D for the circuit in Figure 4.12. What value of current flows in R if it has a value of (a) 10 Ω, (b) 15 Ω?

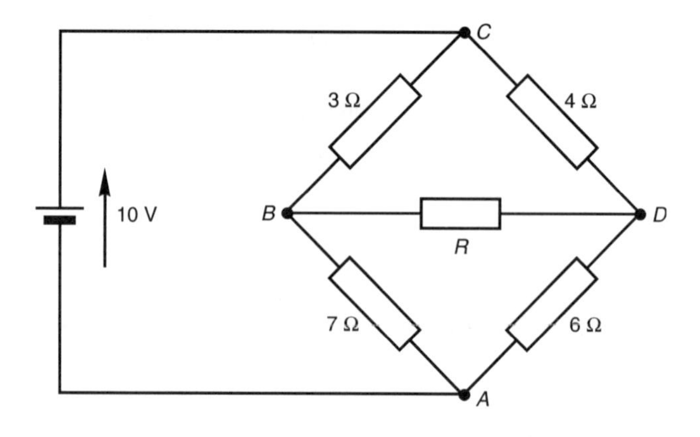

Figure 4.12 Circuit for Worked Example 4.6

Solution

The circuit is similar to a Wheatstone bridge configuration but, since it is not at balance, a current flows in resistor R.

To obtain the Thévenin source voltage, E_T, we remove resistor R from the circuit, and determine the voltage between nodes B and D. Since we do not know which of the nodes is the most positive, we will assume that node B is positive with respect to node D. That is

$$E_T = V_{BD}$$

Having removed resistor R, the circuit is redrawn in Figure 4.13; in the opinion of the author, this is a circuit which is relatively easy to work with compared with Figure 4.12. The current, I_1, in the 3 Ω and 7 Ω resistors is

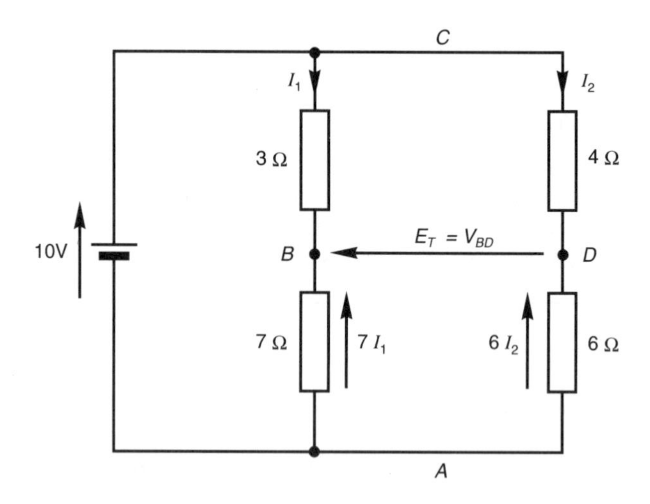

Figure 4.13 Calculation of E_T for Worked Example 4.6

$$I_1 = \frac{10\ \text{V}}{(3+7)\ \Omega} = 1\ \text{A}$$

and the current, I_2, in the 4 Ω and 6 Ω resistors is

$$I_2 = \frac{10\ \text{V}}{(4+6)\ \Omega} = 1\ \text{A}$$

The potential of node *B* with respect to node *A* is

$$V_{BA} = 7I_1 = 7 \times 1 = 7\ \text{V}$$

and the potential of node *D* with respect to node *A* is

$$V_{DA} = 6I_2 = 6 \times 1 = 6\ \text{V}$$

Applying KVL to the path *ABDA* gives

$$7I_1 - E_T - 6I_2 = 0$$

or

$$E_T = 7I_1 - 6I_2 = 7 - 6 = +1\ \text{V}$$

That is, the Thévenin source voltage is +1 V.

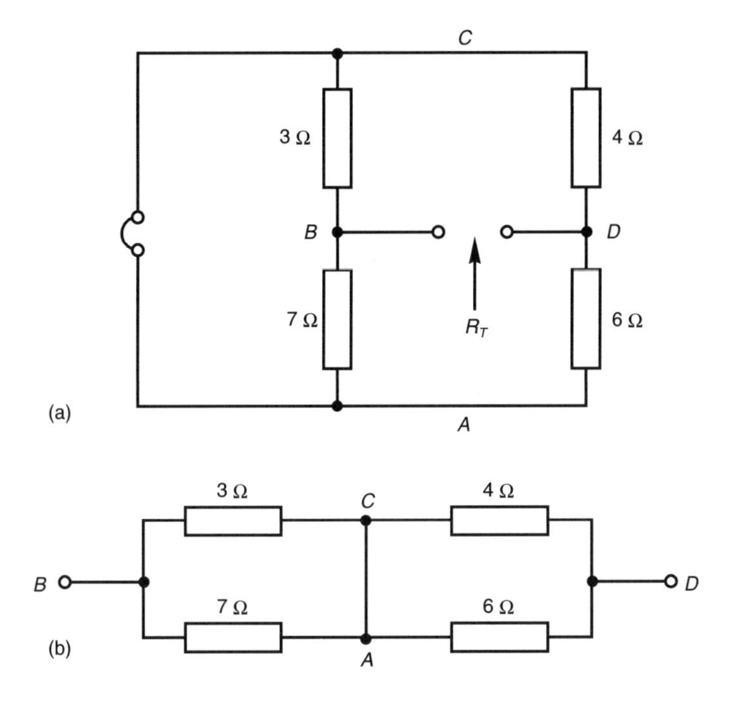

Figure 4.14 Determination of R_T for Worked Example 4.6: (a) Thévenin source resistance with resistance R removed; (b) re-drawn circuit

Next, we need to determine the Thévenin source resistance between the terminals *B* and *D* with resistance *R* removed, as shown in Figure 4.14(a). To do this, we replace the battery by its internal resistance (which we assume to be zero), and re-draw the circuit as shown in Figure 4.14(b). The value of R_T is

$$R_T = \frac{3 \times 7}{3 + 7} + \frac{4 \times 6}{4 + 6} = 2.1 + 2.4 = 4.5\ \Omega$$

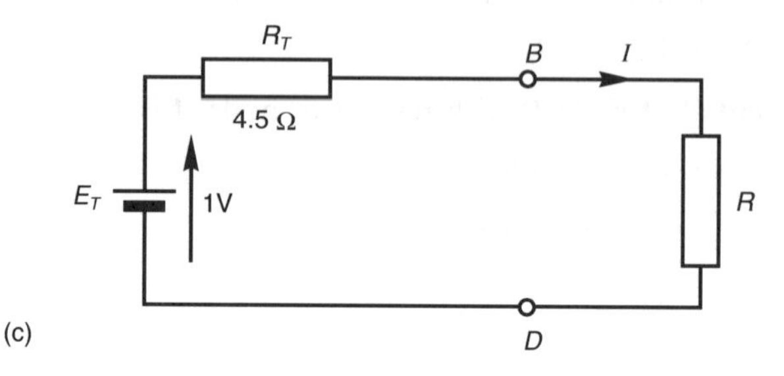

Figure 4.15 The Thévenin equivalent circuit between nodes B and D for the circuit in Figure 4.12

The complete circuit is shown in Figure 4.15.

(a) When $R = 10\ \Omega$, the current in R is

$$I = 1\ \text{V}/(4.5 + 10)\ \Omega = 0.069\ \text{A or } 69\ \text{mA}$$

(b) When $R = 15\ \Omega$, the current in R is

$$I = 1\ \text{V}/(4.5 + 15)\ \Omega = 0.0513\ \text{A or } 51.3\ \text{mA}$$

4.4 Norton's theorem

Whilst Thévenin's theorem says that any complex active network can be replaced by a constant voltage source in series with a source resistance, **Norton's theorem** says that **any complex active network can be replaced by an ideal current source which is shunted by an internal resistance or source resistance**.

At this stage we must ask ourselves 'what do we mean' by an 'ideal current source'? It is, in fact, a source which will 'drive' a constant current into any value of load resistance which can be connected to its terminals. The value of load resistance can range from a short-circuit (zero resistance) to an open-circuit (infinite resistance). In fact we are saying that a constant-current source is equivalent to a variable voltage source, the terminal voltage depending on the load resistance which is connected between its terminals. Since the resistance of the connected load is immaterial, it follows that the *internal resistance of an ideal current source in infinity*!

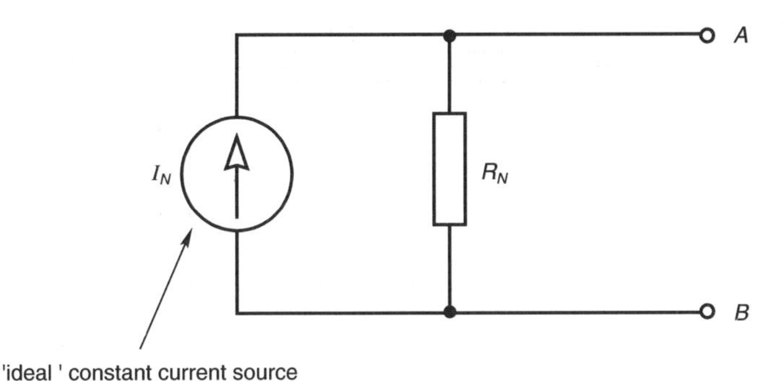

Figure 4.16 Circuit representation of a practical current source (the Norton equivalent circuit)

The circuit in Figure 4.7(a) can therefore be represented by the Norton equivalent circuit in Figure 4.16, in which R_N is the internal resistance of the 'practical' Norton source, and I_N is an 'ideal' Norton source. The arrow inside the Norton source points in the direction in which the current flows in the actual circuit.

Strictly speaking, Thévenin's theorem and Norton's theorem are two sides of the same coin, because we can represent any complex active network using either type of equivalent circuit. Which one we use depends on the application we are considering.

The value of I_N and R_N in the Norton equivalent circuit (see Figure 4.16) are determined for any circuit as follows:

1. **With any external load between terminals *A* and *B* disconnected, the magnitude of the Norton equivalent circuit current source, I_N, is equal to the current which flows in a short-circuit between terminals *A* and *B*.**
2. **With any external load between terminals *A* and *B* disconnected, and with all sources within the network being replaced by their internal resistance (zero in the case of an ideal voltage source, and infinity in the case of an ideal current source), the internal resistance of a 'practical' Norton source, R_N, is equal to the resistance measured between the terminals *A* and *B*.**

Worked Example 4.7

Determine the Norton equivalent circuit of the network in Figure 4.17(a). What current flows in a resistance of (a) 5 Ω, (b) 10 Ω connected between terminals *A* and *B*?

Solution

Initially we set about evaluating the Norton source current, I_N, and its source resistance, R_N, as shown in diagrams (b) and (c), respectively, of Figure 4.17.

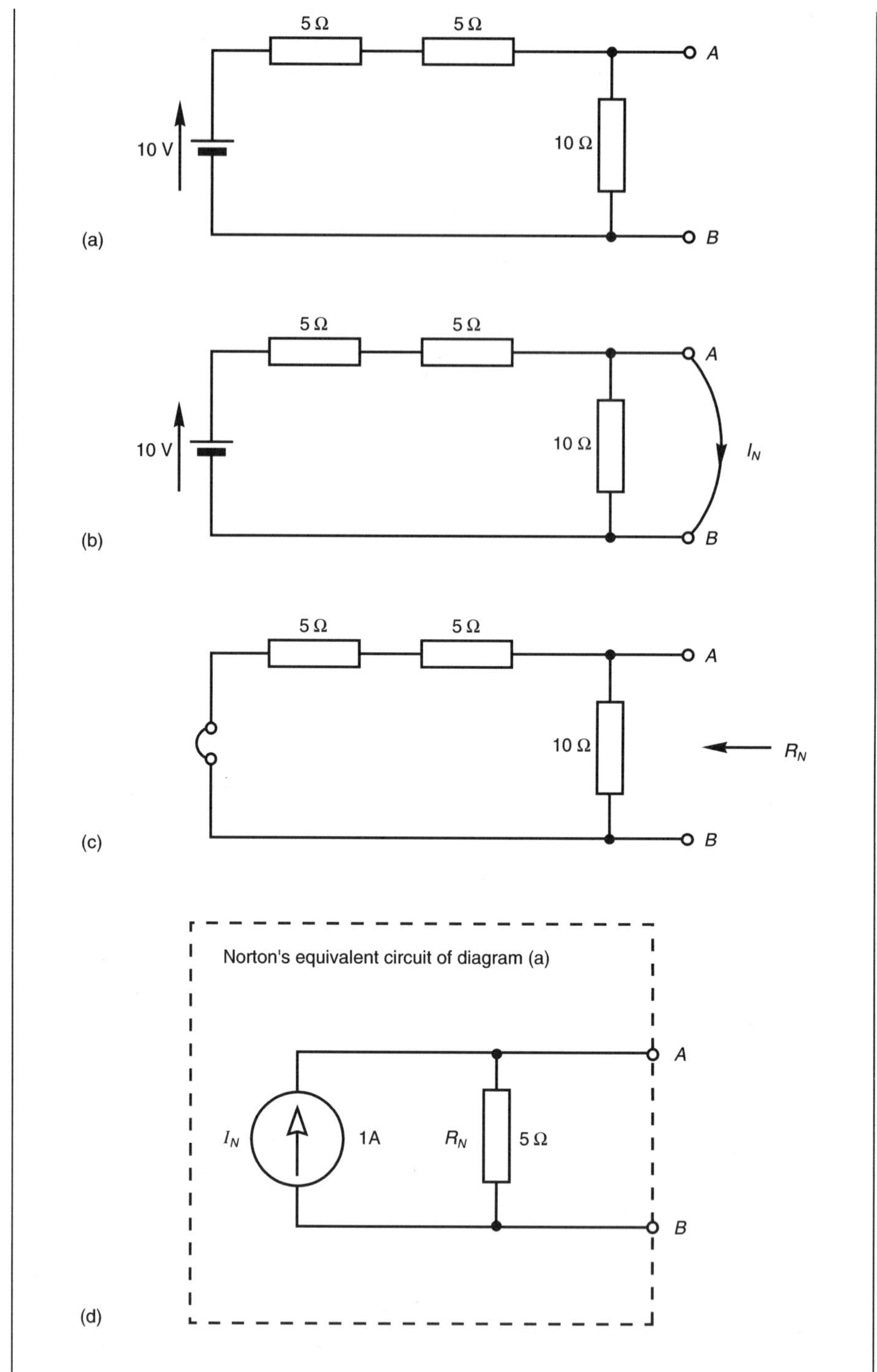

Figure 4.17 Circuit for Worked Example 4.7: (a) actual circuit; (b) determination of I_N; (c) evaluation of R_N; (d) Norton equivalent circuit of the actual circuit in diagram (a).

To determine the value of I_N, we simply short-circuit terminals A and B, and calculate the current flowing in the short-circuit. Since the 10 Ω resistance in diagram (b) is short-circuited, the value of I_N is

$$I_N = 10\ \text{V}/(5+5)\ \Omega = 1\text{A}$$

To determine R_N, the battery in the circuit is replaced by its internal resistance (which is zero), as shown in diagram (c) of Figure 4.17. The resistance between A and B is

$$\begin{aligned} R_N &= 10\ \Omega \text{ in parallel with } (5+5)\ \Omega \\ &= \frac{10 \times (5+5)}{10+(5+5)} = 5\ \Omega \end{aligned}$$

The Norton theorem equivalent circuit of Figure 4.17(a) is shown in Figure 4.17(d). When a load is connected to terminals A and B, the circuit appears as shown in Figure 4.18.

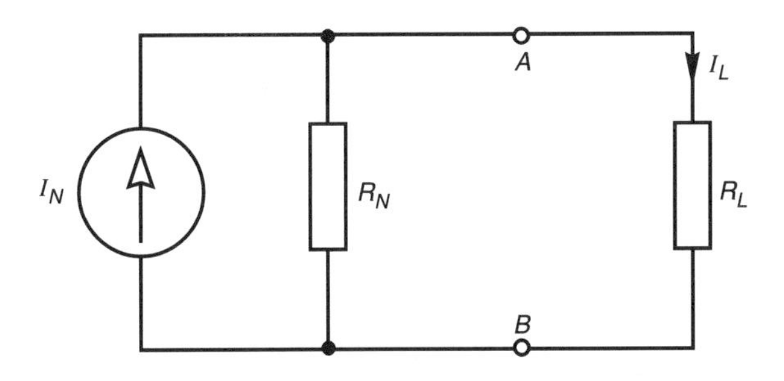

Figure 4.18 Determination of the current in resistor R_L

From what was said in section 2.3 of Chapter 2, the current I_N divides between R_N and R_L, as shown in Figure 4.18. The current in the load resistor is

$$I_L = \frac{I_N R_N}{R_N + R_L}$$

(a) When $R_L = 5\ \Omega$

$$I_L = 1 \times 5/(5+5) = 0.5\ \text{A}$$

(b) When $R_L = 10\ \Omega$, then

$$I_L = 1 \times 5/(5+10) = 0.333\ \text{A}$$

The reader should compare the results with those of parts (a) and (b) of Worked Example 4.4, where the same circuit was solved using Thévenin's theorem.

Worked Example 4.8

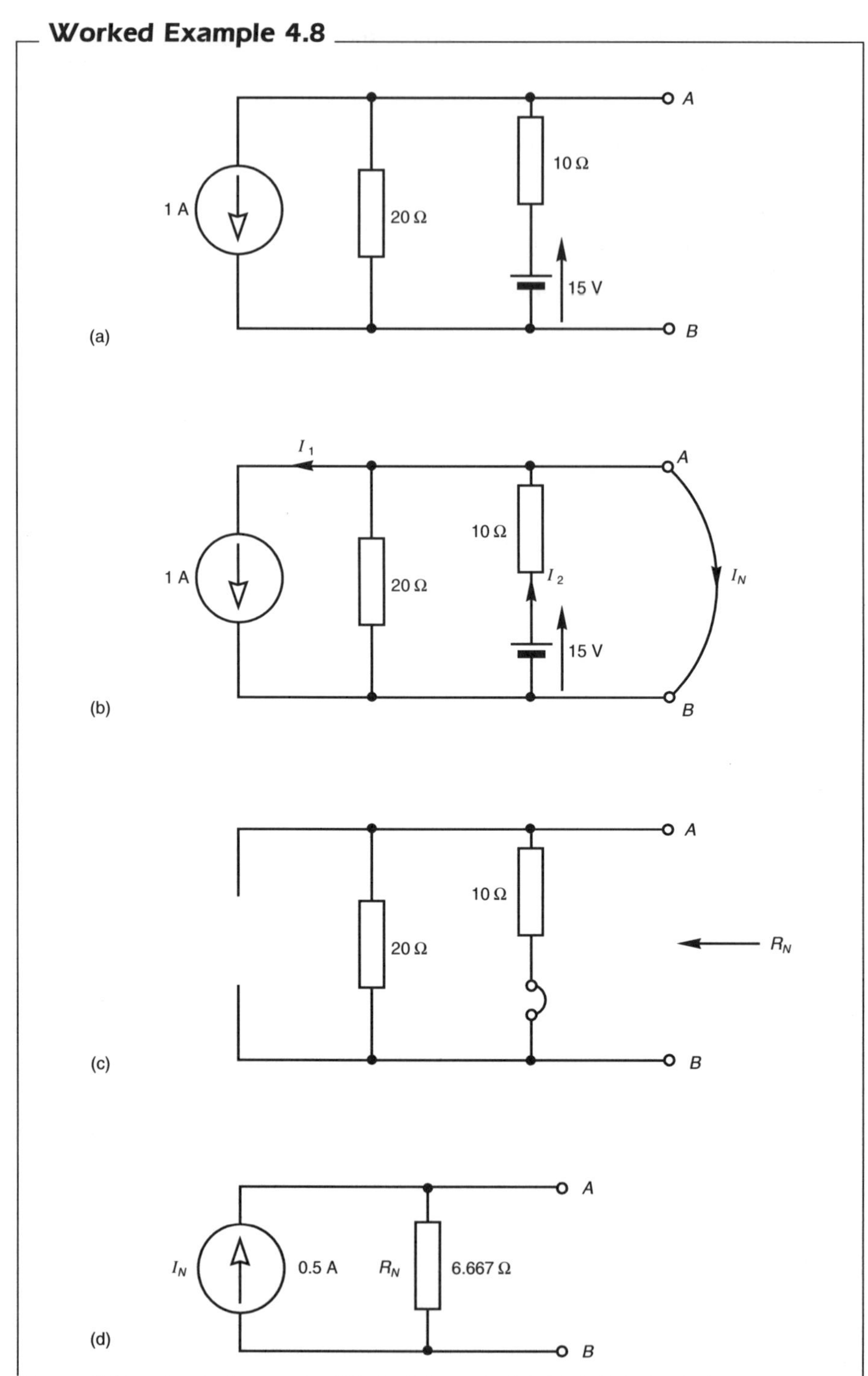

Figure 4.19 Circuit diagram for Worked Example 4.8: (a) initial circuit; (b) evaluation of I_N; (c) determination of R_N; (d) the Norton equivalent circuit of (a)

Determine the Norton equivalent circuit with respect to terminals A and B of Figure 4.19(a). If a load resistance of (a) 40 Ω and (b) 100 Ω is connected between A and B, calculate the current in the load.

Solution

This circuit is interesting because it includes not only a voltage source, but also a current source. Initially, we calculate the current I_N flowing in the short-circuit between terminals A and B in Figure 4.19(b). Since the 20 Ω resistor is short-circuited, it is not involved in the calculations here.

The current source causes a current of $I_1 = 1$ A to flow *away* from terminal A, and the 15 V source provides a current of $I_2 = 15\,\text{V}/10\,\Omega = 1.5$ A into terminal A. Applying KCL to node A in diagram (b) gives

$$I_2 = I_1 + I_N$$

or

$$I_N = I_2 - I_1 = 1.5 - 1 = 0.5\text{ A}$$

Next we determine R_N as shown in Figure 4.19(c). The ideal current source is replaced by its internal resistance (which is an open-circuit), and the ideal voltage source is replaced by its internal resistance (which is a short-circuit). The resistance between A and B is

$$\begin{aligned} R_N &= 10\Omega \text{ in parallel with } 20\Omega \\ &= 10 \times 20/(10 + 20) = 6.667\Omega \end{aligned}$$

The resulting Norton equivalent circuit for Figure 4.19(a) is shown in Figure 4.19(d).

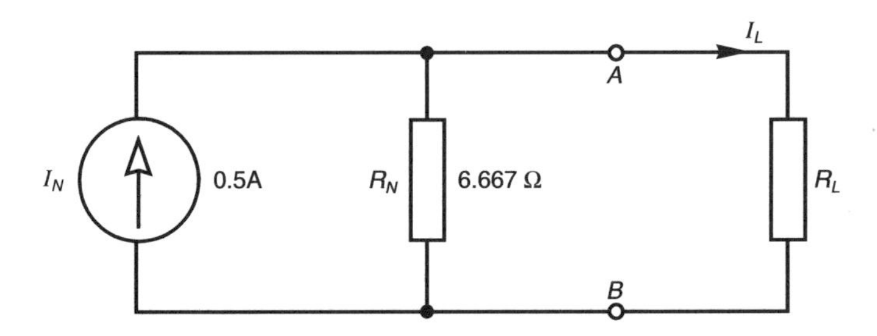

Figure 4.20 Circuit for the final solution of Worked Example 4.8

When a load resistor R_L is connected between terminals A and B, the resulting circuit is shown in Figure 4.20. From the work on current division in parallel circuits in Chapter 2, the current in R_L is calculated from

$$\begin{aligned} I_L &= 0.5 \times 6.667/(6.667 + R_L) \\ &= 3.3335/(6.667 + R_L) \end{aligned}$$

(a) When $R_L = 40\ \Omega$, then

$$I_L = 3.3335/(6.667 + 40) = 0.0714 \text{ A or } 71.4 \text{ mA}$$

(b) When $R_L = 100\ \Omega$, we have

$$I_L = 3.3335/(6.667 + 100) = 0.0312 \text{ A or } 31.2 \text{ mA}$$

Although it is not obvious at this stage, the circuit in Figure 4.18(a) is identical to the circuit in Worked Example 4.5 (Figure 4.9), so that answers (a) and (b) obtained above are the same as those obtained in Worked Example 4.5.

4.5 Conversion between Thévenin and Norton circuits

A Thévenin equivalent circuit has a direct equivalent Norton circuit, and vice versa. These are related by the following equations.

Thévenin to Norton conversion

$$I_N = \frac{E_T}{R_T} \quad \text{and} \quad R_N = R_T$$

Norton to Thévenin conversion

$$E_T = I_N R_N \quad \text{and} \quad R_T = R_N.$$

Worked Example 4.9

(a) Convert a Thévenin equivalent circuit for which $E_T = 5$ V and $R_T = 5\ \Omega$ into its equivalent Norton theorem version.
(b) Convert a Norton equivalent circuit for which $I_N = 0.5$ A and $R_N = 6.667\ \Omega$ into its Thévenin equivalent circuit.

Solution

(a) The value of the Norton current source is

$$I_N = E_T/R_T = 5 \text{ V}/5\ \Omega = 1 \text{ A}$$

and

$$R_N = R_T = 5\ \Omega$$

The results should be compared with those of Worked Examples 4.4 and 4.7.

(b) In this case

$$E_T = I_N R_N = 0.5 \times 6.667 = 3.333 \text{ V}$$

and

$$R_T = R_N = 6.667\ \Omega$$

These results should be compared with those of Worked Examples 4.5 and 4.8.

4.6 The maximum power transfer theorem

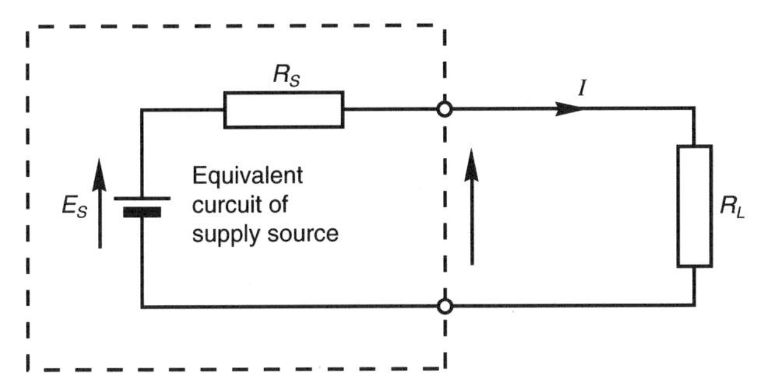

Figure 4.21 The maximum power transfer theorem

When a supply source of the type enclosed in broken lines in Figure 4.21 is connected to the load resistor, R_L, maximum power is transferred to the load when

$$R_L = R_S$$

That is, **maximum power is transferred to the load when the load resistance has the same value as the source resistance.**

Clearly, when R_L is zero (a short-circuit), no voltage appears across the load, and no power is dissipated in it! When R_L is infinite (an open-circuit), no current flows in the load, and no power is consumed by the load is zero.

The reader should note very carefully that, under conditions of maximum power transfer (when $R_L = R_S$), the voltage across the load is equal to the voltage dropped across the internal resistance, so that only one-half of the no-load voltage appears across the load! The reader should also note that, if we are dealing with a current source, maximum power is transferred to the load when the load resistance is equal to the internal resistance of the source; in this case, only one-half of the current produced by the source flows in the load under conditions of maximum power transfer.

Worked Example 4.10

A 10 V source has an internal resistance of 10 Ω. What load resistance causes maximum power transfer? Under maximum power transfer conditions, calculate the current in the load and the power dissipated in it.

For a load of (a) 8 Ω, (b) 12 Ω, determine the current in the load and the power dissipated.

Solution

For maximum power transfer

$$R_L = R_S = 10\ \Omega$$

The current flowing in the load under maximum power transfer conditions is

$$I = V_S/(R_S + R_L) = 10/(10 + 10) = 0.5\ \text{A}$$

and the power dissipated by the load at this time is

$$P = I^2 R_L = 0.52 \times 10 = 2.5\ \text{W}$$

(a) The current in a load of 8 Ω is

$$I = V_S/(R_S + R_L) = 10/(10 + 8) = 0.556\ \text{A}$$

and the power dissipated by the load is

$$P = I^2 R_L = 0.556^2 \times 8 = 2.47\ \text{W}$$

(b) The current in a load of 12 Ω is

$$I = V_S/(R_S + R_L) = 10/(10 + 12) = 0.455\ \text{A}$$

and the power dissipated by the load is

$$P = I^2 R_L = 0.455^2 \times 12 = 2.48\ \text{W}$$

4.7 Computer listings

Listing 4.1 Solution of two simultaneous equations

```
10 CLS '**** ch4-1 ****
20 PRINT "SOLUTION OF TWO SIMULTANEOUS EQUATIONS"
30 PRINT "OF THE FORM:": PRINT
40 PRINT "V1 = A*I1 + B*I2"
50 PRINT "V2 = C*I1 * D*I2": PRINT
60 PRINT "Where V1 and V2 are voltages (or e.m.f.s),"
70 PRINT "A, B, C and D are resistance values,"
80 PRINT "and I1 and I2 are currents.": PRINT
90 INPUT "V1 (volts) = ", V1
100 INPUT "A (ohms) = ", A
110 INPUT "B (ohms) = ", B: PRINT
120 INPUT "V2 (volts) = ", V2
```

```
130 INPUT "C (ohms) = ", C
140 INPUT "D (ohms) = ", D: PRINT
150 Det = (A * D) - (B * C)
160 REM **** There is no solution if DET = 0 ****
170 IF Det = 0 THEN PRINT "The equation cannot be solved.": END
180 DetI1 = (V1 * D) - (V2 * B): DetI2 = (A * V2) - (C * V1)
190 PRINT "I1 = "; DetI1 / Det; "A"
200 PRINT "I2 = "; DetI2 / Det; "A"
210 END
```

We look here at two listings. Computer listing 4.1 deals with the solution of two simultaneous equations of the form

$$V_1 = AI_1 + BI_2$$
$$V_2 = CI_1 + DI_2$$

where $V_1, V_2, A, B, C,$ and D are constants, and I_1 and I_2 are unknowns. The solution uses the method of **determinants** to solve the equations, which the reader can study in *Mastering Mathematics for Electrical and Electronic Engineers*, written by Noel M. Morris and published by Macmillan.

Lines 90 to 140 ask the user to supply the value of $V_1, V_2, A, B, C,$ and D, and lines 150 to 200 calculate the results. Under certain circumstances the equations cannot be solved and, when this occurs, line 170 terminates the program.

Listing 4.2 Relationship between Thévenin and Norton

```
10 CLS '**** CH4-2 ****
20 PRINT "RELATIONSHIP BETWEEN THEVENIN AND NORTON"
30 PRINT : VT = 0: RT = 0: IN = 0: RN = 0
40 PRINT "Choose ONE of the following:": PRINT
50 PRINT "1. Convert a Thevenin source to a Norton source."
60 PRINT "2. Convert a Norton source to a Thevenin source."
70 INPUT "Please enter your selection (1 or 2): ", S: PRINT
80 IF S < 1 OR S > 2 THEN GOTO 10
90 IF S = 2 THEN GOTO 150
100 PRINT "THEVENIN to NORTON source conversion:": PRINT
110 INPUT "Thevenin source voltage (volts) = ", VT
120 INPUT "Thevenin source resistance (ohms) = ", RT: PRINT
130 PRINT "Norton source current = "; VT / RT; "A"
140 PRINT "Norton source resistance = "; RT; "ohms": GOTO 200
150 PRINT "NORTON to THEVENIN source conversion:": PRINT
160 INPUT "Norton source current (A) = ", IN
170 INPUT "Norton source resistance (ohms) = ", RN: PRINT
180 PRINT "Thevenin source voltage = "; IN * RN; "A"
190 PRINT "Thevenin source resistance = "; RN; "ohms"
```

```
200 PRINT
210 PRINT "Do you wish to:": PRINT
220 PRINT "1. return to the main menu, or"
230 PRINT "2. exit from the program?": PRINT
240 INPUT "Enter your selection here: ", S
250 IF S < 1 OR S > 2 THEN GOTO 240
260 IF S = 1 THEN GOTO 10
270 END
```

Listing 4.2 enables the user to convert between a Thévenin equivalent circuit and a Norton equivalent circuit, or vice versa. Lines 10 to 90 allow you to select which of the two conversions you wish to make. Lines 100 to 140 convert a Thévenin equivalent circuit into a Norton equivalent circuit, and lines 150 to 190 convert a Norton equivalent circuit into a Thévenin equivalent circuit. Lines 200 to 260 give a choice either of repeating the program or of leaving the program.

Exercises

4.1 Determine the current flowing out of the positive pole of each battery in Figure 4.22. What is the potential of node B with respect to node A?

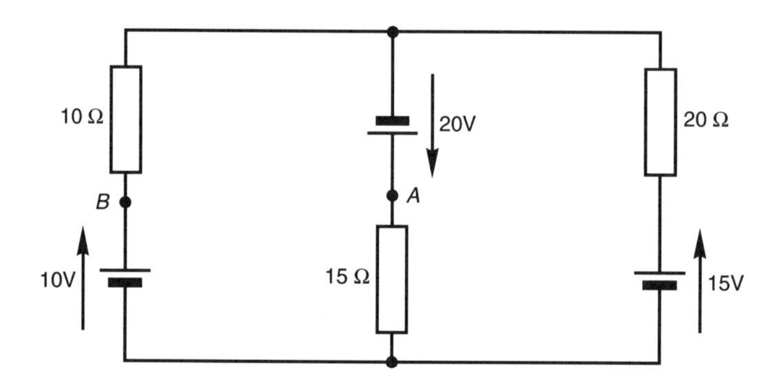

Figure 4.22 Circuit for Exercise 4.1

4.2 What current flows in the 30 Ω resistor in Figure 4.23, and what is the p.d. across the 20 Ω resistor?

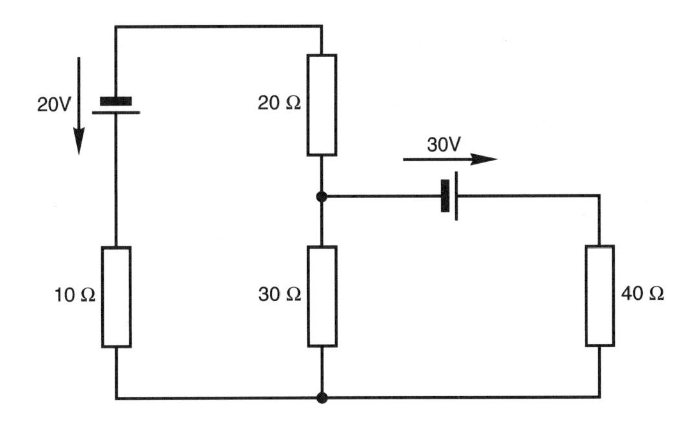

Figure 4.23 Circuit for Exercise 4.2

4.3 What is the potential of node A with respect to node B in Figure 4.24?

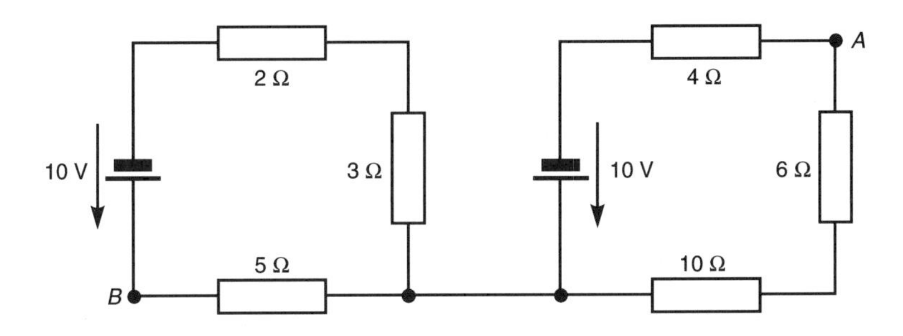

Figure 4.24 Circuit for Exercise 4.3

4.4 Determine the Thévenin equivalent circuit with respect to terminals A and B of the circuit in Figure 4.25.

What current flows in a resistor of (a) 15 Ω, (b) 30 Ω connected between A and B?

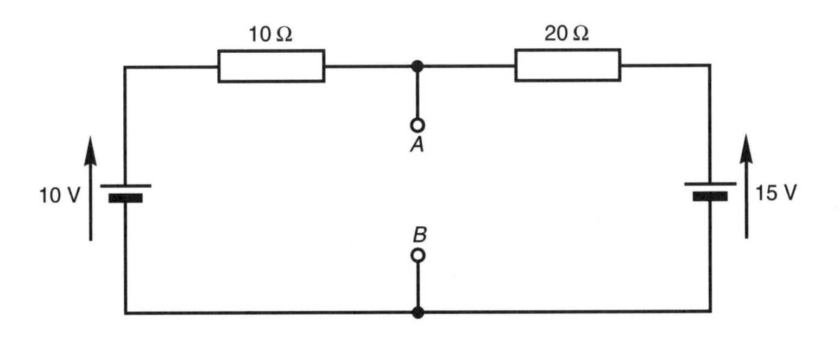

Figure 4.25 Circuit for Exercise 4.4

4.5 Deduce the Thévenin equivalent circuit with respect to terminals A and B of the circuit in Figure 4.26

What current flows in a resistor of (a) 2 Ω, (b) 3 Ω connected between A and B?

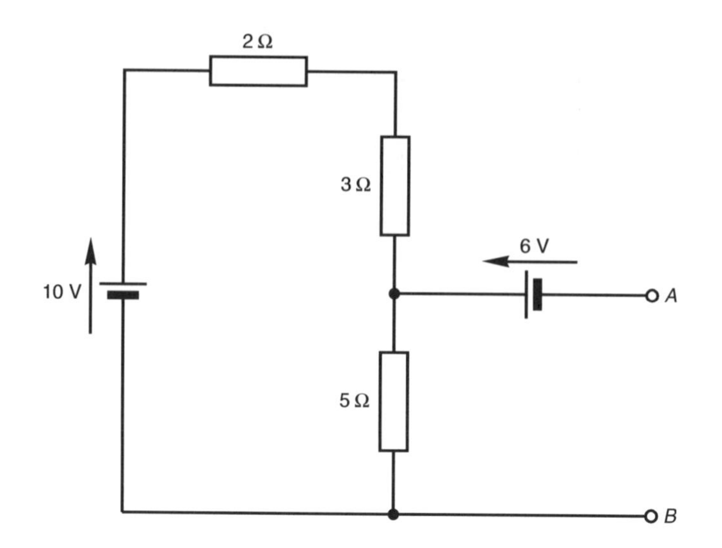

Figure 4.26 Circuit for Exercise 4.5

4.6 If the 10 Ω resistor in Figure 4.25 is replaced by a 20 Ω resistor, determine the Norton equivalent circuit with respect to terminals A and B. Use Norton's theorem to determine what current flows in a resistor of (a) 15 Ω, (b) 20 Ω connected between A and B.

4.7 If the 5 Ω resistor in Figure 4.26 is replaced by a 15 Ω resistor, determine the Norton equivalent circuit with respect to terminals A and B. What current flows in a resistor of (a) 5 Ω, (b) 10 Ω connected between A and B?

4.8 What resistance should be connected between terminals A and B in Exercises 4.4 to 4.7 in order that it consumes maximum power, and what power is consumed by the resistor in each case?

4.9 What resistance connected between A and B in Figure 4.22 consumes maximum power from the circuit?

Summary of important facts

A **network theorem** is the combination of circuit laws for the solution of a particular type of problem.

Kirchhoff's Current Law (KCL) states

1. **The current entering a node or in a circuit is equal to the current flowing away from the node**, or
2. **The sum of the current flowing towards any node in a circuit is zero.**

Kirchhoff's Voltage Law (KVL) states

1. **The algebraic sum of the e.m.fs and p.ds around any closed path is zero**, or
2. **In any closed path, the sum of the e.m.fs is equal to the sum of the p.ds.**

An **ideal voltage source** is one which will theoretically supply a constant voltage to any load. *The internal resistance or source resistance of an 'ideal' voltage source is zero.*

An **ideal current source** is one which will theoretically supply a constant current to any load. *The internal resistance or source resistance of an 'ideal' current source is infinity.*

Thévenin's and Norton's theorem refer to practical voltage and current sources, respectively, which have an internal resistance.

Thévenin's theorem states that any active network, no matter how complex, **can be replaced by an ideal voltage source, E_T, in series with a source resistance, R_T.**

The value of E_T is the voltage which appears between the terminals of the network when the load is disconnected. R_T is the resistance between the terminals of the network when the load is disconnected, and all sources within the network have been replaced by their internal resistance.

Norton's theorem states that any active network, no matter how complex, **can be replaced by an ideal current source, I_N, shunted by a source resistance, R_N.**

The value of I_N is the current which flows into a short-circuit between the terminals of the network. R_N is the resistance between the terminals of the network when the load is disconnected, and all sources within the network have been replaced by their internal resistance.

A Thévenin source and a Norton source are equivalent to one another if

$$E_T = I_N R_N \qquad \text{and} \qquad R_T = R_N$$

or

$$I_N = E_T / R_T \qquad \text{and} \qquad R_N = R_T$$

Maximum power is transferred from a source to a load *if the resistance of the load is equal to the internal resistance of the source.*

Capacitors and capacitor circuits

5.1 Introduction

The capacitor is one of the few electrical components which can store energy, and in this chapter we look at the way in which the capacitor is used in electrical and electronic circuits. By the end of this chapter, the reader will be able to

- Calculate the capacitance of a parallel-plate capacitor.
- Evaluate the electric field strength and electric flux density in a capacitor.
- Determine the charge and energy stored in a capacitor.
- Obtain the capacitance of series-connected, parallel-connected, and series-parallel connected capacitors.
- Determine the voltage distribution between series-connected capacitors.
- Use BASIC programs to deal with the problems in this chapter.

5.2 Capacitance of a parallel-plate capacitor

The **capacitance** of a two-plate parallel-plate capacitor is given by

$$C = \epsilon A/d$$

where ϵ is the *absolute permittivity of the dielectric* in farads per m (F/m), A is the *area of the dielectric* in square metres, and d is the *distance between the plates* (or thickness of the dielectric) in metres, and the capacitance is in farads (or an SI multiple of farads).

The absolute permittivity of the dielectric is also given by

$$\epsilon = \epsilon_0 \epsilon_r$$

where ϵ_0 is the *permittivity of free space*, which has the value 8.854×10^{-12} F/m, and ϵ_r is the relative permittivity of the dielectric material. For free space (a vacuum) the value of the latter is unity (which is also practically true for air), and for other materials its value is typically in the range from 2 (paper) to 10 (glass). Hence

$$C = \frac{\epsilon_0 \epsilon_r A}{d}$$

Practical capacitor values lie in the range from a few picofarads (pF or 10^{-12} F) to several microfarads (μF or 10^{-6} F).

Charge and energy is stored in the dielectric of the capacitor (**Note**: it is NOT stored in the plates), and the capacitance of an n-plate capacitor is

$$C = \frac{\epsilon_0 \epsilon_r (n-1) A}{d}.$$

Worked Example 5.1

A capacitor has 5 metal plates which are 0.35 mm apart, the dielectric being of bakelite which has a relative permittivity of 4.75. If the area of each plate is 200 cm^2, calculate the capacitance of the capacitor.

Solution

The capacitance of the capacitor is

$$\begin{aligned} C &= \frac{\epsilon_0 \epsilon_r (n-1) A}{d} \\ &= \frac{8.854 \times 10^{-12} \times 4.75 \times (5-1) \times 200 \times 10^{-4}}{0.35 \times 10^{-3}} \\ &= 9.613 \times 10^{-9}\ \text{F} = 0.009613 \times 10^{-6}\ \text{F} \\ &= 0.009613\ \mu\text{F or } 9.613\ \text{nF}. \end{aligned}$$

(**Note**: the reader should refer to Chapter 1 for information about changing SI dimensions from one size to another.)

Worked Example 5.2

A 0.2 μF capacitor has five plates. If the relative permittivity of the dielectric is 6.5, and the area of each plate is 1000 cm^2, calculate the relative permittivity of the dielectric.

Solution

From the equation $C = \epsilon_0 \epsilon_r (n-1) A/d$, we see that

$$\begin{aligned} d &= \epsilon_0 \epsilon_r (n-1) A/C \\ &= \frac{8.854 \times 10^{-12} \times 6.5 \times (5-1) \times 1000 \times 10^{-4}}{0.2 \times 10^{-6}} \\ &= 1.15 \times 10^{-4}\ \text{m} = 0.115 \times 10^{-3}\ \text{m} = 0.115\ \text{mm}. \end{aligned}$$

Worked Example 5.3

A capacitor has 5 plates, each of area 300 cm^2, which are separated by 0.3 mm. If its capacitance is 0.02 μF, determine the relative permittivity of the dielectric.

Solution

From the equation $C = \epsilon_0\epsilon_r(n-1)A/d$, we see that

$$\epsilon_r = Cd/(\epsilon_0(n-1)A)$$
$$= \frac{0.02 \times 10^{-6} \times 0.3 \times 10^{-3}}{8.854 \times 10^{-12} \times (5-1) \times 300 \times 10^{-4}}$$
$$= 5.65.$$

5.3 Electric field strength and electric flux density

The **electric field strength** (also known as the **electric stress**, the **electric field intensity**, and the **potential gradient**), symbol E, in a capacitor is given by

$$\boldsymbol{E = V/d} \text{ volts per metre (V/m)}$$

where V is the voltage across the dielectric, and d is the distance between the plates.

(**Note**: do not confuse the symbol E for electric field strength with E for e.m.f.)

The maximum field strength a dielectric can sustain without breakdown is known as the **electric strength** of the dielectric.

The **electric flux density** in a dielectric, symbol D, is given by

$$\boldsymbol{D = Q/A} \text{ coulomb per metre}^2 \text{ (C/m}^2\text{)}$$

where Q is the *charge stored by the capacitor* (see section 5.4), and A is the area of one plate of the capacitor.

The relationship between the flux density, D, and the electric field strength is

$$\boldsymbol{D = E\epsilon = E\epsilon_0\epsilon_r} \quad \text{C/m}^2.$$

Worked Example 5.4

If a voltage of 10 V is maintained between the plates of a capacitor, calculate the electric field intensity in the dielectric if the plates are separated by 0.1 mm.

If the relative permittivity of the dielectric is 6, determine the electric flux density in the dielectric, and the charge stored if the area of the dielectric is 500 cm^2.

Solution

From the equation $E = V/d$, we have

$$E = 10/0.1 \times 10^{-3} = 100\,000 \text{ V/m or } 100 \text{ kV/m}$$

Since

$$D = E\epsilon_0\epsilon_r = 100\,000 \times 8.85410^{-12} \times 6$$
$$= 5.31 \times 10^{-6} \text{ C/m}^2 \text{ or } 5.31\mu \text{ C/m}^2.$$

Now $D = Q/A$, or

$$Q = DA = 5.31 \times 10^{-6} \times 500 \times 10^{-4}$$
$$= 2.655 \times 10^{-7} \text{ coulomb } = 0.2655 \text{ μC}.$$

Worked Example 5.5

A capacitor has a total plate area of 2500 mm^2, and stores a charge of 0.025 μC. What is the electric flux density in the dielectric?

If the electric field strength in the dielectric is 200 kV/m, calculate the relative permittivity of the dielectric.

Solution

The plate area in m^2 is $A = 2500 \times 10^{-6}$ m^2, hence

$$D = Q/A = 0.025 \times 10^{-6}/2500 \times 10^{-6}$$
$$= 1 \times 10^{-5} \text{ C/m}^2 \text{ or } 0.01 \text{ mC/m}^2$$

Also since $D = E\epsilon_0\epsilon_r$, then

$$\epsilon_r = \frac{D}{E\epsilon_0} = \frac{1 \times 10^{-5}}{200\,000 \times 8.854 \times 10^{-12}}$$
$$= 5.64.$$

5.4 Charge and energy stores by a capacitor

The charge, Q coulombs, stored by a capacitor is

$$Q = CV$$

where C is the capacitance in farads, and V is the voltage across the capacitor in volts.

The energy, W joules, stored in a capacitor is

$$W = \frac{1}{2}CV^2$$

Worked Example 5.6

If a 0.5 μF capacitor stores 0.2 mJ of energy, what is its terminal voltage, and what charge is stored?

Solution

From the equation $W = \frac{1}{2}CV^2$, then

$$V = \surd(2W/C) = \surd(2 \times 0.2 \times 10^{-3}/0.5 \times 10^{-6})$$
$$= 28.28 \text{ V}$$

hence

$$Q = CV = 0.5 \times 10^{-6} \times 28.28$$
$$= 1.414 \times 10^{-5} \text{ coulombs or } 14.14 \text{ μC.}$$

Worked Example 5.7

When a voltage of 20 V is applied to a capacitor, it is found to store 0.05 J. Determine the capacitance of the capacitor.

Solution

Since $W = \frac{1}{2}CV^2$, then

$$C = 2W/V^2 = 2 \times 0.05/20^2 = 2.5 \times 10^{-4} \text{ F or } 250 \text{ μF}$$

5.5 Series-connected capacitors

The *reciprocal of the effective capacitance* of the series-connected capacitor circuit in Figure 5.1 is

$$\frac{1}{C_E} + \frac{1}{C_1} + \frac{1}{C_2} + \frac{1}{C_3} \cdots + \frac{1}{C_n}$$

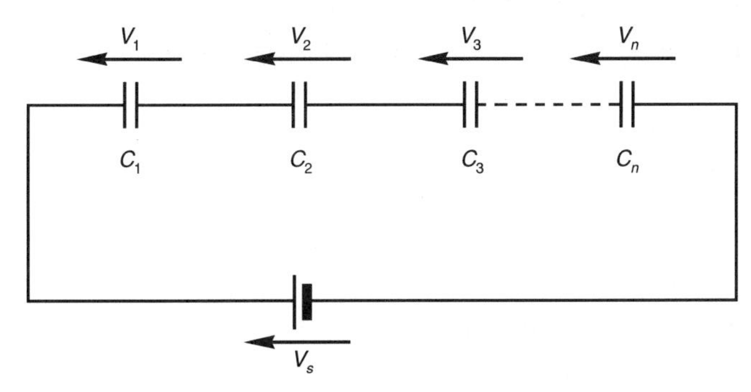

Figure 5.1 Series-connected capacitors

where C_E is the **effective capacitance** of the circuit (sometimes known as a *string of capacitors*), and C_n is the capacitance of the nth capacitor.

The capacitance of series-connected capacitors is always less than the lowest value of capacitance in the circuit.

In the special case of *two series-connected capacitors*, the effective capacitance is

$$C_E = C_1C_2/(C_1 + C_2)$$

If the voltage applied to the circuit containing n series-connected capacitors is V_S, the potential across the nth capacitor is

$$V_n = \frac{C_E V_S}{C_n}$$

that is

$$V_1 = C_E V_S/C_1$$
$$V_2 = C_E V_S/C_2, \text{etc.}$$

In the case of two series-connected capacitors, these equations become

$$V_1 = C_2 V_S/(C_1 + C_2)$$
$$V_2 = C_1 V_S/(C_1 + C_2)$$

The smallest value of capacitance in the circuit always supports the largest voltage across it.

Worked Example 5.8

Capacitors of 5.2 μF and 0.8 μF are connected in series. Determine the effective capacitance of the circuit.

Solution

The reciprocal of the effective capacitance is

$$\frac{1}{C_E} = \frac{1}{5.2 \times 10^{-6}} + \frac{1}{0.8 \times 10^{-6}} = 1.442 \times 10^6$$

hence

$$C_E = 1/1.442 \times 10^6 = 0.693 \times 10^{-6} \text{ F or } 0.693\ \mu\text{F}$$

Alternatively

$$\begin{aligned} C_E &= C_1 C_2/(C_1 + C_2) \\ &= \frac{5.2 \times 10^{-6} \times 0.8 \times 10^{-6}}{(5.2 + 0.8) \times 10^{-6}} \\ &= 0.693 \times 10^{-6} \text{ F or } 0.693\ \mu\text{F} \end{aligned}$$

It is of interest to note that the effective capacitance is less than the lower value of capacitance in the circuit.

Worked Example 5.9

If 10 V d.c. is applied to the series circuit in Worked Example 5.8, determine the voltage across each capacitor in the circuit, and the energy stored by each capacitor.

Solution

The voltage across the nth capacitor is

$$V_n = C_E V_S / C_n$$

In the case of C_1, the voltage is

$$\begin{aligned} V_1 &= C_E V_S / C_1 = 0.693 \times 10^{-6} \times 10/5.2 \times 10^{-6} \\ &= 1.33 \text{ V} \end{aligned}$$

and in the case of C_2, the voltage is

$$\begin{aligned} V_2 &= C_E V_S / C_2 = 0.693 \times 10^{-6} \times 10/0.8 \times 10^{-6} \\ &= 8.67 \text{ V} \end{aligned}$$

(**Note**: $V_1 + V_2 = 10 \text{ V} = V_S$)

Alternatively we see that

$$\begin{aligned} V_1 &= C_2 V_S/(C_1 + C_2) \\ &= 0.8 \times 10^{-6} \times 10/(5.2 + 0.8) \times 10^{-6} = 1.33 \text{ V} \end{aligned}$$

and

$$\begin{aligned} V_2 &= C_1 V_S/(C_1 + C_2) \\ &= 5.2 \times 10^{-6} \times 10/(5.2 + 0.8) \times 10^{-6} = 8.67 \text{ V} \end{aligned}$$

The energy stored by C_1 is

$$W_1 = \frac{1}{2}C_1{V_1}^2 = \frac{1}{2} \times 5.2 \times 10^{-6} \times 1.33^2$$
$$= 4.6 \times 10^{-6} \text{ J or } 4.6 \ \mu\text{J}$$

and the energy stored by C_2 is

$$W_2 = \frac{1}{2}C_2{V_2}^2 = \frac{1}{2} \times 0.8 \times 10^{-6} \times 8.67^2$$
$$= 30 \times 10^{-6} \text{ J or } 30 \ \mu\text{J}$$

The total energy stored by the circuit is

$$W_T = W_1 + W_2 = (4.6 + 30) \ \mu\text{J} = 34.6 \ \mu\text{J}$$

(**Note**:

$$W_T = \frac{1}{2}C_E{V_S}^2 = \frac{1}{2} \times 0.693 \times 10^{-6} \times 10^2$$
$$= 34.6 \times 10^{-6} \text{ J or } 34.6 \ \mu\text{J.)}$$

Worked Example 5.10

Capacitors of 1 μF, 2 μF and 4 μF are connected in series to a 20 V d.c. supply. Determine (a) the effective capacitance of the circuit, (b) the voltage across each capacitor, (c) the energy stored by the 4 μF capacitor, (d) the charge stored by each capacitor.

Solution

(a) The effective capacitance is calculated from

$$\frac{1}{C_E} = \frac{1}{C_1} + \frac{1}{C_2} + \frac{1}{C_3}$$
$$= \frac{1}{1 \times 10^{-6}} + \frac{1}{2 \times 10^{-6}} + \frac{1}{4 \times 10^{-6}} = 1.75 \times 10^6$$

or

$$C_E = 1/1.75 \times 10^6 = 0.571 \times 10^{-6}\text{F or } 0.571 \ \mu\text{F}$$

(b) The voltage across the capacitors are

$$V_1 = C_E V_S / C_1 = 0.571 \times 10^{-6} \times 20/1 \times 10^{-6} = 11.42 \text{ V}$$
$$V_2 = C_E V_S / C_2 = 0.571 \times 10^{-6} \times 20/2 \times 10^{-6} = 5.71 \text{ V}$$
$$V_3 = C_E V_S / C_3 = 0.571 \times 10^{-6} \times 20/4 \times 10^{-6} = 2.86 \text{ V}$$

(**Note**: $V_1 + V_2 + V_3 = 19.99$ V which is very close to 20 V.)

(c) The energy stored by the 4 μF capacitor is

$$W_3 = \frac{1}{2}CV_3{}^2 = \frac{1}{2} \times 4 \times 10^{-6} \times 2.862 = 16.36\ \mu J$$

(d) The charge stored by each capacitor is

$$Q_1 = C_1V_1 = 1 \times 10^{-6} \times 11.42$$
$$= 11.42 \times 10^{-6}\ C \text{ or } 11.42\ \mu C$$
$$Q_2 = C_2V_2 = 2 \times 10^{-6} \times 5.71$$
$$= 11.42 \times 10^{-6}\ C \text{ or } 11.42\ \mu C$$
$$Q_3 = C_3V_3 = 4 \times 10^{-6} \times 2.86$$
$$= 11.44 \times 10^{-6}\ C \text{ or } 11.44\ \mu C$$

The reader will note that the result of the stored charge is practically the same in each case! In fact, each result should be the same. The only reason for the slight difference is due to 'rounding' errors in the calculation.

When capacitors are connected in series, each capacitor stores the same charge, which is also equal to the total charge stored by the circuit.

The total charge stored by the circuit is

$$Q_T = C_EV_S = 0.571 \times 10^{-6} \times 20 = 11.42 \times 10^{-6}\ C \text{ or } 11.42\ \mu C$$

The reader will find it an interesting exercise to discuss why the total charge stored is equal to the charge stored by one capacitor in the circuit.

5.6 Parallel-connected capacitors

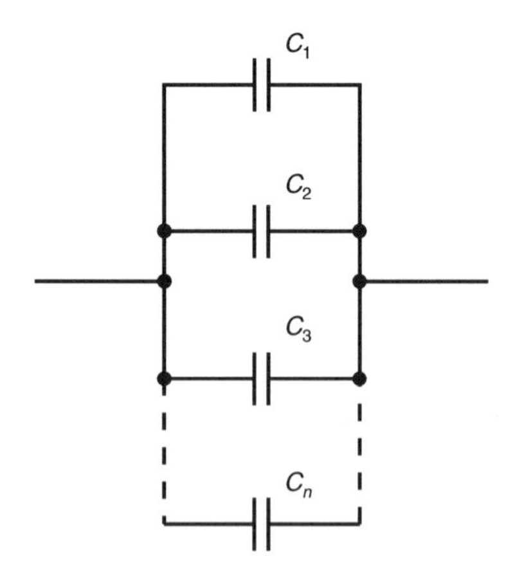

Figure 5.2 Parallel-connected capacitors

The *effective capacitance* of n capacitors connected in parallel (see Figure 5.2) is

$$C_E = C_1 + C_2 + C_3 + \ldots + C_n$$

where C_n is the capacitance of the nth capacitor.

Worked Example 5.11

A two-branch parallel capacitor circuit has a 0.8 μF capacitor in one branch and a 1.2 μF capacitor in the second branch, the supply voltage being 15 V. Determine (a) the effective capacitance of the circuit, (b) the charge stored by each capacitor and the total charge stored by the circuit, and (c) the energy stored by each capacitor and the total energy stored.

Solution

(a) The effective capacitance of the circuit is

$$\begin{aligned} C_E &= C_1 + C_2 \\ &= (0.8 \times 10^{-6}) + (1.2 \times 10^{-6}) \\ &= 2 \times 10^{-6} \text{ F or } 2\ \mu\text{F} \end{aligned}$$

(b) The charge stored by each capacitor is

$$Q_1 = C_1 V_1 = 0.8 \times 10^{-6} \times 15 = 12 \times 10^{-6} \text{ or } 12\ \mu\text{C}$$

$$Q_2 = C_2 V_2 = 1.2 \times 10^{-6} \times 15 = 18 \times 10^{-6} \text{ or } 18\ \mu\text{C}$$

and the total charge stored is

$$Q_T = C_E V_S = 2 \times 10^{-6} \times 15 = 30 \times 10^{-6} \text{ or } 30\ \mu\text{C}$$

Also

$$Q_T = Q_1 + Q_2 = (12 + 18) \times 10^{-6} = 30 \times 10^{-6} \text{ C or } 30\ \mu C$$

That is

$$\begin{aligned} Q_T &= Q_1 + Q_2 \\ &= \textbf{sum of the charge stored on all the capacitors} \end{aligned}$$

(c) The energy stored on each capacitor is

$$\begin{aligned} W_1 &= \frac{1}{2} C_1 V_1{}^2 = \frac{1}{2} \times 0.8 \times 10^{-6} \times 15^2 \\ &= 90 \times 10^{-6} \text{ J or } 90\ \mu\text{J} \end{aligned}$$

$$\begin{aligned} W_2 &= \frac{1}{2} C_2 V_2{}^2 = \frac{1}{2} \times 1.2 \times 10^{-6} \times 15^2 \\ &= 135 \times 10^{-6} \text{ J or } 135\ \mu\text{J} \end{aligned}$$

The total energy stored is

$$W_T = \frac{1}{2}C_E V_S{}^2 = \frac{1}{2} \times 2 \times 10^{-6} \times 15^2$$
$$= 225 \times 10^{-6} \text{ J or } 225\ \mu\text{J}$$

Also

$$W_1 + W_2 = (90 + 135) \times 10^{-6} = 225 \times 10^{-6} \text{ J or } 225\ \mu\text{J}$$

That is

$$W_T = W_1 + W_2$$
$$= \textbf{sum of the energy stored by all the capacitors.}$$

Worked Example 5.12

3 capacitors are connected in parallel to a 10 V d.c. supply. If the total energy stored is 5 μJ, and the capacitance of C_1 and C_2 are respectively 0.01 μF and 0.02 μF, determine the capacitance of the third capacitor.

Solution

The total energy stored by the circuit is

$$W_T = \frac{1}{2}C_E V_S{}^2$$

where C_E is the effective capacitance and V_S is the supply voltage, then

$$C_E = 2W_T/V_S{}^2 = 2 \times 5 \times 10^{-6}/10^2$$
$$= 0.1 \times 10^{-6} \text{ F or } 0.1\ \mu\text{F}$$

Now, for the parallel circuit

$$C_E = C_1 + C_2 + C_3$$

hence

$$C_3 = C_E - (C_1 + C_2) = (0.1 - (0.01 + 0.02)) \times 10^{-6}$$
$$= 0.07 \times 10^{-6} \text{ F or } 0.07\ \mu\text{F}$$

5.7 Series-parallel capacitor circuits

Every series-parallel capacitor circuit must be treated on its merits, and no two are alike. We look at several problems in this section.

For example, a series-parallel capacitor circuit, such as that in Figure 5.3, consists of a combination of series- and parallel-connected capacitors. In general the complete combination can be reduced either to a simple series circuit or a simple parallel circuit. However, we must note that if it can be reduced to a series-type circuit (as in the case of Figure 5.3), the charge stored by the complete circuit is equal to the charge on *any one* of the series elements (that is either C_3 in Figure 5.3, or the capacitor which is equivalent to the parallel combination of C_1 and C_2).

Worked Example 5.13

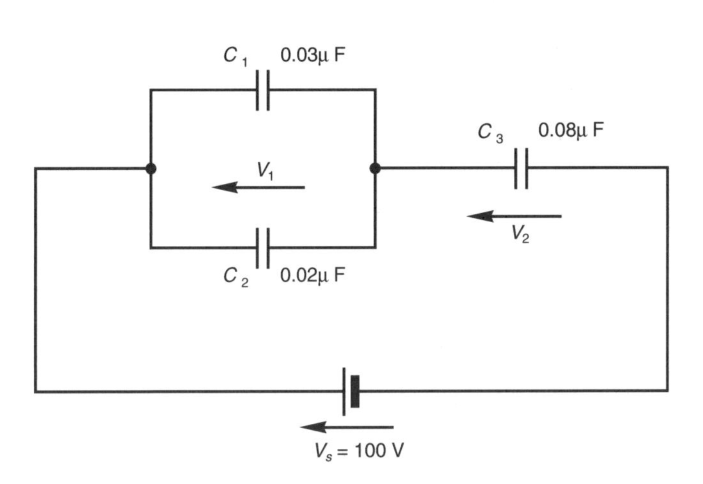

Figure 5.3 Circuit for Worked Example 5.13

For the circuit in Figure 5.3 determine (a) its effective capacitance, (b) the value of V_1 and V_2, (c) the charge stored by each capacitor and the total charge stored, (d) the energy stored by the complete circuit.

Solution

(a) The effective capacitance of the parallel section of the circuit is

$$C_P = C_1 + C_2 = (0.03 + 0.02) \times 10^{-6}$$
$$= 0.05 \times 10^{-6} \text{ F}$$

C_P is connected in series with C_3, and the effective capacitance of the complete circuit is calculated from

$$\frac{1}{C_E} = \frac{1}{C_P} + \frac{1}{C_3} = \frac{1}{0.05 \times 10^{-6}} + \frac{1}{0.08 \times 10^{-6}} = 32.5 \times 10^6$$

and the effective capacitance of the complete circuit is

$$C_E = 1/32.5 \times 10^6 = 0.0308 \times 10^{-6} \text{ F or } 0.0308 \text{ μF}$$

(b) From section 5.5, the voltage V_1 is given by

$$V_1 = C_E V_S / C_P = 0.0308 \times 10^{-6} \times 100/0.05 \times 10^{-6}$$
$$= 61.6 \text{ V}$$

and the voltage V_2 is

$$V_2 = C_E V_S / C_3 = 0.0308 \times 10^{-6} \times 100/0.08 \times 10^{-6}$$
$$= 38.5 \text{ V}$$

(**Note**: $V_1 + V_2 = 100.1$ V, which is within 0.1 per cent of the value of V_S.)

(c) The charge stored by each capacitor is

$$Q_1 = C_1 V_1 = 0.03 \times 10^{-6} \times 61.6$$
$$= 1.85 \times 10^{-6} \text{ C or } 1.85 \text{ µC}$$
$$Q_2 = C_2 V_1 = 0.02 \times 10^{-6} \times 61.6$$
$$= 1.23 \times 10^{-6} \text{ C or } 1.23 \text{ µC}$$
$$Q_3 = C_3 V_2 = 0.08 \times 10^{-6} \times 38.5$$
$$= 3.08 \times 10^{-6} \text{ C or } 3.08 \text{ µC}$$

Also

$$Q_T = C_E V_S = 0.0308 \times 10^{-6} \times 100$$
$$= 3.08 \times 10^{-6} \text{ C or } 3.08 \text{ µC}$$

(d) The total energy stored by the circuit is

$$W_T = \frac{1}{2} C_E {V_S}^2 = \frac{1}{2} \times 0.0308 \times 10^{-6} \times 100^2$$
$$= 154 \times 10^{-6} \text{ J or } 154 \text{ µJ}$$

Worked Example 5.14

For the circuit in Figure 5.4 calculate (a) the effective capacitance of the circuit, (b) the value of V_1 and V_2, (c) the charge stored by each capacitor and the total charge stored, (d) the energy stored by each capacitor and the total energy stored.

Solution

(a) The effective capacitance of the upper branch of Figure 5.4 is

$$C_S = C_1 C_2 / (C_1 + C_2)$$
$$= 5 \times 10^{-6} \times 3 \times 10^{-6} / (5 + 3) \times 10^{-6}$$
$$= 1.875 \times 10^{-6} \text{ F or } 1.875 \text{ µF}$$

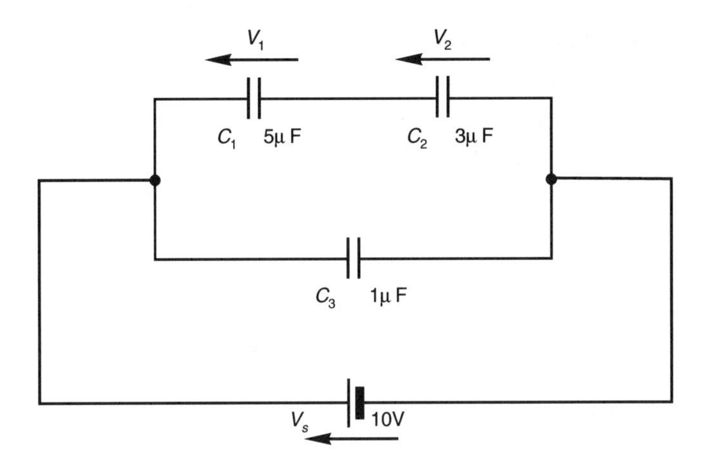

Figure 5.4 Circuit for Worked Example 5.14

This capacitance is connected in parallel with C_3, and the effective capacitance of the complete circuit is

$$\begin{aligned} C_E &= C_S + C_3 = (1.875 + 1) \times 10^{-6} \\ &= 2.875 \times 10^{-6} \text{ F or } 2.875 \text{ μF} \end{aligned}$$

(b) The value of V_1 is calculated from

$$V_1 = C_S V_S / C_1 = 1.875 \times 10^{-6} \times 10/5 \times 10^{-6} = 3.75 \text{ V}$$

and V_2 is

$$V_2 = C_S V_S / C_2 = 1.875 \times 10^{-6} \times 10/3 \times 10^{-6} = 6.25 \text{ V}$$

(**Note**: $V_1 + V_2 = 10 \text{ V} = V_S$)

(c) The charge stored by each capacitor is

$$\begin{aligned} Q_1 &= C_1 V_1 = 5 \times 10^{-6} \times 3.75 \\ &= 18.75 \times 10^{-6} \text{ C or } 18.75 \text{ μC} \\ Q_2 &= C_2 V_2 = 3 \times 10^{-6} \times 6.25 \\ &= 18.75 \times 10^{-6} \text{ C or } 18.75 \text{ μC} \\ Q_3 &= C_3 VS = 1 \times 10^{-6} \times 10 \\ &= 10 \times 10^{-6} \text{ C or } 10 \text{ μC} \end{aligned}$$

and the total charge stored by the circuit is

$$\begin{aligned} Q_T &= C_E V_S = 2.875 \times 10^{-6} \times 10 \\ &= 28.75 \times 10^{-6} \text{ C or } 28.75 \text{ μC} \end{aligned}$$

(**Note**: Since C_1 and C_2 are in series, this branch only stores 18.75 μC and, since this is in parallel with C_3, the total charge stores is

$$Q_T = 18.75 + 10 = 28.75\ \mu\text{C}.$$

(d) The energy stored in capacitors C_1, C_2 and C_3 is

$$W_1 = \frac{1}{2}C_1V_1{}^2 = \frac{1}{2} \times 5 \times 10^{-6} \times 3.75^2$$
$$= 35.16 \times 10^{-6}\ \text{J or } 35.16\ \mu\text{J}$$

$$W_2 = \frac{1}{2}C_2V_2{}^2 = \frac{1}{2} \times 3 \times 10^{-6} \times 6.25^2$$
$$= 58.59 \times 10^{-6}\ \text{J or } 58.59\ \mu\text{J}$$

$$W_3 = \frac{1}{2}C_3V_3{}^2 = \frac{1}{2} \times 1 \times 10^{-6} \times 10^2$$
$$= 50 \times 10^{-6}\ \text{J or } 50\ \mu\text{J}$$

and the total energy stored is

$$W_T = \frac{1}{2}C_EV_S{}^2 = \frac{1}{2} \times 2.875 \times 10^{-6} \times 10^2$$
$$= 143.75 \times 10^{-6}\ \text{J or } 143.75\ \mu\text{J}$$

(**Note**: $W_1 + W_2 + W_3 = (35.16 + 58.59 + 50)\ \mu\text{J} = 143.75\ \mu\text{J}$.)

Worked Example 5.15

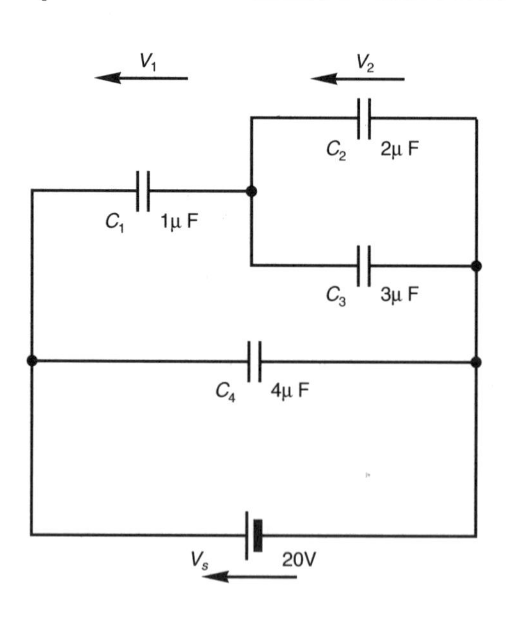

Figure 5.5 Circuit for Worked Example 5.15

For the circuit in Figure 5.5 calculate (a) the effective capacitance of the circuit, (b) the value of V_1 and V_2, (c) the charge and energy stored by C_2.

Solution

(a) Initially we calculate the effective capacitance of C_2 and C_3 in parallel, which is

$$C_{P1} = C_2 + C_3 = (2 + 3) \times 10^{-6} = 5 \times 10^{-6}\ \text{F}$$

Since the combination is in series with C_1, the effective capacitance of the series-parallel combination in the top branch is

$$\begin{aligned} C_S &= C_1 C_{P1}/(C_1 + C_{P1}) \\ &= 1 \times 10^{-6} \times 5 \times 10^{-6}/(1 + 5) \times 10^{-6} \\ &= 0.833 \times 10^{-6}\ \text{F or } 0.833\ \mu\text{F} \end{aligned}$$

The upper branch is connected in parallel with C_4, so that the effective capacitance of the circuit is

$$\begin{aligned} C_E &= C_S + C_4 = (0.833 + 4) \times 10^{-6} \\ &= 4.833 \times 10^{-6}\ \text{F or } 4.833\ \mu\text{F} \end{aligned}$$

(b) Since $V_S = 20$ V is applied to the upper branch, the value of V_1 and V_2 is

$$\begin{aligned} V_1 &= C_S V_S / C_1 = 0.833 \times 10^{-6} \times 20/1 \times 10^{-6} = 16.66\ \text{V} \\ V_2 &= C_S V_S / C_{P1} = 0.833 \times 10^{-6} \times 20/5 \times 10^{-6} = 3.33\ \text{V} \end{aligned}$$

(**Note**: $V_1 + V_2 = 19.99\ \text{V} \approx V_S$. The reader should also note that, since the capacitance of C_1 is less than C_{P1}, C_1 supports the largest voltage.)

(c) The charge stored by C_2 is

$$\begin{aligned} Q_2 &= C_2 V_2 = 2 \times 10^{-6} \times 3.33 \\ &= 6.66 \times 10^{-6}\ \text{C or } 6.66\ \mu\text{C} \end{aligned}$$

and the energy stored is

$$\begin{aligned} W_2 &= \frac{1}{2} C_2 {V_2}^2 = \frac{1}{2} \times 2 \times 10^{-6} \times 3.33^2 \\ &= 11.09 \times 10^{-6}\ \text{J or } 11.09\ \mu\text{J} \end{aligned}$$

5.8 Computer listings

Listing 5.1 Capacitance of a parallel-plate capacitor, and charge and energy stored in a capacitor

```
10 CLS '**** ch5-1 ****
20 PRINT "CAPACITANCE OF A PARALLEL-PLATE CAPACITOR,"
30 PRINT "CHARGE AND ENERGY STORED IN A CAPACITOR."
40 PRINT
50 PRINT "Select one of the following:": PRINT
60 PRINT "1. Capacitance of a parallel-plate capacitor."
70 PRINT "2. Charge and energy stored in a capacitor.": PRINT
80 INPUT "Enter your selection here: ", S
90 IF S < 1 OR S > 2 THEN GOTO 10
100 CLS
110 IF S = 2 THEN GOTO 270
120 PRINT "The equation is:": PRINT
130 PRINT "C = Er * Eo * (N - 1) * A/D: PRINT"
140 PRINT "    where Er = relative permittivity,"
150 PRINT "    Eo = permittivity of free space"
160 PRINT "       = 8.854E-12,"
170 PRINT "    N = number of plates,"
180 PRINT "    A = area of one plate (m^2)"
190 PRINT "    D = distance between plates (m).": PRINT
200 INPUT "Er = ", Er
210 INPUT "N = ", N
220 INPUT "A (m^2) = ", A
230 INPUT "D (m) = ", D: PRINT
240 PRINT "C = "; Er * 8.854E-12 * (N - 1) * A / D; "farads"
250 PRINT "  = "; Er * 8.854E-06 * (N - 1) * A / D; "microfarads"
260 GOTO 330
270 PRINT "The equations used are"
280 PRINT "Q = C*V and W = C*V^2/2": PRINT
290 INPUT "C (farads) = ", C
300 INPUT "V (volts) = ", V: PRINT
310 PRINT "Q = "; C * V; "coulombs"
320 PRINT "W = "; C * V ^ 2 / 2; "joules": PRINT
330 PRINT : PRINT "Do you wish to:": PRINT
340 PRINT "1. return to the main menu, or"
350 PRINT "2. exit from the program?:": PRINT
360 INPUT "Enter your selection here (1 or 2): ", S
370 IF S < 1 OR S > 2 THEN GOTO 360
380 IF S = 1 THEN GOTO 10
390 END
```

Three listings are included in this section. Computer listing 5.1 offers two options; the first is to calculate the capacitance of an n-plate parallel-plate capacitor, and the second determines the charge and energy stored by a capacitor. Both options use equations given in this chapter, and can be used either to calculate the answer to a specific problem, or to check an answer which has been solved by hand.

Listing 5.2 Capacitance of series- and parallel-connected capacitors

```
10 CLS '**** CH5-2 ****
20 PRINT "CAPACITANCE OF CAPACITOR NETWORKS"
30 PRINT : Ze = 0: Ce = 0
40 DIM C(5): FOR N = 0 TO 5: C(N) = 0: NEXT N
50 PRINT "Select ONE of the following.:": PRINT
60 PRINT "1. Determine the equivalent capacitance"
70 PRINT "   of series-connected capacitors."
80 PRINT "2. Determine the equivalent capacitance"
90 PRINT "   of parallel-connected capacitors.": PRINT
100 INPUT "Enter your selection here: ", S
110 IF S < 1 OR S > 2 THEN GOTO 10
120 IF S = 2 THEN GOTO 250
130 CLS
140 INPUT "Number of capacitors in series (max. 5) = ", N
150 IF N < 2 OR N > 5 THEN GOTO 130
160 PRINT
170 FOR Y = 1 TO N
180     PRINT "For capacitor "; Y
190     INPUT " capacitance (F) = ", C(Y)
200     Ze = Ze + 1 / C(Y)
210 NEXT Y
220 PRINT
230 PRINT "Equivalent capacitance = "; 1 / Ze; "F"
240 GOTO 360
250 CLS
260 INPUT "Number of capacitors in parallel (max. 5) = ", N
270 IF N < 2 OR N > 5 THEN GOTO 250
280 PRINT
290 FOR Y = 1 TO N
300     PRINT "For capacitor "; Y
310     INPUT " capacitance (F) = ", C(Y)
320     Ce = Ce + C(Y)
330 NEXT Y
340 PRINT
350 PRINT "Equivalent capacitance = "; Ce; "F"
360 PRINT : PRINT "Do you wish to:": PRINT
370 PRINT "1. return to the main menu, or"
380 PRINT "2. leave the program?: PRINT"
390 INPUT "Enter your selection here (1 or 2): ", S
```

```
400 IF S < 1 OR S > 2 THEN GOTO 390
410 IF S = 1 THEN GOTO 10
420 END
```

Computer listing 5.2 enables you to determine the equivalent capacitance of up to 5 capacitors which are either connected in series or in parallel. The capacitance of a series-parallel circuit can be determined using a combination of the options in this listing.

Listing 5.3 Voltage distribution between series-connected capacitors

```
10 CLS '**** CH5-3 ****
20 PRINT "VOLTAGE DISTRIBUTION BETWEEN"
30 PRINT "SERIES-CONNECTED CAPACITORS"
40 DIM C(5): FOR N = 0 TO 5: C(N) = 0: NEXT N
50 PRINT : Ze = 0
60 INPUT "Number of capacitors in series (max. 5) = ", N
70 IF N < 2 OR N > 5 THEN GOTO 10
80 FOR Y = 1 TO N
90     PRINT "For capacitor "; Y
100    INPUT " capacitance (F) = ", C(Y)
110    Ze = Ze + 1 / C(Y)
120 NEXT Y
130 PRINT : Ce = 1 / Ze
140 INPUT "Voltage (d.c.) across capacitors = ", V: PRINT
150 PRINT "Effective capacitance = "; Ce; "F"
160 PRINT "                      = "; Ce * 1000000; "microfarads"
170 PRINT
180 FOR Y = 1 TO N
190     PRINT "Voltage across C"; Y; " = "; V * Ce / C(Y); "V"
200 NEXT Y
210 END
```

Finally, listing 5.3 allows you to evaluate the voltage distribution in a series-connected string of up to five capacitors.

Exercises

5.1 The plates of a 5-plate capacitor are separated by a dielectric of relative permittivity 7, which is 0.4 mm thick. If the area of each plate is 250 cm^2, what is the capacitance of the capacitor?

5.2 A capacitor of capacitance 20.7 nF has a dielectric of relative permittivity 6.5. If it has 7 plates which are separated by 0.5 mm, what is the area of each plate?

5.3 A 0.33 μF capacitor has an effective plate area of 1200 cm^2, and employs a dielectric with a relative permittivity of 8. If the capacitor has 9 plates, what is the thickness of the dielectric?

5.4 Determine the relative permittivity of the dielectric of a capacitor of capacitance 0.2 μF, having 7 plates of total area 900 cm^2 which are separated by 0.12 mm.

5.5 Determine the electric flux density in a dielectric of area 400 cm^2 which stores a charge of 0.5 μC.

5.6 What is the potential gradient in a dielectric 0.2 mm thick which has 15 V applied to it?

5.7 Calculate the relative permittivity of a dielectric having an electric field strength of 90 kV/m, and an electric flux density of 5 $\mu C/m^2$.

5.8 A 10μF capacitor stores 0.02 J. Determine the voltage between the terminals of the capacitor, and the charge it stores.

5.9 Capacitors of 0.1μF and 0.02 μF are connected (a) in series, (b) in parallel. In each case calculate (i) the effective capacitance of the circuit, (ii) the energy stored when 10 V is applied to the circuit.

5.10 Determine the effective capacitance of a circuit containing capacitors of 0.1 μF, 100 nF and 0.3 μF which are connected (a) in series, (b) in parallel.

5.11 What charge and energy is stored by each of the circuits in Problem 5.10 when 10 V is applied to them?

5.12 A capacitance of 0.01 μF is connected in series with a parallel circuit comprising a 0.02 μF and a 0.04 μF capacitor. What is the effective capacitance of the circuit?

5.13 What voltage appears across the 0.01 μF capacitor in Problem 5.12 when 10 V is applied to the circuit? What charge is stored by the 0.02 μF capacitor?

5.14 Two capacitors are connected in series to a 10 V supply. One of the capacitors has a capacitance of 0.04 μF, and supports 2 V between its terminals. What is the capacitance of the other capacitor, and what is the equivalent capacitance of the circuit?

5.15 Three capacitors are connected in series to a 100 V supply, the equivalent capacitance of the circuit being 57.14 nF. If the voltage across C_1 is 57.14 V, and its capacitance is 0.1 μF, what charge is stored by each capacitor, and what voltage appears across each of the other capacitors?

5.16 If, in Problem 5.15, the energy stored by the largest value of capacitance is 40.84 μJ, what is the capacitance of each capacitor?

5.17 If 50 V d.c. is applied to the circuit in Problem 5.12, what voltage appears (a) across the 0.01 μF capacitor, (b) across the parallel circuit? What is the final value of energy stored by the complete circuit?

Summary of important facts

The **capacitance** of an n-plate parallel-plate capacitor is given by

$$C = \frac{\epsilon(n-1)A}{d} = \frac{\epsilon_0\epsilon_r(n-1)A}{d}$$

where

C is the capacitance in farads (F)
ϵ is the absolute permittivity of the dielectric (F/m)
ϵ_0 is the permittivity of free space (8.854×10^{-12} F/m)
ϵ_r is the relative permittivity of the dielectric (dimensionless)
A is the cross-sectional area of the dielectric (m^2)
d is the distance between the plates.

The **electric field strength** (or **potential gradient**) in the dielectric of a capacitor is

$$\mathbf{E} = \mathbf{V}/\mathbf{d} \text{ volts per metre}$$

where V is the potential between the plates, and d is the distance between them.

The **charge**, Q coulombs, stored by a capacitor is

$$Q = CV$$

where C is in farads and V is in volts.

The **electric flux density**, D, in a dielectric is

$$D = Q/A \text{ coulombs/m}^2$$

where Q is the charge stored in coulombs, and A is the area of the dielectric in m^2. Also

$$D = \epsilon E = \epsilon_0\epsilon_r E$$

where E is the potential gradient in the dielectric.

The **energy**, W, stored by a capacitor is

$$W = \frac{1}{2}CV^2 \text{ joules}$$

where C is in farads and V is in volts.

The reciprocal of the equivalent capacitance of n series-connected capacitors is given by

$$\frac{1}{C_E} = \frac{1}{C_1} + \frac{1}{C_2} + \ldots + \frac{1}{C_n}$$

In the *special case of two series-connected capacitors*

$$C_E = \frac{C_1 C_2}{C_1 + C_2}$$

The equivalent capacitance of a series circuit is always less than the lowest individual value of capacitance in the circuit.

If V_S is the d.c. voltage applied to n capacitors in series, the voltage V_n across the nth capacitor is

$$V_n = C_E V_S / C_n$$

that is

$$V_1 = C_E V_S / C_1$$

$$V_2 = C_E V_S / C_2, \text{ etc.}$$

In the special case of two series-connected capacitors

$$V_1 = V_S C_2 / (C_1 + C_2)$$

$$V_2 = V_S C_1 / (C_1 + C_2)$$

The smallest value of capacitance in a series circuit supports the highest voltage across it.

The **effective capacitance of a parallel-connected circuit** is

$$C_E = C_1 + C_2 + \ldots + C_n.$$

Inductance and mutual inductance

6.1 Introduction

The inductor is one of the basic building blocks upon which electrical and electronic circuits are founded, and form a springboard to important branches of the subject. The reader will gain confidence in dealing with problems relating to inductive circuits as he/she reads through this chapter. By the end of this chapter, the reader will be able to

- Understand the units involved in inductive calculations.
- Calculate the value of the e.m.f. induced in a conductor.
- Determine the self-inductance of a coil.
- Evaluate the energy stored in a magnetic circuit.
- Determine the inductance of series-connected and parallel-connected inductive circuits.
- Understand mutual inductance.
- Calculate the effective inductance of series-connected inductive circuits having mutual inductance between the inductors.
- Use BASIC software to solve the equations in this chapter.

6.2 Inductance

When current flows in a conductor it produces a **magnetic flux**. The name given to this property is **self-inductance** (symbol L); the unit of inductance is the **henry** (H). *When one ampere of current flows in a coil of inductance one henry, it produces a magnetic flux of one* **Weber** (Wb).

The greater the number of turns on the coil, the greater the magnetic flux produced.

Conversely, *when the magnetic flux linking with a conductor changes, an e.m.f. is induced in it.*

It is the above property which allows us to construct and use electrical generators, motors and transformers. A circuit symbol for an inductor is shown in Figure 6.1.

Figure 6.1 Symbol for an inductor

The equation relating inductance to the change in flux and the current is

$$\boldsymbol{L} = \textbf{number of turns} \times \frac{\textbf{change in flux}}{\textbf{change in current}}$$

$$= \boldsymbol{N\frac{d\Phi}{dI}}$$

In engineering, we refer to the product

number of turns $\times$ flux

as the **number of flux linkages**, hence

$$\boldsymbol{L} = \frac{\textbf{change in flux linkages}}{\textbf{change in current}}$$

Another relationship which is important in inductive circuits is

$$\boldsymbol{L} = \frac{(\text{number of turns})^2}{\text{reluctance of the magnetic circuit}}$$

$$= \frac{\boldsymbol{N^2}}{\boldsymbol{S}} = \frac{\boldsymbol{N^2}}{\ell/(\mu_r\mu_0 \boldsymbol{a})}$$

where

S = reluctance of the magnetic circuit
μ_r = relative permeability of the magnetic circuit
μ_0 = absolute permeability of free space
= $4\pi \times 10^{-7}$ henry per metre.

The reluctance of magnetic circuits, and calculations using it, is investigated in detail in Chapter 7.

The reader should note here is that the above relationships are correct only for a coil which does not have an iron core, i.e. a coil with an air core. In a coil with an iron core, the flux and current do not change proportionally.

Worked Example 6.1

If a current of 5 A causes a 2000 turn coil to produce a flux of 8 μWb, what is the inductance of the coil?

Solution

The solution is obtained from the equation

$$L = N\,d\Phi/dI$$
$$= 2000 \times (8 \times 10^{-6})/5 = 3.2 \times 10^{-3} \text{ H or } 3.2 \text{ mH.}$$

Worked Example 6.2

A current of 2.5 A flows in the 1200-turn winding of an air-cored coil. If the inductance of the coil is 0.5 H, determine the magnetic flux produced by the coil.

Solution

From the equation

$$L = N\,d\Phi/dI$$

we see that

$$N\,d\Phi = L\,dI$$

or

$$d\Phi = L\,dI/N = 0.5 \times 2.5/1200$$
$$= 1.042 \times 10^{-3} \text{ Wb or } 1.042 \text{ mWb}$$

Worked Example 6.3

A coil of 2500 turns is uniformly wound on an air core of mean diameter 35 cm and cross-sectional area of 5 cm^2. Determine the self-inductance of the coil.

Solution

We will first convert the essential dimensions in SI units as follows:

Diameter of ring $= 35 \text{ cm} = 0.35 \text{ m}$

Cross-sectional area of magnetic circuit $= 5 \text{ cm}^2 = 5 \times 10^{-4} \text{ m}^2$

The reluctance, S, of the magnetic circuit is

$$S = \ell/(\mu_r \mu_0 A) = \pi \times 0.35/(1 \times (4\pi \times 10^{-7}) \times 5 \times 10^{-4})$$
$$= 1.75 \times 10^9 \text{ ampere turns/Wb}$$

The inductance of the coil is

$$L = N^2/S = 2500^2/1.75 \times 10^9$$
$$= 3.57 \times 10^{-3} \text{ H or } 3.57 \text{ mH.}$$

6.3 e.m.f. induced in an inductor

There are two basic equations which give the e.m.f. induced in an inductor, and the first of these is

$$\boldsymbol{e = N \times \textbf{ rate of change of flux } = N\, d\Phi/dt}$$

where

e is the induced e.m.f.

N is the number of turns on the coil

$d\Phi$ is the change in magnetic flux in Wb which occurs in dt seconds.

The second equation is

$$\boldsymbol{e = L \times \textbf{ rate of change of current } = L\frac{di}{dt}}$$

where

L is the *self-inductance* of the coil in henrys (H)

di is the change in current (A) which occurs in dt seconds.

Worked Example 6.4

The magnetic flux linking the 2000-turn coil of an electromagnet changes uniformly from 0.7 mWb to 0.8 mWb in 50 ms. Determine the average value of the e.m.f. induced in the coil.

Solution

In this case

$$N = 2000 \text{ turns}$$
$$d\Phi = 0.8 - 0.7 = 0.1 \text{ mWb} = 0.1 \times 10^{-3} \text{ Wb}$$
$$dt = 50 \text{ ms} = 50 \times 10^{-3} \text{ s}$$

hence

$$e = N = \frac{d\Phi}{dt} = 2000 \times \frac{0.1 \times 10^{-3}}{50 \times 10^{-3}} = 4 \text{ V}.$$

Worked Example 6.5

If the average voltage induced in a coil of 1500 turns is 3 V when the flux linking it changes by 0.2 mWb, determine the time taken for the change of flux to occur.

Solution

From the equation

$$e = Nd\Phi/dt$$

we see that

$$e\,dt = N\,d\Phi$$

or $$dt = N\,d\Phi/e = 1500 \times 0.2 \times 10^{-3}/3 = 0.1 \text{ s}$$

Worked Example 6.6

When the flux linking with a coil changes by 4 mWb in 144 ms, it is found that the average value of the induced e.m.f. in the coil is 25 V. How many turns of wire are there on the coil?

Solution

From the equation

$$e = N\,d\Phi/dt$$

then $$N = e\,dt/d\Phi = 25 \times (144 \times 10^{-3})/4 \times 10^{-3} = 900$$

(**Note**: in practice, every coil has an *exact number of turns*. However, the result of a practical experiment may appear to show that the coil does not have an exact number of turns, and this may be due to many factors, including the accuracy of the instrument, the accuracy of reading, leakage flux, etc.)

6.4 Energy stored in the magnetic field of an inductor

When current flows in an inductor, energy is stored in its magnetic field. When the current reduces in value, energy is returned from the magnetic field to the circuit. The energy, W, stored in the magnetic field is

$$W = \frac{1}{2} LI^2$$

If L is in henrys and I is in amperes, then W is in joules.

Worked Example 6.7

The inductance of each field coil of a four-pole d.c. machine is 2.25 H, the coils being connected in series. What energy is stored in the magnetic field system when the field current is 1.5 A?

Solution

Since $W = \frac{1}{2}LI^2$, then the energy stored is

$$W = \frac{1}{2} \times (4 \times 2.25) \times 1.5^2 = 10.125 \text{ J}$$

Worked Example 6.8

If a 0.2 H inductor stores 10 J in its magnetic field, calculate the current flowing in the field coil.

Solution

From the equation $W = \frac{1}{2}LI^2$, it follows that

$$I^2 = 2W/L$$

or

$$I = \sqrt{(2W/L)} = \sqrt{(2 \times 10/0.2)} = 10 \text{ A.}$$

6.5 Mutual inductance

When magnetic flux links with a wire or a coil, an e.m.f. is induced in it. If the magnetic flux is produced by another coil or circuit, the e.m.f. is said to be **mutually induced**, and a state of *mutual inductance* or *mutual coupling* exists between the two circuits. It is possible for more than two circuits to be mutually coupled.

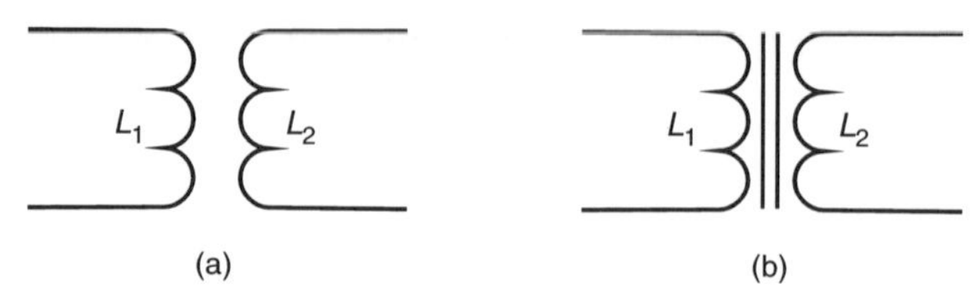

Figure 6.2 Mutually coupled coils with: (a) an air core; (b) an iron core

Symbols representing coils which are mutually coupled on (a) an air core and (b) an iron core are shown in Figure 6.2 (a) and (b), respectively. The current which produces the magnetic flux is known as the **primary current**, and flows in the **primary winding**. The coil which has an e.m.f. induced in it is the **secondary winding**, the induced e.m.f. being the **secondary induced e.m.f.**

If only a relatively small proportion of the flux which leaves the primary winding reaches the secondary winding, the coils are said to be **loosely coupled**. If the majority of the flux leaving the primary winding links with the secondary winding, as in the case of a *power transformer*, the coils are said to be **closely coupled** or **tightly coupled**.

The **mutual inductance**, M, between two coils of inductance L_1 and L_2, respectively, is given by

$$\boldsymbol{M = k\sqrt{(L_1 L_2)}}$$

where k is the **magnetic coupling coefficient**, whose value lies between zero (loose coupling) and unity (close coupling). Also

$$M = \frac{\text{change in flux linkages with the secondary}}{\text{change in primary current}}$$

$$= N_2 \times \frac{\text{change in flux linking with the secondary}}{\text{change in primary current}}$$

$$= N_2 \times \frac{k\,d\Phi_2}{dI_1}$$

and

$$M = N_1 \times \frac{k\,d\Phi_1}{dI_2}$$

Worked Example 6.9

Two coils, A and B, are wound on an iron core, and have perfect coupling between them. If the self-inductance of A is 0.5 H, and the mutual inductance between A and B is 0.671 H, determine the self-inductance of coil B.

Assume that there is no magnetic leakage, and the coupling coefficient is unity.

Solution

Applying the above values into the equation $M = k\sqrt{(L_A L_B)}$, we get

$$0.671 = 1 \times \sqrt{(0.5 \times L_B)}$$

hence

$$0.671^2 = 0.5 \times L_B$$

or

$$L_B = 0.671^2/0.5 = 0.9 \text{ H}$$

Worked Example 6.10

If the self-inductance of L_1 is 0.2 H, and that of L_2 is 0.25 H, determine the coefficient of coupling between them if the mutual inductance is 111.8 mH.

Solution

From the equation

$$M = k\sqrt{(L_1 L_2)}$$

then

$$k = M/\sqrt{(L_1 L_2)} = 111.8 \times 10^{-3}/\sqrt{(0.2 \times 0.25)}$$
$$= 0.5$$

Worked Example 6.11

Two identical 1200-turn coils are wound so that 60 per cent of the flux produced by one coil links with the other coil. A current of 6 A in either coil produced a flux of 0.25 mWb in its own coil. Determine (a) the self-inductance of each coil, (b) the mutual inductance between the coils.

Solution

(a) For coil L_1, the self-inductance is given by

$$L_1 = N_1 \frac{d\Phi}{dI} = 1200 \times \frac{0.25 \times 10^{-3}}{6}$$
$$= 0.05 \text{ H or } 50 \text{ mH}$$

Also, L_2 has the same inductance.

(b) Since 60 per cent of the flux links one coil to the other, then

$$M = k\sqrt{(L_1 L_2)} = 0.6 \times \sqrt{(50 \times 10^{-3} \times 50 \times 10^{-3})}$$
$$= 0.03 \text{ H or } 30 \text{ mH.}$$

Worked Example 6.12

A current of 5 A flows in a coil of 800 turns, and produces a flux of 10 mWb. If the current is reduced to zero in a time of 10 ms, determine the average value of the e.m.f. induced in the coil.

80 per cent of the flux produced by the coil links with a second coil of 600 turns. Determine (a) the mutual inductance between the coils, (b) the average value of the e.m.f. induced in the second coil when the primary current is reduced to zero.

Solution

The inductance of the primary coil is

$$L_1 = N_1 \frac{d\Phi_1}{dI_1} = 800 \times \frac{10 \times 10^{-3}}{5} = 1.6 \text{ H}$$

The self-induced e.m.f. in the primary when the current is reduced to zero is

$$e_1 = L_1 \frac{dI_1}{dt} = 1.6 \times \frac{5}{10 \times 10^{-3}} = 800 \text{ V}$$

The mutual inductance between the coils is

$$M = N_2 \frac{k\, d\Phi_2}{dI_1} = 600 \times \frac{0.8 \times 10 \times 10^{-3}}{5} = 0.96 \text{ H}$$

and the e.m.f. induced in the secondary winding when the flux collapses is

$$e_2 = N_2 \frac{d\Phi_2}{dt} = 600 \times \frac{0.8 \times 10 \times 10^{-3}}{10 \times 10^{-3}} = 480 \text{ V.}$$

6.6 Series-connected inductors

Figure 6.3 Series-connected inductors having no mutual coupling between them.

In the case where *no magnetic coupling exists between series-connected coils* (see Figure 6.3), the effective inductance of the circuit is

$$L_e = L_1 + L_2 + L_3 + \ldots + L_n$$

Where *mutual inductance exists* between two series-connected coils (see Figure 6.4), it can be one of two kinds, namely:

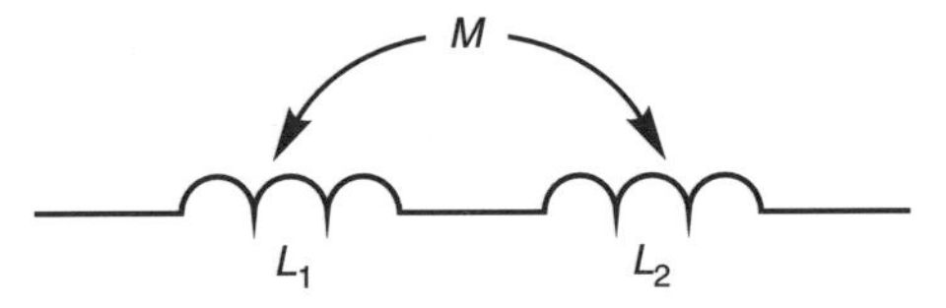

Figure 6.4 Two series-connected coils which are mutually coupled

(a) *series-aiding*, in which the mutual flux assists the flux in the mutually-coupled coil, or
(b) *series-opposing*, in which the mutual flux opposes the flux in the mutually-coupled coil.

For *two series-aiding coils*, the effective inductance is

$$L_e = L_1 + L_2 + 2M$$

and for two series-opposing coils, the effective inductance is

$$L_e = L_1 + L_2 - 2M$$

Worked Example 6.13

Determine the effective inductance of four series-connected coils of inductance 10 mH, 15 mH, 5 mH and 8 mH which do not have any mutual inductance between them.

Solution

The effective inductance is

$$L_e = L_1 + L_2 + L_3 + L_4 = 10 + 15 + 5 + 8 = 38 \text{ mH}.$$

Worked Example 6.14

When two mutually-coupled coils are connected in series-aiding, the effective inductance is 70 mH. When they are connected in series-opposing it is 30 mH. Determine the value of the mutual inductance between the coils.

Solution

For the two coils connected in series-aiding, the equation for the circuit is

$$L_1 + L_2 + 2M = 70 \text{ mH}$$

and for series-opposing it is

$$L_1 + L_2 - 2M = 30 \text{ mH}$$

Subtracting the second equation from the first gives

$$\begin{aligned} L_1 + L_2 + 2M &= 70 \\ L_1 + L_2 - 2M &= 30 \\ \hline \end{aligned}$$

SUBTRACT $\quad 4M = 40$

or $\quad M = 10 \text{ mH}$

It also follows that $L_1 + L_2 = 70 - 20 = 50$ mH

6.7 Parallel-connected inductors

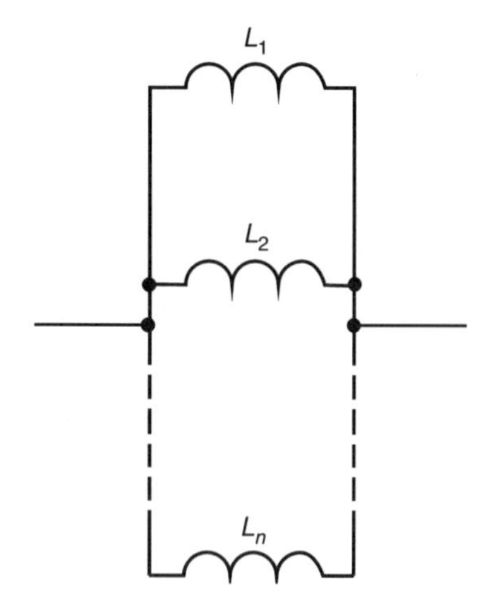

Figure 6.5 Parallel-connected inductors

If there is no mutual inductance between the parallel-connected inductors (see Figure 6.5), the reciprocal of the effective inductance of the circuit is

$$\frac{1}{L_e} = \frac{1}{L_1} + \frac{1}{L_2} + \ldots + \frac{1}{L_n}$$

In the special case of *two parallel-connected inductors*, the effective inductance is

$$L_e = \frac{L_1 L_2}{L_1 + L_2}$$

Worked Example 6.15

Determine the effective inductance of the circuit in Figure 6.6(a). No mutual coupling exists between the inductors.

Solution

In this case we commence at the most remote point from the input terminals, and work our way towards the input. Initially we have L_1 and L_2 in parallel, giving

$$\begin{aligned} L_{P1} &= \frac{L_1 L_2}{L_1 + L_2} = \frac{2 \times 2}{2 + 2} \\ &= 1 \text{ H} \end{aligned}$$

This leaves us with the circuit in Figure 6.6(b). Next, we combine L_3 and L_{P1} as a series circuit to give

$$\begin{aligned} L_S &= L_3 + L_{P1} = 3 + 1 \\ &= 4 \text{ H} \end{aligned}$$

which is shown in Figure 6.13(c). Finally, we combine L_S and L_4 in parallel to give

$$\begin{aligned} L_e &= \frac{L_4 L_S}{L_4 + L_S} = \frac{1 \times 4}{1 + 4} \\ &= 0.8 \text{ H} \end{aligned}$$

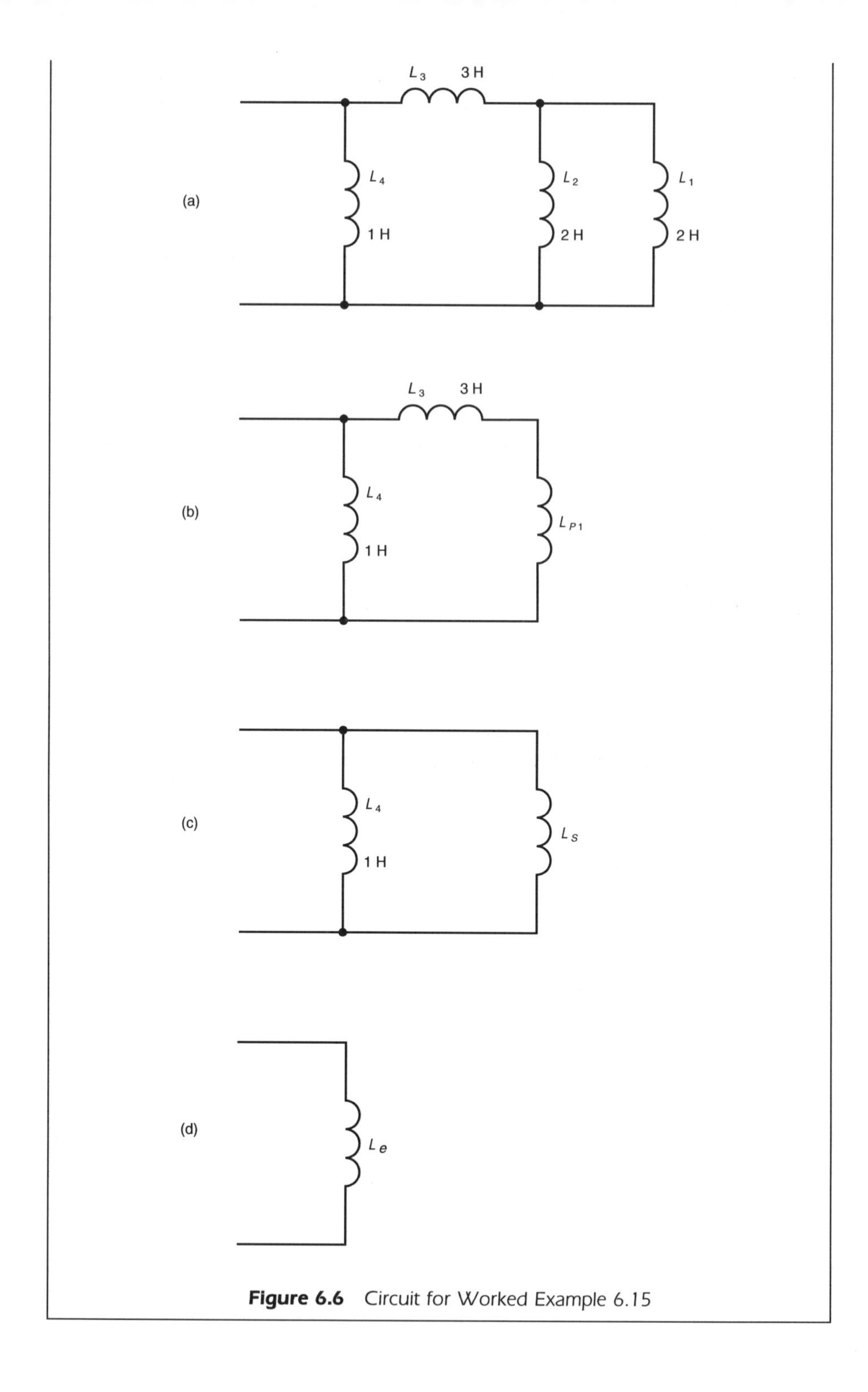

Figure 6.6 Circuit for Worked Example 6.15

6.8 Computer listing

Listing 6.1 Inductance and mutual inductance

```
10 CLS '**** CH6-1 ****
20 PRINT "INDUCTANCE AND MUTUAL INDUCTANCE": PRINT
30 L = 0: L1 = 0: L2 = 0: M = 0: N = 0: dPHI = 0: dt = 0: dI = 0
40 A = 0: MUr = 0: PI = 3.14159: MUo = PI * .0000004: S = 0: k = 0
50 LENGTH = 0
60 PRINT "Select ONE of the following:": PRINT
70 PRINT "1. Inductance = turns * dPHI/ dI": PRINT
80 PRINT "2. Reluctance = length/(MUr * MUo * area)": PRINT
90 PRINT "3. Inductance = (turns)^2/reluctance": PRINT
100 PRINT "4. e.m.f. = turns * dPHI/dt": PRINT
110 PRINT "5. e.m.f. = inductance * dI/dt": PRINT
120 PRINT "6. mutual inductance = k * SQR(L1 * L2)": PRINT
130 INPUT "Please enter your selection (1 - 6): ", Sel
140 IF Sel < 1 OR Sel > 6 THEN GOTO 10
150 IF Sel = 1 THEN GOTO 300
160 IF Sel = 2 THEN GOTO 500
170 IF Sel = 3 THEN GOTO 700
180 IF Sel = 4 THEN GOTO 900
190 IF Sel = 5 THEN GOTO 1100
200 IF Sel = 6 THEN GOTO 1300
300 CLS '*** Option 1 ***
310 PRINT "The equation is: Inductance = turns * dPHI/ dI"
320 PRINT : INPUT "Number of turns = ", N
330 INPUT "Change in flux (Wb) = ", dPHI
340 INPUT "Change in current (A) = ", dI: PRINT
350 IF dI <= 0 THEN GOSUB 3000: GOTO 310
360 PRINT "Inductance (H) = "; N * dPHI / dI: GOTO 1500
500 CLS '*** Option 2 ***
510 PRINT "The equation is: Reluctance = length/(MUr * MUo * area)"
520 PRINT : INPUT "Length (m) = ", LENGTH
530 INPUT "Relative permeability = ", MUr
540 INPUT "Area (m^2) = ", A: PRINT
550 IF (MUr * MUo * A) <= 0 THEN GOSUB 3000: GOTO 510
560 PRINT "Reluctance (A/Wb) = "; LENGTH / (MUr * MUo * A): GOTO 1500
700 CLS '*** Option 3 ***
710 PRINT "The equation is: Inductance = (turns)^2/reluctance"
720 PRINT : INPUT "Number of turns = ", N
730 INPUT "Reluctance (A/Wb) = ", S: PRINT
740 IF S <= 0 THEN GOSUB 3000: GOTO 710
750 PRINT "Inductance (H) = "; N ^ 2 / S: GOTO 1500
900 CLS '*** Option 4 ***
```

```
910 PRINT "The equation is: Induced e.m.f. = N * dPHI/dt": PRINT
920 INPUT "Number of turns = ", N
930 INPUT "Change in flux (Wb) = ", dPHI
940 INPUT "Change in time (s) = ", dt
950 IF dt <= 0 THEN GOSUB 3000: GOTO 910
960 PRINT : PRINT "Induced e.m.f. (V) = "; N * dPHI / dt: GOTO 1500
1100 CLS '*** Option 5 ***
1110 PRINT "The equation is: Induced e.m.f. = L * dI/dt": PRINT
1120 INPUT "Inductance (H) = ", L
1130 INPUT "Change of current (A) = ", dI
1140 INPUT "Change of time (s) = ", dt
1150 IF dt <= 0 THEN GOSUB 3000: GOTO 1110
1160 PRINT : PRINT "Induced e.m.f. (V) = "; L * dI / dt: GOTO 1500
1300 CLS '*** Option 6 ***
1310 PRINT "The equation is: M = k * SQR(L1 * L2)": PRINT
1320 INPUT "Magnetic coupling coefficient = ", k
1330 IF k < 0 OR k > 1 THEN GOSUB 3100: GOTO 1310
1340 INPUT "Inductance L1 (H) = ", L1
1350 INPUT "Inductance L2 (H) = ", L2
1360 IF (L1 * L2) < 0 OR L1 < 0 OR L2 < 0 THEN GOSUB 3200: GOTO 1310
1370 PRINT : PRINT "Mutual inductance (H) = "; k * SQR(L1 * L2)
1500 PRINT
1510 PRINT "Do you wish to: ": PRINT
1520 PRINT "1. Return to the menu, or"
1530 PRINT "2. exit from the program.": PRINT
1540 INPUT "Enter your selection here (1 or 2): ", Sel
1550 IF Sel < 1 OR Sel > 2 THEN CLS : GOTO 1500
1560 IF Sel = 1 THEN GOTO 10
1570 END
3000 CLS : PRINT "ERROR: DENOMINATOR IS NEGATIVE OR ZERO"
3010 PRINT "--------------------------------------": PRINT
3020 RETURN
3100 CLS : PRINT "ERROR: k < 0 OR k > 1"
3110 PRINT "-------------------": PRINT
3120 RETURN
3200 CLS : PRINT "ERROR: INDUCTANCE CANNOT BE NEGATIVE"
3210 PRINT "------------------------------------": PRINT
3220 RETURN
```

In this case Computer listing 6.1 deals with five important equations used in this chapter, together with an equation used to determine the reluctance of a magnetic circuit. The latter equation is also very useful in association with Chapter 7.

There are three principal reasons for the introduction of errors when supplying data to these equations. These are:

1. A 'division by zero error', which is produced if we attempt to divide a number by zero
2. Use of an unsuitable value of coupling coefficient in the mutual inductance equation $M = k\sqrt{(L_1 L_2)}$
3. Use of an unsuitable value of inductance in the mutual inductance equation.

The first of these errors is dealt with by the subroutine in lines 3000 to 3020, the second is handled by lines 3100 to 3120, and the final one by lines 3200 to 3220.

Once again, lines 1510 to 1560 give you the option of returning to the main menu.

Exercises

6.1 Express 7.5 mH in (a) H, (b) μH.

6.2 A coil of 200 turns wound on a non-magnetic core carries a current of 2 A. If the flux produced by the coil is 600 μWb, what is the inductance of the coil?

6.3 A coil of inductance 200 mH produces a magnetic flux of 800 μWb, the coil having 1000 turns and is wound on a non-magnetic former. What current flows in the coil?

6.4 Determine the flux produced by a 1500-turn air-cored coil of inductance 100 mH, which carries a current of 10 A.

6.5 How many turns of wire are there on an air-cored coil of inductance 150 mH, which produces a flux of 500 μWb when carrying a current of 4 A?

6.6 What is the inductance of an air-cored coil having 500 turns, which is wound on a former of length 10 cm and 1 cm diameter.

6.7 An iron core has a length of 10 cm and cross-sectional area of 3×10^{-5} m^2, which is wound with 200 turns of wire.
If the relative permeability of the iron is 10 000, calculate the inductance of the coil.

6.8 Determine the e.m.f. induced in a coil of 500 turns, if a flux of 70 mWb which is acting through the coil is reduced to zero in 100 ms.

6.9 What is the e.m.f. induced in the coil in Exercise 6.8 if the flux is reversed?

6.10 If a voltage of 1200 V is induced in a coil when a flux of 30 mWb is reversed in 0.05 s, calculate the number of turns on the coil.

6.11 If an e.m.f. of 500 V is induced in a coil of 800 turns when a flux change takes place over 1.6 ms, determine the flux change.

6.12 If a flux of 50 mWb is reversed in a coil of 100 turns, the average value of the induced e.m.f. is 200 V. Determine the time taken to reverse the flux.

6.13 Calculate the voltage induced in a 10 H inductor if the current flowing through it changes by 10 A in 20 ms.

6.14 If the current in an inductor changes from 20 A to 4 A in 0.5 s, and the magnitude of the induced e.m.f. is 80 V, calculate the inductance of the coil.

6.15 The current flowing through a coil of inductance 12.5 mH is reversed in 25 ms, and the magnitude of the induced e.m.f. is found to be 50 V. Determine the original value of the current.

6.16 What energy is stored in the magnetic field of a 0.2 H inductor when it carries a current of 2 A.

6.17 A current of 10 A flows in the coil of inductance 100 mH. Calculate the energy stored in its magnetic field. If the circuit is opened in 20 ms, calculate the power which must be dissipated.

6.18 Inductors of inductance 0.1 H, 0.4 H and 0.5 H are connected (a) in series, (b) in parallel. If no mutual coupling exists between the coils, what is the effective inductance in each case?

6.19 Coil X and coil Y have respective self-inductances of 5 H and 4 H, are wound on the same iron core. When they are connected in series-aiding, the effective inductance is found to be 11 H. Determine (a) the mutual inductance between them, (b) the effective inductance of the two if they are connected in series-opposing.

6.20 What energy is stored in the combined magnetic field of the two coils in Exercise 6.19 when they each carry a current of 10 A and are connected in (a) series-aiding, (b) series-opposing?

6.21 Coil A and coil B are connected in series with one another. Coil C and coil D are also connected in series with one another. The two circuits are connected in parallel with each other, there being no mutual coupling between the coils. If the inductances are

$$\text{coil } A = 5 \text{ H}, \text{ coil } B = 4 \text{ H},$$
$$\text{coil } C = 3 \text{ H}, \text{ coil } D = 10 \text{ H}$$

Calculate the effective inductance of the circuit.

6.22 Amend the software in Computer listing 6.1 so that, in option 1, you can determine either (a) the inductance, (b) the number of turns, (c) *dPHI* or (d) *dI* (given all other data).

6.23 Amend the software in Computer listing 6.1 so that it includes an option for calculating the energy stored in the magnetic field of an inductor.

Summary of important facts

When current flows in a conductor it produces a **magnetic flux**, and when the conductor is wound in form of a coil it intensifies the flux. The coil has the property of **self-inductance** (L), whose unit is the **henry** (H).

When a current of 1 A flows in an inductance of 1 H, it produces a magnetic flux of 1 weber (Wb).

Conversely, **when the magnetic flux linking with a coil or conductor changes, an e.m.f. is induced in it**.

The self-inductance of a coil is given by

$$L = \frac{\text{change in flux linkages}}{\text{change in current}} = N\frac{d\Phi}{dI}$$

also

$$L = \frac{N^2}{S} = \frac{N^2}{\ell/\mu a} = \frac{N^2}{l/(\mu_0\mu_r\mu a)}$$

where N is the number of turns on the coil, S is the reluctance of the magnetic circuit, ℓ is the length of the magnetic path, μ is the permeability of the magnetic circuit $(= \mu_0\mu_r = 4\pi \times 10^{-7} \times \mu_r)$, and a is the cross-sectional area of the magnetic circuit.

The e.m.f. induced in an inductor is

$$e = N\frac{d\Phi}{dt} = L\frac{di}{dt}$$

where dt is change in time.

The **energy stored in a magnetic field** is given by

$$W = \frac{1}{2}LI^2$$

where W is the energy in joules, L the inductance in henrys, and I the current in amperes.

Mutual inductance (M) exists between coils or circuits when the magnetic flux produced by one coil or circuit links with the other coil or circuit. The mutual inductance between two coils L_1 and L_2 is

$$M = k\sqrt{(L_1L_2)}$$

where k is the *magnetic coupling coefficient* between the coils (whose value lies between zero and unity), and is the proportion of flux leaving one coil which reaches the other coil. If k has a value in the region of unity, the coils are said to be **closely coupled** (as in the case of a power transformer), and if k has a low value the coils are said to be **loosely coupled**.

The mutual inductance is also given by

$$M = kN_2\frac{d\Phi_2}{dI_1} = kN_1\frac{d\Phi_1}{dI_2}$$

where I_1, N_1 and Φ_1 refer to coil L_1, and I_2, N_2 and Φ_2 refer to coil L_2.

The effective inductance, L_e, of series-connected inductors which are not mutually coupled is

$$L_e = L_1 + L_2 + \ldots + L_n$$

If mutual inductance exists between the coils so that the mutual flux between the coils aids one another, the two are said to be connected in **series-aiding**. The effective inductance of the two coils connected in series-aiding is

$$L_e = L_1 + L_2 + 2M$$

where M is the mutual inductance between the coils.

If the mutual flux produced by one coil opposes that produced by the other, the two are said to be connected in **series-opposing**. The effective inductance in this case is

$$L_e = L_1 + L_2 - 2M$$

The reciprocal of the effective inductance of n parallel-connected inductors which have no mutual coupling between them is

$$\frac{1}{L_e} = \frac{1}{L_1} + \frac{1}{L_2} + \ldots + \frac{1}{L_n}$$

7 Electromagnetism

7.1 Introduction

We look in this chapter at the subject of electromagnetism, which is a very important area in electronic and electrical engineering. By the end of this chapter, the reader will be able to

- Calculate the e.m.f. induced in a conductor which is in motion in a magnetic field.
- Determine the force acting on a current-carrying conductor in a magnetic field.
- Evaluate the flux density in a magnetic field.
- Calculate the magnetomotive force and magnetising force produced by a conductor or coil.
- Understand and calculate the reluctance of a magnetic circuit.
- Determine the reluctance of series and parallel magnetic circuits.
- Calculate the magnetic flux in a complex magnetic circuit.
- Solve problems involving magnetic leakage.
- Use software to solve many of the equations and problems in this chapter.

7.2 e.m.f. induced in a conductor

When a conductor moves relative to a magnetic field (or vice versa), an e.m.f. is induced in the conductor. The equation for the induced e.m.f., e, when *the movement of the conductor is perpendicular to the magnetic field* is

$$\boldsymbol{e = B\ell v}$$

where e = induced e.m.f. (in volts)

B = flux density of the magnetic field (tesla)
ℓ = active length of the conductor (m) in the magnetic field
v = velocity of the conductor in m/s *perpendicular* to the magnetic field

If the conductor moves at an angle θ *relative to the magnetic field* (see Figure 7.1), the velocity of the conductor perpendicular to the magnetic field is $v \sin\theta$, and the e.m.f. equation becomes

$$\boldsymbol{e = B\ell v \sin\theta}$$

(**Note**: If the conductor moves perpendicular to the magnetic field, $\theta = 90°$.)

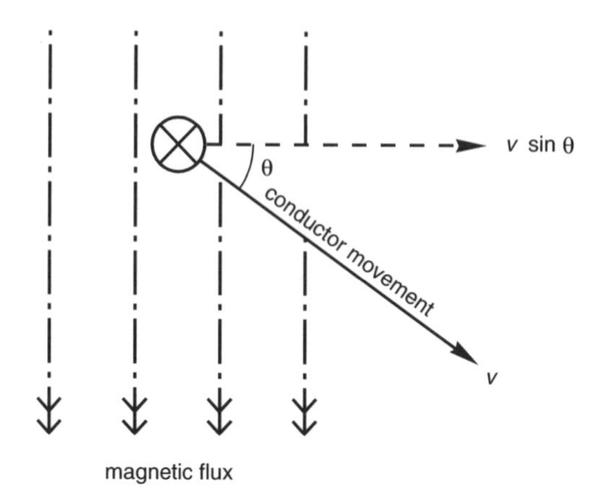

Figure 7.1 e.m.f induced in a conductor when it moves at an angle to the magnetic field

7.2.1 Angular measurement in radians

In engineering, we often use the radian angular measurement, where

$$360° \equiv 2\pi \text{ radians (rad)}$$

or $$1 \text{ rad} \equiv 57.3°$$

The two angular measurements are related by the equation

$$\frac{\textbf{angle in degrees}}{\textbf{180}} = \frac{\textbf{angle in radians}}{\boldsymbol{\pi}}$$

or

$$\textbf{angle in degrees} = \textbf{angle in radians} \times \mathbf{180}/\boldsymbol{\pi}$$

and

$$\textbf{angle in radians} = \textbf{angle in degrees} \times \boldsymbol{\pi}/\mathbf{180}$$

Hence the radian equivalent of 30° is

$$30 \times \pi/180 = 0.5236 \text{ rad}$$

and the degree equivalent of 1.5 rad is

$$1.5 \times 180/\pi = 85.94°.$$

Worked Example 7.1

Calculate the value of the e.m.f. induced in a conductor of active length 20 cm when it moves with a velocity of 15 m/s (a) perpendicular to a magnetic field, (b) at an angle of 30° relative to a magnetic field of flux density 200 mT.

Solution

(a) In this case, the movement of the conductor is perpendicular to the field, so that

$$e = B\ell v = (200 \times 10^{-3}) \times 0.2 \times 15 = 0.6 \text{ V}$$

or, alternatively

$$e = B\ell v \sin\theta = (200 \times 10^{-3}) \times 0.2 \times 15 \times \sin 90^\circ$$
$$= 0.6 \text{ V}$$

(b) Here $\theta = 30^\circ$, so that

$$e = B\ell v \sin\theta = (200 \times 10^{-3}) \times 0.2 \times 15 \times \sin 30^\circ$$
$$= 0.3 \text{ V}.$$

Worked Example 7.2

If 10 V is induced in a conductor moving at 20 m/s at right-angles to a magnetic field of flux density 0.5 T, determine the active length of the conductor.

Solution

From the equation e = Blv, we see that

$$\ell = e/Bv = 10/(0.5 \times 20) = 1 \text{ m}.$$

7.3 Force acting on a current-carrying conductor

The force acting on a current-carrying conductor in a magnetic field is

$$F = BI\ell \text{ newtons (N)}$$

where B is the flux density in tesla
I is the current flowing in the conductor in amperes

ℓ is the active length of the conductor in metres.

The force acts mutually at right-angles to the flux and the current.

Worked Example 7.3

A conductor which carries a current of 50 A in a flux density of 0.25 T, has an active length of 0.6 m. Calculate the force acting on the conductor.

Solution

The force is given by

$$F = BI\ell = 0.25 \times 50 \times 0.6 = 7.5 \text{ N}.$$

Worked Example 7.4

Determine the active length of a conductor which experiences a force of 10 N when it carries a current of 25 A in a magnetic field of flux density 0.5 T.

Solution

From the equation $F = BI\ell$, it follows that

$$\ell = F/BI = 10/(0.5 \times 25) = 0.8 \text{ m}$$

7.4 Flux and flux density

Magnetic flux is simply a *condition of space*, and we say that the direction of the magnetic field at a point is the direction of the force experienced by an isolated *N*-pole placed at that point. The unit of magnetic flux is the **weber** (Wb), and the symbol for flux is Φ.

The **magnetic flux density**, symbol B, is the amount of magnetic flux passing through unit area perpendicular to the flux, and is defined as follows:

$$\mathbf{B} = \frac{\textbf{flux}}{\textbf{area perpendicular to flux}} = \frac{\mathbf{\Phi}}{\mathbf{a}} \text{ tesla (T)}$$

One tesla is equivalent to one Wb/m^2.

Worked Example 7.5

The flux density in a magnetic circuit of cross-sectional area 300 cm^2 is 0.01 T. Determine the value of the magnetic flux in the circuit.

Solution

The reader may like to refer to Chapter 1 to understand the process of unit size conversion. From the equation $B = \Phi/a$, then

$$\Phi = aB = (300 \times (10^{-2})^2) \times 0.01 = 300 \times 10^{-6}$$
$$= 300\ \mu\text{Wb or } 0.3\ \text{mWb}$$

7.5 Magnetomotive force and magnetising force

The **magnetomotive force** (m.m.f., symbol F) in a magnetic circuit is equivalent to e.m.f. in an electrical circuit. The m.m.f. is responsible for the production of the magnetic flux in the magnetic circuit, and is given by the equation

$$F = IN \text{ ampere turns or amperes}$$

where I is the current in amperes in the coil, and
N is the number of turns on the coil.

Since the number of turns on the coil is dimensionless, the product (amperes × turns) is, dimensionally, equal to amperes.

The **magnetising force** (also known as the **magnetic field intensity** or the **magnetic field strength**), symbol H, is the m.m.f. per unit length of the magnetic circuit as follows:

$$H = \frac{F}{\ell} = \frac{IN}{\ell} \text{ ampere turns/m or amperes/m}$$

where ℓ is the length of the magnetic circuit in metres.

Worked Example 7.6

A 200-turn coil carries a current of 5 A. If the length of the coil is 0.1 m, calculate (a) the m.m.f. produced by the coil, (b) the magnetising force inside the coil.

Solution

(a) The m.m.f. is given by

$$F = IN = 5 \times 200 = 1000 \text{ ampere turns or amperes}$$

(b) The magnetising force inside the coil is

$$H = F/\ell = 1000/0.1 = 10\,000 \text{ ampere turns/m or A/m.}$$

7.6 Permeability

In a magnetic material, the flux density, B, is related to the magnetising force, H, which produces it by the equation

$$B = \mu H$$

where μ is the **absolute permeability** of the material, and has dimensions of henrys per metre (H/m). The *permeability of free space* of a vacuum is

$$\mu_0 = 4\pi \times 10^{-7}\ \text{H/m}$$

For all practical purposes, the permeability of air is equal to μ_0. For a magnetic material, the absolute permeability is

$$\mu = \mu_0 \mu_r$$

where μ_r is the **relative permeability** of the material, which may have a value between unity (for a non-magnetic material) and well in excess of 20 000.

The value of μ_r depends on the point on the B–H curve of the material at which it is operating; that is to say, μ_r varies with the value of the magnetising current. Table 7.1 lists three values of B, H and μ_r for a sample of cast steel. The reader will note that, in this case, the value of μ_r passes through a peak as the value of H increases.

Table 7.1 ***B***, ***H*** and μ_r for cast steel

B (T)	0.4	0.8	1.2
H (A/m)	470	720	1230
μ_r	677	833	777

Worked Example 7.7

A rectangular-shaped iron circuit is built up from laminations, and has a mean magnetic length of 40 cm, and a uniform cross-sectional area of 5 cm^2. A coil wound uniformly on the iron produces a magnetomotive force of 150 ampere turns, and produces a flux of 0.25 mWb in the iron. Determine (a) the flux density in the iron, (b) the relative permeability of the iron.

Solution

(a) The flux density in the iron is

$$B = \text{magnetic flux/area} = 0.25 \times 10^{-3}/5 \times (10^{-2})^2$$
$$= 0.5\ \text{T}$$

(b) The magnetising force is

$$H = F/\ell = 150/40 \times 10^{-2} = 375 \text{ A/m}$$

and since $B = \mu_0\mu_r H$, then

$$\mu_r = B/(\mu_0 H) = 0.5/(4\pi \times 10^{-7} \times 375) = 1061.$$

7.7 Reluctance

The reluctance of a magnetic circuit is the magnetic equivalent of resistance in an electric circuit. It is the total opposition of the circuit to the establishment of magnetic flux. The reluctance, S, of a magnetic circuit is

$$S = \frac{\ell}{\mu_0\mu_r a} \text{ A/Wb}$$

where ℓ is the length of the magnetic path, and
a is the cross-sectional area of the magnetic circuit.
The relationship between m.m.f., flux and reluctance (known as Ohm's law for the magnetic circuit) is

$$F = \Phi S.$$

Worked Example 7.8

A mild steel ring has a mean circumference of 0.25 m and a cross-sectional area of 0.001 m^2. If the flux in the core is 1.5 mWb, and the relative permeability of the ring at the operating flux density is 625, calculate the current needed in a coil of 1000 turns wound uniformly around the ring to produce the flux.

Solution

The reluctance of the core is

$$S = \ell/(\mu_0\mu_r a) = 0.25/(625 \times 4\pi \times 10^{-7} \times 0.001) = 0.3183 \times 10^6 \text{ A/Wb}$$

The m.m.f. is calculated from

$$F = \Phi S = 1.5 \times 10^{-3} \times 0.3183 \times 10^6$$
$$= 478 \text{ ampere turns}$$

Since $F = NI$, then

$$I = F/N = 478/1000 = 0.478 \text{ A}.$$

7.8 Series-connected magnetic circuits

The analogy between electrical and magnetic circuits is sufficiently close to allow us to use methods in magnetic circuits which are similar to those used in electrical circuits.

If we have n magnetic elements in series with one another, the *effective reluctance*, S_e, of the circuit is

$$S_e = S_1 + S_2 + S_3 + \ldots + S_n$$

where S_1 is the reluctance of element 1 of the circuit, S_2 is the reluctance of element 2 of the circuit, etc.

Worked Example 7.9

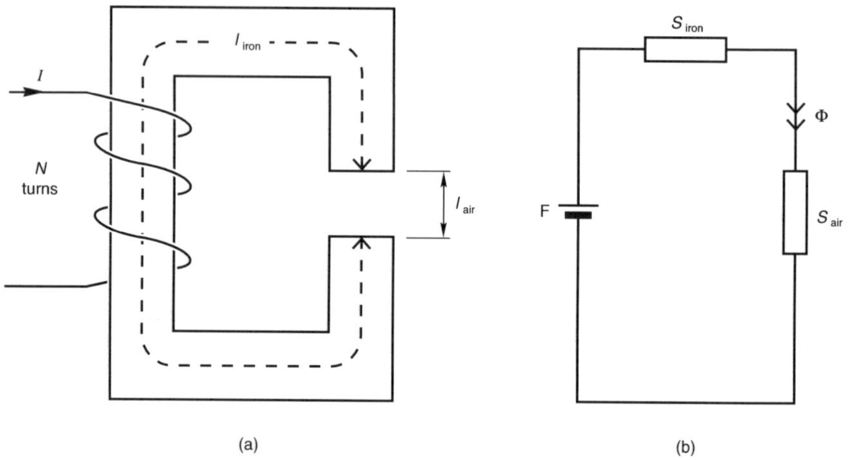

Figure 7.2 Circuit for Worked Example 7.9: (a) the series magnetic field; (b) its equivalent circuit

The iron section of the magnetic circuit in Figure 7.2(a) has a mean length of 0.25 m and a cross-sectional area of 0.001 m^2. The air gap has a length of 1 mm and the same cross-sectional area as the iron path; all the flux in the iron may be assumed to cross the air gap, i.e. there is no magnetic leakage or fringing.

If the magnetic flux in the circuit is 1.5 mWb, and the relative permeability of the iron at the operating flux density is 625, calculate the current required in the 1000-turn coil to produce the flux.

Solution

The 'electrical' equivalent circuit of the magnetic circuit is shown in Figure 7.2(b), and we see that the equivalent reluctance of the circuit is

$$S_e = S_{iron} + S_{air}$$

The reluctance of the iron path is calculated from the equation

$$\begin{aligned} S_{iron} &= \ell_{iron}/(\mu_0 \mu_r a) \\ &= 0.25/(625 \times 4\pi \times 10^{-7} \times 0.001) \\ &= 0.318 \times 10^6 \text{ A/Wb} \end{aligned}$$

and the reluctance of the air gap is

$$\begin{aligned} S_{air} &= \ell_{air}/(\mu_0 a) = 10^{-3}/(4\pi \times 10^{-7} \times 0.001) \\ &= 0.8 \times 10^6 \text{ A/Wb} \end{aligned}$$

The reader should note that, even though the length of the air gap is only 1 mm, the reluctance of the air gap is 2.5 times greater that of the iron. The total reluctance of the magnetic circuit is

$$S_e = S_{iron} + S_{air} = 1.118 \times 10^6 \text{ A/Wb}$$

The m.m.f. required to produce the flux is

$$F = \Phi S = 1.5 \times 10^{-3} \times 1.118 \times 10^6 = 1677 \text{ ampere turns}$$

Since $F = IN$, the current in the coil is

$$I = F/N = 1677/1000 = 1.677 \text{ A}$$

(**Note**: had the air gap been replaced by iron, a current of only 0.477 A would be needed to establish the flux iron.)

Worked Example 7.10

A magnetic circuit has two iron parts A and B, together with an air gap C, all connected in series. The dimensions of the three parts are given below. The B–H curve for the iron sections is shown in Figure 7.3.

Part	Length (cm)	Area (cm^2)
A	35	4.5
B	10	3.0
C	0.1	3.5

Determine the current required in a coil of 1200 turns wound uniformly on the iron circuit to produce a magnetic flux of 0.35 mWb in the air gap. Assume that there is no magnetic leakage.

Solution

Since parts A and B have different cross-sectional areas, each has a different flux density in it. For part A, the flux density is

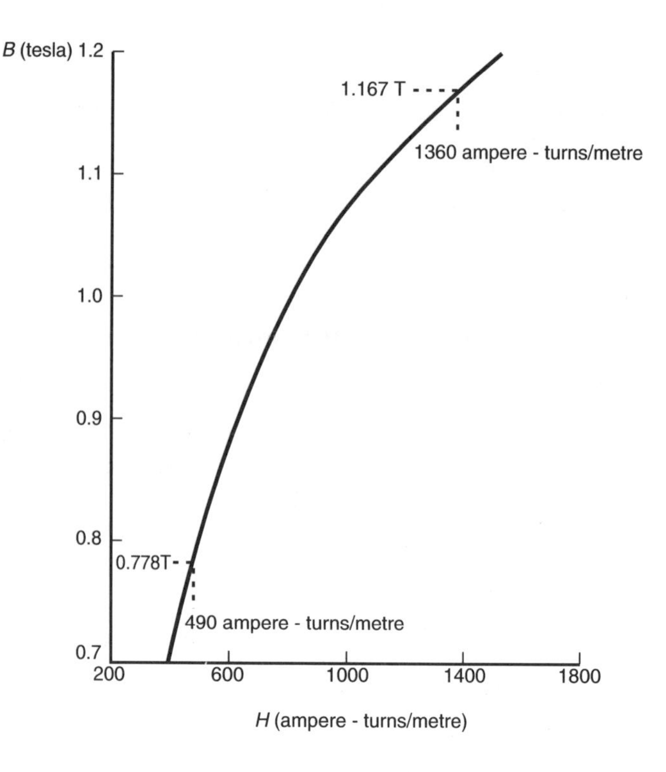

Figure 7.3 B–H curve for Worked Example 7.10

$$B_A = \Phi/\text{area} = 0.35 \times 10^{-3}/(4.5 \times (10^{-2})^2)$$
$$= 0.778 \text{ T}$$

and for part B, the flux density is

$$B_B = \Phi/\text{area} = 0.35 \times 10^{-3}/(3 \times (10^{-2})^2)$$
$$= 1.167 \text{ T}$$

From the B–H curve in Figure 7.3, we see that the relevant magnetising force values are

$$H_A = 490 \text{ ampere turns/m}$$
$$H_B = 1360 \text{ ampere turns/m}$$

It follows that the respective magnetomotive force values are

$$F_A = H_A \ell_A = 490 \times 35 \times 10^{-2} = 171.5 \text{ ampere turns}$$
$$F_B = H_B \ell_B = 1360 \times 10 \times 10^{-2} = 136 \text{ ampere turns}$$

so that the iron circuit requires an m.m.f. of

$$F_{\text{iron}} = F_A + F_B = 171.5 + 136 = 307.5 \text{ ampere turns}$$

The reluctance of the air gap is

$$S_C = \ell_C/\mu_0 a$$
$$= 0.1 \times 10^{-2}/(4\pi \times 10^{-7} \times 3.5 \times 10^{-4})$$
$$= 2.274 \times 10^6 \text{ ampere turns/Wb}$$

hence the m.m.f. required for the air gap is

$$F_{air} = \Phi S_C = 0.35 \times 10^{-3} \times 2.274 \times 10^6$$
$$= 795.9 \text{ ampere turns}$$

The total m.m.f. requirement for the complete circuit is

$$F = F_{iron} + F_{air} = 307.5 + 795.9$$
$$= 1103.4 \text{ ampere turns}$$

Since $F = IN$, the current in the coil is

$$I = F/N = 1103.4/1200 = 0.92\text{A}.$$

7.9 Parallel magnetic circuits

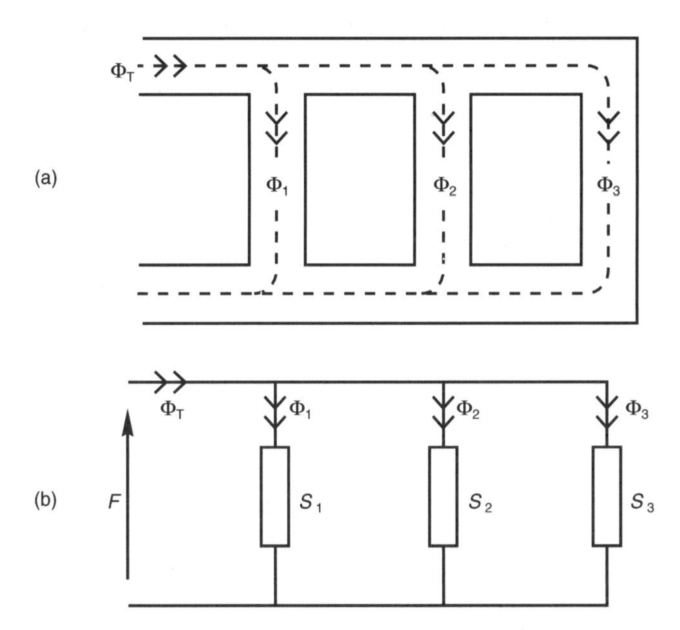

Figure 7.4 Parallel magnetic circuits: (a) a magnetic circuit with parallel branches; (b) its equivalent circuit

When a magnetic circuit splits into several parallel branches, as shown in Figure 7.4(a), it has the equivalent magnetic circuit in Figure 7.4(b), in which

$$\Phi 1 = \text{flux in branch 1}$$
$$\Phi 2 = \text{flux in branch 2}$$
$$\Phi 3 = \text{flux in branch 3}$$
$$\Phi T = \text{total flux} = \Phi 1 + \Phi 2 + \Phi 3$$
$$S_1 = \text{reluctance of branch 1}$$
$$S_2 = \text{reluctance of branch 2}$$
$$S_3 = \text{reluctance of branch 3}$$

The m.m.f. *across each branch* of the circuit is

$$\mathbf{F} = \mathbf{\Phi_1\, S_1 + \Phi_2\, S_2 + \Phi_3\, S_3 = \Phi_T\, S_e}$$

where S_e is the equivalent reluctance of the parallel circuit, whose *reciprocal* is given by

$$\frac{\mathbf{1}}{\mathbf{S_e}} = \frac{\mathbf{1}}{\mathbf{S_1}} + \frac{\mathbf{1}}{\mathbf{S_2}} + \frac{\mathbf{1}}{\mathbf{S_3}}$$

The magnetic flux in each of branch is

$$\Phi_1 = \Phi_T S_e / S_1$$
$$\Phi_2 = \Phi_T S_e / S_2$$
$$\Phi_3 = \Phi_T S_e / S_3$$

In the *special case of two parallel branches*, the effective reluctance is

$$\mathbf{S_e = S_1 S_2 / (S_1 + S_2)}.$$

Worked Example 7.11

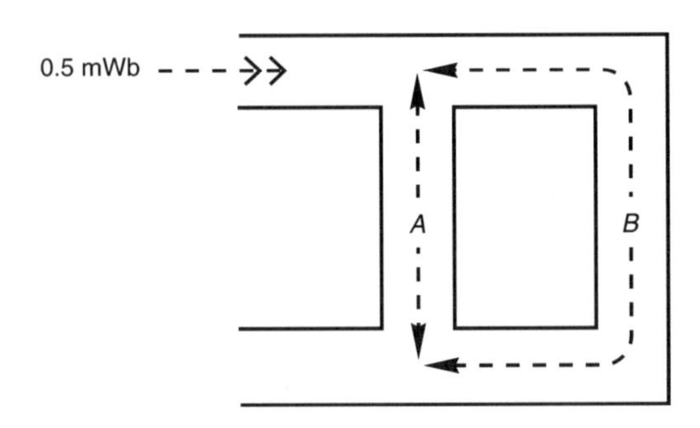

Figure 7.5 Magnetic circuit for Worked Example 7.11

The two branches, *A* and *B*, of a magnetic circuit are shown in Figure 7.5, the details being:

Branch	Length (cm)	Area (cm^2)	μ_r
A	7.5	4	1000
B	10	8	800

Determine (a) the reluctance of each branch, (b) the magnetic flux in each branch, (c) the total m.m.f. across the two branches.

Solution

(a) The reluctance of branch A of the iron circuit is

$$\begin{aligned} S_A &= \ell_A/(\mu_A a_A) \\ &= 7.5 \times 10^{-2}/(1000 \times 4\pi \times 10^{-7} \times 4 \times 10^{-4}) \\ &= 0.149 \times 10^6 \text{ A/Wb} \end{aligned}$$

For branch B the reluctance is

$$\begin{aligned} S_B &= \ell_B/(\mu_B a_B) \\ &= 10 \times 10^{-2}/(800 \times 4\pi \times 10^{-7} \times 8 \times 10^{-4}) \\ &= 0.124 \times 10^6 \text{ A/Wb} \end{aligned}$$

The effective reluctance of the two branches in parallel is

$$\begin{aligned} Se &= \frac{S_A S_B}{S_A + S_B} = \frac{(0.149 \times 0.124) \times 10^6}{(0.149 + 0.124) \times 10^6} \\ &= 67.68 \times 10^3 \text{ A/Wb} \end{aligned}$$

(b) The magnetic flux in branch A is calculated from

$$\begin{aligned} \Phi_A &= \Phi_T S_e/S_A \\ &= 0.5 \times 10^{-3} \times 67.68 \times 103/0.149 \times 10^6 \\ &= 0.227 \times 10^{-3} \text{ Wb or } 0.227 \text{ mWb} \end{aligned}$$

and

$$\begin{aligned} \Phi_B &= \Phi_T S_e/S_B \\ &= 0.5 \times 10^{-3} \times 67.68 \times 10^3/0.124 \times 10^6 \\ &= 0.273 \times 10^{-3} \text{ Wb or } 0.273 \text{ mWb} \end{aligned}$$

(**Note**: $\Phi_A + \Phi_B = \Phi_T = 0.5$ mWb)

(c) The total m.m.f. requirement for the two branches is

$$\begin{aligned} F &= \Phi_T S_e = 0.5 \times 10^{-3} \times 67.68 \times 10^3 \\ &= 33.84 \text{ ampere turns or A.} \end{aligned}$$

7.10 Magnetic leakage and fringing

Magnetic flux which fails to follow the 'useful' path is known as **magnetic leakage** or **fringing flux**, and is accounted for in calculations by means of a **magnetic leakage coefficient** as follows:

$$\textbf{leakage coefficient} = \frac{\textbf{total magnetic flux produced}}{\textbf{useful magnetic flux}}$$

The value of this coefficient is greater than unity and, in a well-designed magnetic system such as an electrical machine, its value may lie in the range 1.15 to 1.25.

Worked Example 7.12

Suppose in Worked Example 7.9 (relevant details are repeated below for the convenience of the reader), the air gap flux is to remain at 1.5 mWb, but the magnetic circuit now has a leakage coefficient of 1.2. Determine the current in the coil to produce a flux of 1.5 mWb in the air gap.

Solution

The leakage coefficient of 1.2 means that the current in the coil must produce a flux in the iron of

$$1.2 \times 1.5 = 1.8 \text{ mWb}$$

The net effect is that the current in the coil in this example must be higher than it is in Worked Example 7.9. From the solution of Worked Example 7.9, we know that

$$\begin{aligned}
\text{gap flux} &= 1.5 \times 10^{-3} \text{ Wb} \\
\text{reluctance of the iron circuit} &= 0.318 \times 10^{6} \text{ A/Wb} \\
\text{air gap reluctance} &= 0.8 \times 10^{6} \text{ A/Wb} \\
\text{number of turns on the coil} &= 1000
\end{aligned}$$

Since the iron circuit and the air gap are in series with one another, the total m.m.f. requirement is

$$\begin{aligned}
F &= \Phi_{\text{iron}} S_{\text{iron}} + \Phi_{\text{gap}} S_{\text{gap}} \\
&= 1.2\Phi_{\text{gap}} S_{\text{iron}} + \Phi_{\text{gap}} S_{\text{gap}} \\
&= \Phi - \text{gap}(1.2 S_{\text{iron}} + S_{\text{gap}}) \\
&= 1.5 \times 10^{-3}((1.2 \times 0.318 \times 106) + (0.8 \times 10^{6})) \\
&= 1772 \text{ ampere turns or A}
\end{aligned}$$

hence the current required in the coil is

$$I = F/N = 1772/1000 = 1.772 \text{ A}$$

That is, the current in the coil in this case must be about 6 per cent greater than in Worked Example 7.9 to account for the magnetic leakage.

7.11 Computer listing

Listing 7.1 Equations for electromagnetism

```
10 CLS '**** CH7-1 ****
20 PRINT "ELECTROMAGNETISM": PRINT
30 DIM S(5): FOR Y = 0 TO 5: S(Y) = 0: NEXT Y
40 PRINT : A = 0: B = 0: C = 0: F = 0: M = 0: MUr = 0
45 L = 0: R = 0: S = 0
50 PRINT "Select ONE of the following:": PRINT
60 PRINT "1. Induced e.m.f. = B * L * v * sin THETA"
70 PRINT "2. Force = B * I * L"
80 PRINT "3. Flux density = MUr * MUo * H"
90 PRINT "4. M.M.F. = I * N"
100 PRINT "5. Magnetising force = M.M.F./L"
110 PRINT "6. Reluctance = L/(MUr * MUo * A)"
120 PRINT " where L = length, A = area,"
130 PRINT " MUr = relative permeability, and"
140 PRINT " MUo = permeability of free space."
150 PRINT "7. Calculate the reluctance of a series"
160 PRINT " magnetic circuit."
170 PRINT "8. Calculate the reluctance of a parallel"
180 PRINT " magnetic circuit."
190 PRINT "9. Calculate magnetic flux using FLUX = F/S"
200 PRINT " where F the MMF and S the reluctance.": PRINT
210 INPUT "Enter your selection here (1 - 9): ", Sel
220 IF Sel < 1 OR Sel > 9 THEN GOTO 10
230 IF Sel = 1 THEN GOTO 400
240 IF Sel = 2 THEN GOTO 600
250 IF Sel = 3 THEN GOTO 800
260 IF Sel = 4 THEN GOTO 1000
270 IF Sel = 5 THEN GOTO 1200
280 IF Sel = 6 THEN GOTO 1400
290 IF Sel = 7 THEN GOTO 1600
300 IF Sel = 8 THEN GOTO 1800
310 IF Sel = 9 THEN GOTO 2000
400 CLS : PRINT "INDUCED E.M.F. DUE TO MOVEMENT": PRINT
410 PRINT "The equation is: Induced e.m.f. = BLv sin THETA"
420 PRINT : INPUT "Flux density (Tesla) = ", B
430 INPUT "Length of conductor (m) = ", L
```

```
440 INPUT "Linear velocity of conductor (m/s) = ", v
450 PRINT "Angle of movement of conductor with"
460 INPUT " respect to the magnetic field (deg.) = ", THETA
470 THETA = THETA * 3.142159 / 180: PRINT
480 PRINT "Induced e.m.f. = "; B * L * v * SIN(THETA); "V"
490 PRINT : GOTO 2200
600 CLS : PRINT "FORCE ON A CONDUCTOR": PRINT
610 PRINT "The equation is: Force = BIL": PRINT
620 INPUT "Flux density (Tesla) = ", B
630 INPUT "Current in conductor (A) = ", I
640 INPUT "Length of conductor (m) = ", L: PRINT
650 PRINT "Force on conductor = "; B * I * L; "N": PRINT
660 GOTO 2200
800 CLS : PRINT "FLUX DENSITY": PRINT
810 PRINT "The equations is: Flux density = MUr * MUo * H": PRINT
820 INPUT "Relative permeability = ", MUr
830 INPUT "Magnetising force (A/m) = ", H: PRINT
840 B = MUr * 4 * 3.14159E-07 * H
850 PRINT "Flux density = "; B; "T"
860 PRINT " = "; B * 1000; "mT": PRINT
870 GOTO 2200
1000 CLS : PRINT "CALCULATION OF M.M.F.": PRINT
1010 PRINT "The equation is : M.M.F. = I * N": PRINT
1020 INPUT "Current (A) = ", I
1030 INPUT "Number of conductors = ", N: PRINT
1040 PRINT "Magnetomotive force = "; I * N; "(A or ampere-turns)"
1050 PRINT : GOTO 2200
1200 CLS : PRINT "CALCULATION OF MAGNETISING FORCE": PRINT
1210 PRINT "The equation is: Magnetising force = M.M.F./L": PRINT
1220 INPUT "Magnetomotive force (A or ampere-turns) = ", I
1230 INPUT "Length of magnetic circuit (m) = ", L: PRINT
1240 PRINT "Magnetising force = "; I / L; "A/m": PRINT
1250 GOTO 2200
1400 CLS : PRINT "RELUCTANCE OF IRON PATH": PRINT
1410 PRINT "The formulae is S = L/(MUr * MUo * A)": PRINT
1420 INPUT "Length of iron (m) = ", L
1430 INPUT "Relative permeability of iron = ", MUr
1440 INPUT "Area of iron (m^2) = ", A: PRINT
1450 PRINT "Reluctance = "; L / (MUr * 1.2566E-06 * A); "A/Wb"
1460 PRINT " = "; L / (MUr * 1.2566 * A); "* 10^6 A/Wb"
1470 GOTO 2200
1600 CLS : PRINT "RELUCTANCE OF A SERIES MAGNETIC CIRCUIT": PRINT
1610 INPUT "Number of reluctances in series (max. 5) = ", N: PRINT
1620 S = 0: IF N < 2 OR N > 5 THEN GOTO 1600
1630 FOR Y = 1 TO N
1640 PRINT "For reluctance"; Y
1650 INPUT " value (A/Wb) = ", S(Y)
```

```
1660 S = S + S(Y)
1670 NEXT Y
1680 PRINT : PRINT "Effective reluctance = "; S; "A/Wb"
1690 GOTO 2200
1800 CLS : PRINT "RELUCTANCE OF A PARALLEL MAGNETIC CIRCUIT"
1810 PRINT
1820 INPUT "Number of reluctances in parallel (max. 5) = ", N
1830 PRINT
1840 IF N < 2 OR N > 5 THEN GOTO 1800
1850 R = 0
1860 FOR Y = 1 TO N
1870 PRINT "For reluctance"; Y
1880 INPUT " value (A/Wb) = ", S(Y)
1890 R = R + 1 / S(Y)
1900 NEXT Y
1910 PRINT : PRINT "Effective reluctance = "; 1 / R; "A/Wb"
1920 GOTO 2200
2000 CLS : PRINT "CALCULATION OF FLUX USING FLUX = F/S": PRINT
2010 INPUT "MMF (ampere turns) = ", F
2020 INPUT "Reluctance (A/Wb) = ", S: PRINT
2030 PRINT "Flux = "; F / S; "Wb": PRINT
2200 PRINT "Do you wish to:": PRINT
2210 PRINT "1. return to the main menu, or"
2220 PRINT "2. leave the program?": PRINT
2230 INPUT "Enter your selection here: ", S
2240 IF S < 1 OR S > 2 THEN CLS : GOTO 2200
2250 IF S = 1 THEN GOTO 10
2260 END
```

Once again, a comprehensive set of solutions is presented in Computer listing 7.1, dealing with many of the equations introduced in this chapter, namely

1. induced e.m.f. $= B \sin \theta$
2. force $= BI\ell$
3. flux density $= \mu_r \mu_0 H$
4. magnetomotive force $= IN$
5. magnetising force $= F/\ell$
6. reluctance $= l/(\mu_r \mu_0 a)$
7. reluctance of a series magnetic circuit
8. reluctance of a parallel magnetic circuit
9. $\Phi = F/S$.

In the case of option 1, the reader should note that angle θ is with respect to the direction of the magnetic field. That is, if the conductor *moves across the magnetic field*, then $\theta = 90°$.

Exercises

In the following, the permeability of free space (μ_0) has the value $4\pi \times 10^{-7}$ H/m.

7.1 A straight conductor of active length 50 cm cuts a magnetic field perpendicular to the line of action of the field. If the flux density is 0.2 T and the velocity of the conductor is 8 m/s, calculate the e.m.f. induced in the conductor.

7.2 If the conductor in Problem 7.1 moves at an angle of (a) 60 ° and (b) 0.5 rad to the line of action of the magnetic field, determine the e.m.f. induced in the conductor.

7.3 Calculate the e.m.f. generated between the wing-tips of an aircraft which is flying horizontally at a velocity of 650 km/h if the vertical component of the earth's magnetic field is 40 T. The wing span of the aircraft is 46 m.

7.4 A 250-turn coil is rotated in a uniform magnetic field of flux density 0.1 T. If the active length of each conductor is 0.25 m and the linear velocity of the conductor is 8 m/s, calculate the maximum value of e.m.f. induced in the coil.

7.5 A straight conductor carries a current of 50 A, the length of the conductor being 20 cm. If the conductor is at right-angles to a magnetic field of flux density 0.5 T, determine the force acting on the conductor.

7.6 A galvanometer coil carries a current of 10 mA and has 25 turns of wire. If the active length of each side of the coil is 20 mm, and it moves at right-angles to a uniform magnetic field of flux density 0.12 T, calculate the torque produced by the coil if its radius is 20 mm.

7.7 A coil of 400 turns carries a current of 0.5 A and is wound uniformly on a magnetic circuit which is 20 cm in length. Calculate the m.m.f. produced and the magnetising force.

7.8 A mild steel ring of uniform cross-section, and of mean diameter 15 cm is wound with a coil of 800 turns. If the magnetising force is to be 4500 A/m, calculate the current in the coil.

7.9 A ferromagnetic ring has a uniform cross-sectional area of 550 mm^2 and a mean radius of 10 cm. If the flux density in the core is to be 1.3 T, and the relative permeability of the iron is 2000, calculate the reluctance of the magnetic circuit, and the m.m.f. required.

7.10 A coil carrying a current of 1.25 A is uniformly wound on an iron torroid of cross-sectional area 125 mm^2 and mean diameter 10 cm. The coil has 20 turns, and the relative permeability of the iron is 1000. If the magnetic flux in the iron is 3.5 Wb, calculate the current in the coil.

7.11 A 120-turn coil is uniformly wound on a ferromagentic ring of mean diameter 250 mm and cross-sectional area 1000 mm^2. If the current in the coil is 7 A and the magnetic flux in the core is 1 mWb, determine (a) the m.m.f. and the magnetising force at the mean circumference of the ring, (b) the relative permeability of the iron.

7.12 A coil of 250 turns is uniformly wound over a ferromagnetic ring of cross-sectional area 200 mm^2 and mean diameter 120 mm. The coil has a resistance of 10 Ω and is connected to a 25 V d.c. supply. Determine (a) the steady value of current in the coil, (b) the m.m.f. produced by the coil, (c) the magnetising force at the mean circumference, (d) the reluctance of the ring given that the relative permeability of the ring is 2000, (e) the flux density in the ring.

7.13 If the relative permeability of a steel ring at a flux density of 1.3 T is 800, determine the magnetising force required to produce this flux density if the mean length of the ring is 1.0 m.

What m.m.f. would be required to maintain a flux density of 1.3 T if a radial air gap of 1 mm were introduced into the ring? The effects of fringing and magnetic leakage can be neglected.

7.14 A mild steel ring is wound with 1500 turns of wire, and carries a current of 2 A. If the resulting magnetising force is 2000 A/m, determine the mean diameter of the ring.

7.15 A magnetic circuit has a mean length of 475 mm and a uniform cross-sectional area of 5 cm^2. If the magnetic circuit is uniformly wound with 189 turns of wire which carry a current of 1.0 A, and the flux produced is 0.2 mWb, calculate the flux density in the core and the relative permeability of the iron.

7.16 A magnetic circuit consists of three parts X, Y and Z; parts X and Y are of iron and Z is an air gap. Part X has a mean length of 40 cm and a cross-sectional area of 3 cm^2, part Y has a mean length of 15 cm and a cross-sectional area of 3.5 cm^2, and part Z is an air gap of length 0.1 cm and is 3 cm^2 in area.

(a) If magnetic leakage can be neglected, determine the value of the current which must flow in a coil of 1000 turns of wire wound uniformly around the iron circuit if the flux density in the air gap is to be 1.0 T. The characteristic of the magnetic circuit is given in Table 7.2.

(b) If the magnetic circuit has a leakage coefficient of 1.1, and the flux density in the air gap is to remain at 1.0 T, calculate the new value of current in the coil.

Table 7.2 Table for Exercise 7.16

B (T)	0.79	0.925	1.0	1.09	1.11
H (A/m)	500	700	900	1400	1600

7.17 An iron circuit has three parts in series with one another, the length of each part being 0.3 m. The cross-sectional area of the first part is 300 mm^2, that of the second is 600 mm^2 and that of the third is 1200 mm^2. If the relative permeability of the first part is 400, and that of the second and third parts is 2600, calculate the m.m.f. required to produce a magnetic flux in the core of 0.5 mWb. Neglect the effects of fringing and leakage.

7.18 A series magnetic circuit has a ferromagentic part of length 0.4 m and an air gap of length 1.0 mm. The circuit has a coil of 500 turns wound on it, and the cross-sectional area of the steel is 550 mm^2. The following points lie on the magnetic curve of the steel:

H (A/m)	500	1000	2000	5000
B (T)	1.2	1.35	1.45	1.51

Estimate the current required to produce a flux of 0.7 mWb in the air gap if the leakage coefficient is (a) 1.0 and (b) 1.15.

7.19 A 680-turn coil is wound on the centre limb of the cast steel magnetic circuit in Figure 7.6. If magnetic leakage can be neglected, determine the current required in the coil to produce a flux of 1.6 mWb in the air gap. The details of the $B - H$ curve for the magnetic material are given in Table 7.3.

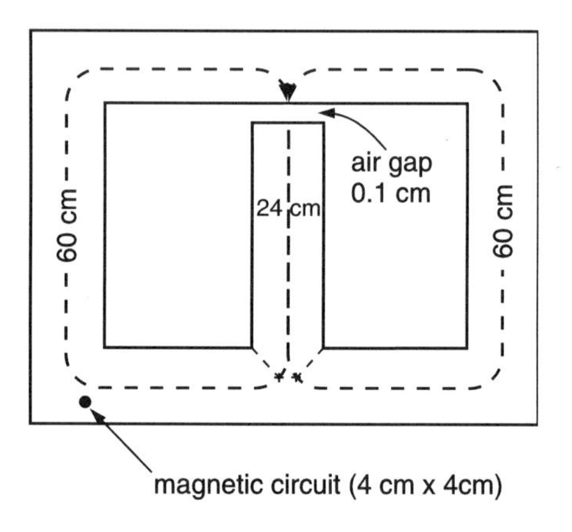

Figure 7.6 Magnetic circuit for Exercise 7.19

Table 7.3 Table for Exercise 7.19

B (T)	0.4	0.6	0.9	1.1	1.9
H (A/m)	470	590	800	1040	1200

7.20 A coil is wound on a non-magnetic ring of length 1.0 m and cross-sectional area 2.5 cm^2. When a certain current passes through the coil, the flux density is found to be 0.06 T. Determine the energy stored in the magnetic circuit.

7.21 Determine the energy stored in the magnetic circuit of a coil of self-inductance 2.5 H when it carries a current of 2.4 A.

Summary of important facts

The **e.m.f. induced in a conductor** when it moves in a perpendicular direction to a magnetic field is

$$\mathbf{e = B\ell v \text{ V}}$$

where e is the induced e.m.f. in volts
B is the flux density in tesla
ℓ is the *active length* of the conductor in metres
v is the velocity of the conductor perpendicular to the magnetic field in metre/s.

If the conductor moves at an angle θ relative to the line of action of the magnetic field, the induced e.m.f. in the conductor is

$$\mathbf{e = B\ell v \sin\theta \text{ V}}$$

The relationship between an angle in degrees and an angle in radians is

$$\theta(\mathbf{deg}) = \theta\,(\mathbf{rad}) \times \mathbf{180/\pi}$$

and

$$\theta(\mathbf{rad}) = \theta\,(\mathbf{deg}) \times \pi\,\mathbf{/180}$$

The force, F, acting on a current-carrying conductor in a magnetic flux density of B tesla is

$$\mathbf{F = BI\ell N}$$

where I is the current in the conductor
ℓ is the active length of the conductor in the magnetic field.

Flux density is given by

$$\mathbf{B = \Phi/a} \text{ tesla}$$

where Φ is the flux in webers
a is the area in m^2 through which the flux passes.
The **magnetomotive force** (m.m.f.), F, produced by a coil is

$$\mathbf{F = NI \text{ ampere turns or amperes}}$$

where N is the number of turns on the coil and I is the current in the coil.
The **magnetomotive force**, H (also known as the **magnetic field intensity**, or the **magnetic field strength**), is given by

$$\mathbf{H = F/\ell = NI/\ell \text{ ampere turns/m or amperes/m}}$$

where ℓ is the length (m) of the magnetic circuit.

The **flux density**, B, and the magnetising force, H, are related by the following equation

$$\mathbf{B = \mu H}$$

where μ is the absolute permeability of the magnetic circuit. The permeability of free space, μ_0, is a constant whose value is

$$\mu_0 = 4\pi \times 10^{-7} \text{ H/m}$$

Also $\mu = \mu_0\mu_r$, where μ_r is the relative permeability of the magnetic material. In the case of air, $\mu_r \approx 1.0$; the value of μ_r can be very high, and its value may vary with the operating flux density.

The **reluctance**, S, of a magnetic circuit is its opposition to magnetic flux, and is given by

$$\boldsymbol{S} = \frac{\ell}{\mu a} = \frac{\ell}{\mu_0\mu_r a} \quad \text{A/Wb}$$

where ℓ is the length (m) of the magnetic circuit

μ is the reluctance of the magnetic circuit
a is the cross-sectional area of the magnetic circuit.

In the case of ***n*** **series-connected magnetic elements**, the effective reluctance of the circuit is

$$\boldsymbol{S_e = S_1 + S_2 + \ldots + S_n} \quad \text{A/Wb}$$

The *reciprocal of the reluctance* of ***n*** **parallel-connected magnetic branches** is

$$\frac{\mathbf{1}}{\boldsymbol{S_e}} = \frac{\mathbf{1}}{\boldsymbol{S_1}} + \frac{\mathbf{1}}{\boldsymbol{S_2}} + \ldots + \frac{\mathbf{1}}{\boldsymbol{S_n}}$$

In the special case of *two parallel-connected magnetic branches* it is

$$\boldsymbol{S_e} = \frac{\boldsymbol{S_1 S_2}}{\boldsymbol{S_1 + S_2}} \quad \text{A/Wb}$$

Magnetic flux which fails to follow the 'useful' path is said to *leak* or to *fringe* from the magnetic circuit, and is accounted for by a **leakage coefficient** as follows:

$$\textbf{leakage coefficient} = \frac{\textbf{total magnetic flux}}{\textbf{useful magnetic flux}}$$

The value of this coefficient is equal to or greater than unity. In a well-designed magnetic circuit, its value lies in the range 1.15 to 1.25.

8 Alternating current

8.1 Introduction

This chapter introduces the reader to the exciting and challenging topic of alternating current. By the end of chapter, the reader will be able to

- Appreciate alternating waveforms.
- Understand the meaning of frequency, angular frequency, periodic time and wavelength of a wave.
- Determine the instantaneous, peak, peak-to-peak, average (mean) and root-mean-square (r.m.s.) value of a wave.
- Calculate the reactance of inductors and capacitors.
- Determine the phase angle of a wave, and the phase angle difference between waves.
- Draw and make calculations with phasors.
- Add and subtract phasors.
- Be introduced to software which handles calculations relating to this chapter.

8.2 Alternating waveforms

In electronic and electrical engineering, the waveform most frequently encountered is the sine wave (or a variant of it, such as the cosine wave). These waves are collectively known as **sinusoidal waves**, several of which are shown in Figure 8.1.

A sinusoidal wave is only one of many **alternating waves**, and a feature of an alternating wave is that *the area under the positive half-cycle of the curve is equal to the area under the negative half cycle*. This is illustrated for the sine wave in Figure 8.1(a).

An alternating wave may be plotted either to a base of time or a base of angle. If the curve is plotted to a base of angle (which may either be in degrees or radians), **one cycle** has been completed after 360° or 2π rad. When we use time as the base of the wave, the time taken for one complete cycle of the wave is known as the **periodic time**.

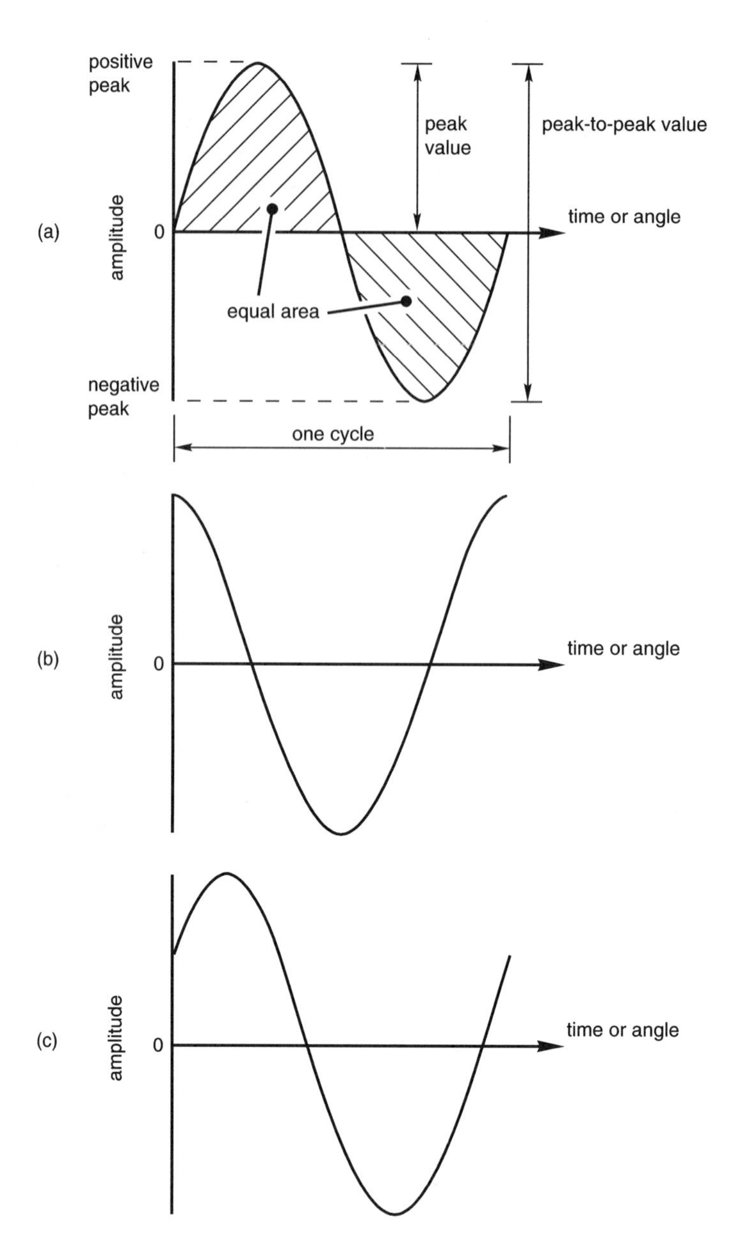

Figure 8.1 Sinusoidal waves: (a) the sine wave; (b) the cosine wave; (c) a sinusoidal wave whose phase shift is between a sine wave and a cosine wave

The maximum positive value of the wave is known as the **positive peak value**, and the maximum negative peak value is the **negative peak value**. In a true alternating wave, these have the same numerical value, and it is known as the **peak value**. The difference between the positive peak and the negative peak is the **peak-to-peak value**. These features are shown in Figure 8.1(a).

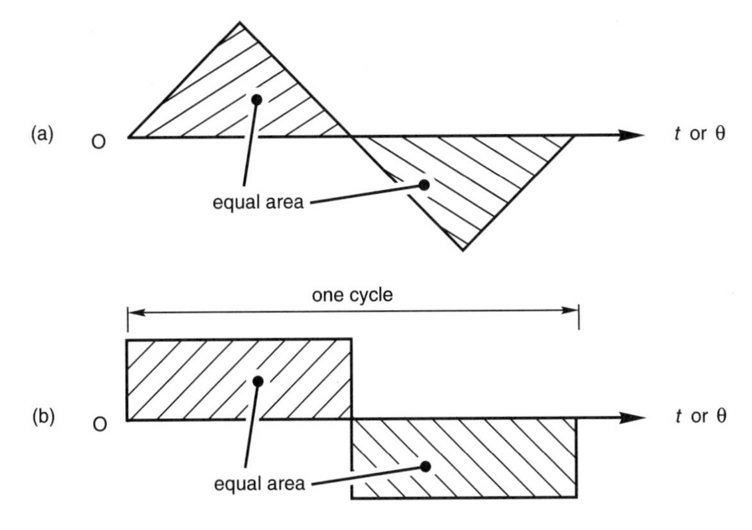

Figure 8.2 Triangular and rectangular waves: (a) one form of triangular wave; (b) a rectangular wave

The sine waves in Figure 8.1 are typical of those produced by an electronic oscillator or an alternator. Figure 8.2 illustrates two **non-sinusoidal waves** and, once again, the area under the positive half-cycle is equal to the area under the negative half-cycle. These waves are typically produced by an electronic oscillator.

Worked Example 8.1

The reader should refer to Chapter 7 for the equations relating to the conversion of degrees to radians, and vice versa.

(a) Convert the following angles into radians: (i) 45°, (ii) 170°, (iii) −95°.
(b) Convert the following angles into degrees: (i) 0.7 rad, (ii) −1.5 rad, (iii) 2π/3 rad.

Solution

(a) (i) $\theta_1 = 45° \times \pi/180 = 0.785$ rad
(ii) $\theta_2 = 170° \times \pi/180 = 2.967$ rad
(iii) $\theta_3 = -95° \times \pi/180 = -1.658$ rad

(b) (i) $\theta_4 = 0.7 \text{ rad} \times 180/\pi = 40.11°$
(ii) $\theta_5 = -1.5 \text{ rad} \times 180/\pi = -85.94°$
(iii) $\theta_6 = (2\pi/3) \text{ rad} \times 180/\pi = 120°$

8.3 Frequency, periodic time, angular frequency and wavelength

The **frequency** of a wave is the number of cycle it completes in one second. If the time taken to complete one cycle is T seconds (which is the **periodic time** of the wave), then

$$\textbf{frequency, } f = \frac{1}{T} \text{ hertz (Hz)}$$

The **angular frequency**, ω, of a wave is given by

$$\omega = 2\pi f \quad \text{rad/s}$$

The **wavelength**, λ, of a wave in metres is related to the frequency by the equation

$$\lambda f = K$$

or

$$\lambda = K/f$$

where K is a constant whose value depends on the medium in which the wave is propagated. If we are dealing with a radio wave in space (or in air), then $K = 3 \times 10^8$. In another medium, such as in a printed circuit board or in a cable, the value of K will be a little less than 3×10^8.

Worked Example 8.2

For a 50 Hz wave, determine (a) its angular frequency, (b) its periodic time, (c) its wavelength if it is propagated in free space (for the moment we will assume that the overhead grid line system is equivalent!).

Solution

(a) The angular frequency of the wave is

$$\omega = 2\pi f = 2\pi \times 50 = 314.2 \text{ rad/s}$$

(b) The periodic time of the wave is

$$T = 1/f = 1/50 = 0.02 \text{ s or } 20 \text{ ms}$$

(c) The wavelength is calculated from

$$\lambda = K/f = 3 \times 10^8/50 = 6 \times 10^6 \text{ m or } 6000 \text{ km}$$

That is, in imperial units, $\lambda = 3729$ miles! From this we see that, by the time the second cycle commences, the beginning of the first cycle would be 6000 km away.

Worked Example 8.3

(a) A radio station transmits on a frequency of 10 MHz. Determine (i) its wavelength and (ii) its angular frequency.
(b) Infra-red waves have a wavelength of around 2×10^{-5} m. What is the frequency of the waves?
(c) What is the wavelength of a transmitting station having a frequency of 1 GHz?

Solution

(a) (i) From the equation $\lambda f = 3 \times 10^8$, we see that

$$\lambda = 3 \times 10^8/f = 3 \times 10^8/10 \times 10^6 = 30 \text{ m}$$

(ii) The angular frequency of the wave is given by

$$\omega = 2\pi f = 2\pi \times 10 \times 10^6$$
$$= 62.84 \times 10^6 \text{ rad/s or } 62.84 \text{ mega rad/s}$$

(b) The equation is

$$f = 3 \times 10^8/\lambda = 3 \times 10^8/2 \times 10^{-5}$$
$$= 1.5 \times 10^{13} \text{ Hz}$$

(c) From $\lambda f = 3 \times 10^8$, we see that

$$\lambda = 3 \times 10^8/f = 3 \times 10^8/1 \times 10^9 = 0.3 \text{ m}$$

8.4 Average value of a wave

The *mathematical average value* of a wave is its average value taken over a complete cycle. It has been pointed out earlier that a true alternating wave has equal positive and negative areas, so that the mathematical average value of an alternating wave is zero!

When electronic and electrical engineers discuss the 'average value' of a wave, it is understood to mean the **rectified average value**, that is the value of either one of the half-cycles of the wave. In the majority of cases, this is the average value of the positive half-cycle.

For a **sinusoidal wave** the (rectified) average value or **mean value** is

$$\textbf{mean value} = \frac{2}{\pi} \times \textbf{maximum value}$$
$$= \mathbf{0.637} \times \textbf{maximum value}$$

The **mean value of any wave** (including a non-sinusoidal wave) which has n mid-ordinates is

$$\textbf{mean value} = \frac{v_1 + v_2 + \ldots + v_n}{n}$$

where $v_1, v_2, \ldots, v_n$ is the value of the mid-ordinates (see Worked Example 8.5).

Worked Example 8.4

If the mean value of a sinusoidal voltage wave is 50 V, determine its maximum value.

Solution

$$\begin{aligned} \text{Since mean value} &= \text{maximum value} \times 2/\pi, \text{ then} \\ \text{maximum value} &= \text{mean value} \times \pi/2 = 50 \times \pi/2 \\ &= 78.5 \text{ V} \end{aligned}$$

Worked Example 8.5

The following values are the *ordinates* of an alternating current wave for the positive half-cycle. Determine the mean value of the wave.

Time (ms)	0	1	2	3	4	5	6	7	8
Current (mA)	0	23	40	50	42	33	25	16	0

Solution

Since we are provided with nine ordinates, there are eight mid-ordinates. Assuming that the points are connected together by straight lines, the *eight mid-ordinates* are as follows:

mid-ordinate 1: $(0 + (23 - 0))/2$	=	11.5 mA
mid-ordinate 2: $(23 + (40 - 23))/2$	=	31.5 mA
mid-ordinate 3: $(40 + (50 - 40))/2$	=	45.0 mA
mid-ordinate 4: $(42 + (50 - 42))/2$	=	46.0 mA
mid-ordinate 5: $(33 + (42 - 33))/2$	=	37.5 mA
mid-ordinate 6: $(25 + (33 - 25))/2$	=	29.0 mA
mid-ordinate 7: $(16 + (25 - 16))/2$	=	20.5 mA
mid-ordinate 8: $(0 + (16 - 0))/2$	=	8.0 mA
sum of mid-ordinates	=	229.0 mA

The 'electrical' mean value of the wave is

$$\frac{\text{sum of mid-ordinates}}{\text{number of mid-ordinates}} = \frac{229}{8} = 28.63 \text{ mA}.$$

8.5 Root-mean-square (r.m.s.) value of a wave

The r.m.s. value of an alternating wave is the **effective value** of the wave, and is equivalent to the d.c. value which produces an equal power in a resistor. The r.m.s. value is calculated *over one complete cycle*. The r.m.s. value of a *sinusoidal wave* is given by

$$\textbf{r.m.s. value} = \frac{\textbf{maximum value}}{\sqrt{2}}$$
$$= \mathbf{0.7071 \times} \textbf{ maximum value}$$

The **r.m.s. value of any wave** (including non-sinusoidal waves) which has n mid-ordinates is

$$\textbf{r.m.s. value} = \sqrt{\left[\frac{v_1^2 + v_2^2 + \ldots + v_n^2}{n}\right]}$$

Worked Example 8.6

What is the maximum value of a sinusoidal wave having an r.m.s. value of 450 V?

Solution

Since the r.m.s. value of a sine wave = maximum value/$\sqrt{2}$, then

$$\text{maximum value} = \sqrt{2} \times \text{ r.m.s. value } = \sqrt{2} \times 450$$
$$= 636.4 \text{ V}.$$

Worked Example 8.7

Determine the r.m.s. value of the waveform in Worked Example 8.5. It may be assumed that the second half-cycle is a mirror image of the first half-cycle.

Solution

For Worked Example 8.5, the mid-ordinate and (mid-ordinate)^2 values for the first half-cycle are as follows:

Mid-ordinate number	Mid-ordinate (mA)	(Mid-ordinate)^2 (mA)^2
1	11.5	132.25
2	31.5	992.25
3	45.0	2025.00
4	46.0	2116.00
5	37.5	1406.25
6	29.0	841.00
7	20.5	420.25
8	8.0	64.00
SUM		7997.00

Since the second half-cycle is a mirror-image of the first half-cycle, the sum of the (mid-ordinate)^2 values for the complete cycle is 15 994 (mA)^2, and there are $(2 \times 8) = 16$ mid-ordinates for the complete cycle. The r.m.s. value of the wave is

$$\sqrt{\left[\frac{\text{sum of (mid-ordinate)}^2\text{ values}}{\text{number of mid-ordinates}}\right]}$$

$$= \sqrt{\left[\frac{15\,944}{16}\right]} = 31.62 \text{ mA.}$$

8.6 Form factor and peak factor (or crest factor)

The **form factor** and **peak factor** (or **crest factor**) of a wave give an indication of its general shape. The factors are described as follows:

$$\textbf{form factor} = \frac{\textbf{r.m.s. value}}{\textbf{average value}}$$

$$\begin{matrix}\textbf{peak factor}\\ \textbf{or}\\ \textbf{crest factor}\end{matrix} = \frac{\textbf{maximum value}}{\textbf{r.m.s. value}}$$

Two differently shapes waves may both, for example, have the same form factor, but the *waves only have the same shape if both the form factor for both of them is the same and the peak factor for both of them is the same.*

For a *sinusoidal wave*, the factors are

form factor = 1.11
peak factor = 1.414.

Worked Example 8.8

Determine the form factor and peak factor for the waveform in Worked Examples 8.5 and 8.7.

Solution

From Worked Examples 8.5 and 8.7 we see that

peak value of the wave = 50 mA
average value of the wave = 28.63 mA
r.m.s. value of the wave = 31.62 mA

Hence

$$\text{form factor} = \frac{\text{r.m.s. value}}{\text{average value}} = \frac{31.62 \times 10^{-3}}{28.63 \times 10^{-3}} = 1.104$$

$$\text{peak factor} = \frac{\text{maximum value}}{\text{r.m.s. value}} = \frac{50 \times 10^{-3}}{31.62 \times 10^{-3}} = 1.58.$$

8.7 Inductive reactance

A pure inductance has no resistance, but when alternating current flows through it, a 'back' e.m.f. is induced in it. This e.m.f. has the effect of opposing the flow of current through it, and the effect is called **inductive reactance**, symbol X_L. The equation for inductive reactance is

$$X_L = 2\pi fL = \omega L$$

If f is in hertz and L in henrys, X_L is in ohms.

Worked Example 8.9

(a) Calculate the inductive reactance of a 10 mH inductor at a frequency of 10 kHz.
(b) If the inductive reactance of a 5 H inductor is 3.2 kΩ, what is the supply frequency?
(c) If the reactance of an inductor is 6 Ω at a frequency of 100 kHz, what is its inductance?

Solution

(a) The equation for inductive reactance is

$$X_L = 2\pi f L = 2\pi \times 10 \times 10^3 \times 10 \times 10^{-3}$$
$$= 6.284\Omega$$

(b) From $X_L = 2\pi f L$, we see that

$$f = X_L/(2\pi L) = 3.2 \times 10^3/(2\pi \times 5)$$
$$= 101.9 \text{ Hz}$$

(c) The equation for L is

$$L = X_L/(2\pi f) = 6/(2\pi \times 100 \times 10^3)$$
$$= 9.55 \times 10^{-6} \text{ H or } 9.55 \ \mu\text{H}.$$

8.8 Alternating current in a pure inductor

We can apply Ohm's law to a pure inductor, and the r.m.s. current in the inductor is given by

$$I_L = \frac{\textbf{r.m.s. voltage across the inductor}}{X_L} = \frac{V_L}{X_L}.$$

Worked Example 8.10

What current flows in a pure inductance of 0.2 H at a frequency of 200 Hz when 50 V r.m.s. is applied to the inductor?

Solution

The reactance of the inductor is

$$X_L = 2\pi f L = 2\pi \times 200 \times 0.2 = 251.3\Omega$$

and the current flowing in it is

$$I_L = V_S/X_L = 50/251.3 = 0.2 \text{ A}.$$

8.9 Capacitive reactance

In an a.c. circuit, the voltage between the terminals of a capacitor is continuously changing. When this occurs, the capacitor charging current also changes; that is, a capacitor draws an alternating current from the supply. The current is limited in value by what is known as the **capacitive reactance**, X_C, which is given by

$$\mathbf{X_C = 1/(2\pi fC) = 1/(\omega C)}$$

If f is in hertz, and C in farads, then X_C is in ohms. The farad is a very large unit of capacitance, and practical capacitors have a value of a few microfarads (μF or 10^{-6} F), nanofarads (nF or 10^{-8} F), or picofarads (pF or 10^{-12} F).

Worked Example 8.11

(a) Calculate the capacitive reactance of a 4 μF capacitor at a frequency of 1 kHz.
(b) A 4.7 nF capacitor has a reactance of 35 Ω. What is (i) the frequency, (ii) the angular frequency of the supply?
(c) A capacitor has a reactance of 0.2 Ω at a frequency of 100 Hz. Determine the capacitance of the capacitor.

Solution

(a) The capacitive reactance is calculated from

$$X_C = 1/(2\pi fC) = 1/(2\pi \times 1 \times 10^3 \times 4 \times 10^{-6})$$
$$= 39.8\Omega$$

(b) From the equation $X_C = 1/(2\pi fC)$, we see that

$$f = 1/(2\pi \times X_C \times C)$$
$$= 1/(2\pi \times 35 \times 4.7 \times 10^{-9})$$
$$= 967\,500 \text{ Hz or } 0.9675 \text{ MHz}$$

(c) The equation for capacitance is

$$C = 1/(2\pi \times X_C \times f)$$
$$= 1/(2\pi \times 0.2 \times 100 \times 10^3)$$
$$= 7.96 \times 10^{-6} \text{ F or } 7.96\ \mu\text{F}$$

8.10 Alternating current in a pure capacitor

As with inductors, we can calculate the r.m.s. current in a pure capacitor using Ohm's law as follows:

$$I_C = \frac{\textbf{r.m.s. voltage across the capacitor}}{\textbf{reactance of the capacitor}} = \frac{V_S}{X_C}.$$

Worked Example 8.12

The sinusoidal current flowing through a pure capacitor is 31.4 mA when 10 V r.m.s. at a frequency of 1 kHz is applied to it. What is the capacitance of the capacitor?

Solution

From $I_C = V_S/X_C$, then

$$X_C = V_S/I_C = 10/31.4 \times 10^{-3} = 318.5\Omega$$

hence

$$C = 1/(2\pi \times X_C \times f) = 1/(2\pi \times 1000 \times 318.5)$$
$$= 0.5 \times 10^{-6} \text{ or } 0.5\ \mu\text{F}.$$

8.11 Phase angle of a sinusoidal wave

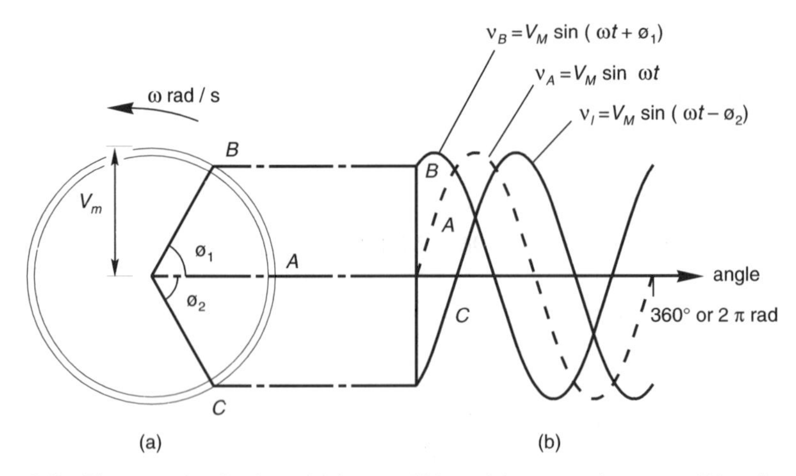

Figure 8.3 Phase angle of a sinusoidal wave. Wave A is a pure sine wave, B is a sine wave which leads wave A by ϕ_1, and wave C lags behind A by ϕ_2

A sine wave is the wave drawn to a base of time or angle which represents the vertical displacement of the tip of a line which rotates at a constant speed in an anti-clockwise direction (see Figure 8.3(a)).

Wave A (see Figure 8.3(b)) is a sine wave representing the displacement of the tip of line A (see Figure 8.3(a), which commences in the horizontal direction or the **reference direction** when $t = 0$. The **instantaneous value**, v_A, of this wave at any angle θ is given by

$$v_A = V_m \sin\theta = V_m \sin\omega t$$

where ω is the angular frequency of the wave, and V_m is the *maximum value* of the wave, which occurs when $\theta = 90°$.

The line B (Figure 8.3(a)) is at angle $+\phi_1$ with respect to the reference direction at $t = 0$, and the graph drawn out by the tip of the line is shown in graph B in Figure 8.3(b). The equation for graph B is

$$v_B = V_m \sin(\theta + \phi_1) = V_m \sin(\omega t + \phi_1)$$

In this case we say that sine wave B *leads* sine wave A by ϕ_1, that is B is in advance of A at all points in time.

The angle ϕ_1 can either be expressed in degrees or in radians. If $\phi_1 = 45°$ (or $\pi/4$ rad), we may say that

$$v_B = V_m \sin(\omega t + 45°) = V_m \sin(\omega t + \pi/4)$$

Line C in Figure 8.3(a) is at angle $-\phi_2$ with respect to the reference direction at $t = 0$, and the corresponding sine wave which is drawn out is shown in graph C in Figure 8.3(b). The equation of graph C is

$$v_C = V_m \sin(\theta - \phi_2) = V_m \sin(\omega t - \phi_2)$$

In this case we say that sine wave C *lags behind* sine wave A, that is graph C is behind graph A at all points in time. If $\phi_1 = 45°$ (or $\pi/4$ rad), we may say that

$$v_C = V_m \sin(\omega t - 45°) = V_m \sin(\omega t - \pi/4).$$

Worked Example 8.13

When $t = 3$ ms, calculate the value of
(a) $v = 100 \sin\omega t$, (b) $i = 20\sin(\omega t + 45°)$, (c) $v = 50\sin(\omega t - 0.6)$.
The frequency is 50 Hz.

Solution

Since $f = 50$ Hz, then

$$\omega = 2\pi f = 2\pi \times 50 = 314.2 \text{ rad/s}$$

The quantity ωt has the dimensions of

$$\frac{\text{rad}}{\text{sec}} \times \text{sec} = \text{rad}$$

We can therefore perform the calculation either in radians or in degrees, and we will do both.

(a) In *radians*

$$\omega t = 314.2 \times 3 \times 10^{-2} \text{ rad} = 0.9425 \text{ rad}$$

and

$$v = 100 \sin 0.9425 = 100 \times 0.809 = 80.9V$$

In *degrees*

$$0.9425 \text{ rad} \equiv (rad \times 57.3)^\circ = 54^\circ$$

and

$$v = 100 \sin 54^\circ = 100 \times 0.809 = 80.9 \text{ V}$$

(b) In *radians*

$$45^\circ \equiv (45/57.3) \text{ rad} = 0.7853 \text{ rad}$$

hence

$$\omega t + 45^\circ \equiv 0.9425 + 0.7853 = 1.7278 \text{ rad}$$

therefore

$$i = 20 \sin 1.7278 \text{ rad} = 20 \times 0.9877 = 19.75 \text{ A}$$

In *degrees*

$$\omega t = 0.9425 \text{ rad} \equiv (0.9425 \times 57.3)^\circ = 54^\circ$$

hence

$$i = 20 \sin(54 + 45)^\circ = 20 \sin 99^\circ$$
$$= 20 \times 0.9877 = 19.75 \text{ A}$$

(**Note**: in the case of example (b), the sine wave has passed through its maximum value.)

(c) In *radians*

$$v = 50 \sin(\omega t - 0.6) = 50 \sin(0.9425 - 0.6)$$
$$= 50 \sin 0.3425 \text{ rad} = 50 \times 0.3358 = 16.79 \text{ V}$$

In *degrees*

$$0.6 \text{ rad} \equiv (0.6 \times 57.3)^\circ = 34.38^\circ$$

and

$$\omega t - 0.6 \text{ rad} \equiv (54 - 34.38)^\circ = 19.62^\circ$$

hence

$$v = 50 \sin 19.62^\circ = 50 \times 0.3358 = 16.79 \text{ V}.$$

Worked Example 8.14

A voltage sine wave has a value of 70.71 V when $t = 1$ ms, the periodic time of the wave being 8 ms. Calculate (a) the frequency and angular frequency of the wave, (b) its value when $t = 4.5$ ms.

Solution

(a) Since $T = 8$ ms, then

$$f = 1/T = 1/8 \times 10^{-3} = 125 \text{ Hz}$$

The angular frequency is

$$\omega = 2\pi f = 2\pi \times 125 = 785.4 \text{ rad/s}$$

(b) The equation of the wave is $v = V_m \sin \omega t$, hence

$$V_m = \frac{v}{\sin \omega t}$$

Now, when $t = 1$ ms

$$\omega t = 785.4 \times 1 \times 10^{-3} = 0.7854 \text{ rad} \equiv (0.7854 \times 57.3)^\circ = 45^\circ$$

therefore

$$V_m = \frac{70.71}{\sin 45^\circ} = 100 \text{ V}$$

Since the wave has completed 360° in 8 ms, then in 4.5 ms we can say that

$$\frac{\text{angle}}{4.5 \times 10^{-3}} = \frac{360^\circ}{8 \times 10^{-3}}$$

or

$$\text{angle} = 360 \times 4.5 \times 10^{-3}/8 \times 10^{-3} = 202.5^\circ$$

The value of the wave at this angle is

$$v = 100 \sin 202.5^\circ = -38.27 \text{ V}.$$

8.12 Introduction to phasors

The magnitude of a sinusoidal quantity continuously changes and, additionally, may have a differing phase relation with other waveforms in the same circuit.

Engineers have devised the concept of the **phasor**, which 'fixes' a sinusoidal quantity both in magnitude and angle. This allows us not only to be able to visualise and 'draw' phasors, but also to add and subtract them using standard graphical and mathematical methods (see also sections 8.13 and 8.14).

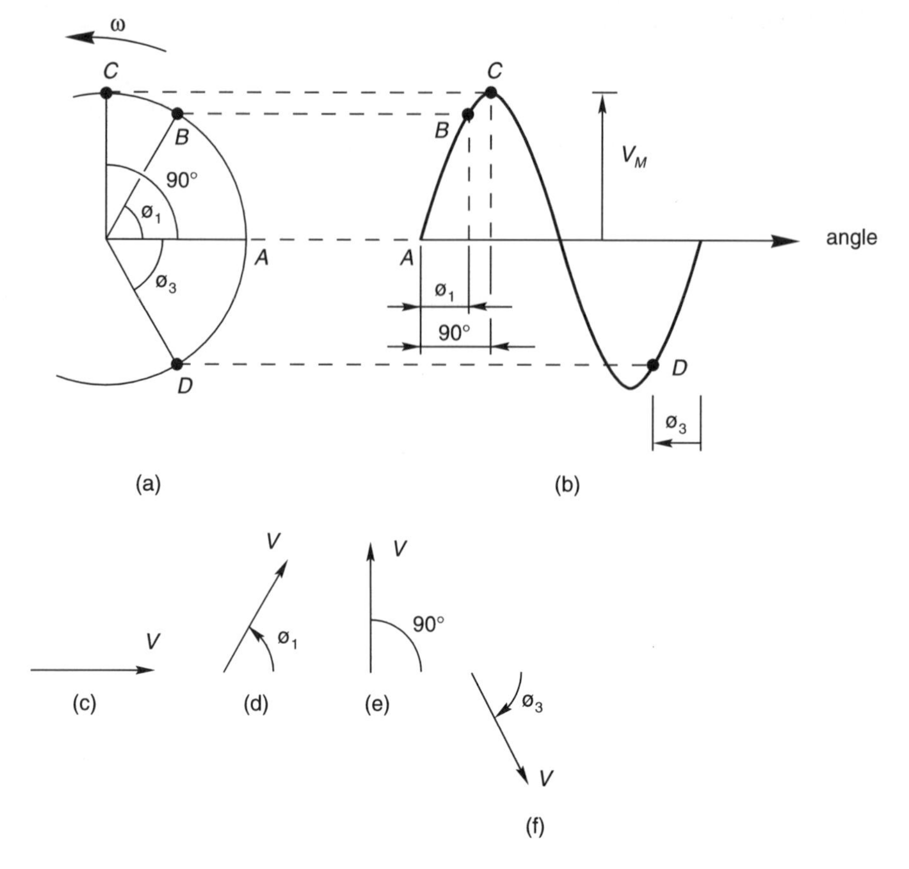

Figure 8.4 Drawing phasors: (a) a rotating line of length V_m which traces out the sine wave in diagram; (b) the phasor representation of the sine wave at (c) 0°, (d) ϕ_1, (e) 90°, (f) ϕ_3

Figure 8.4 illustrates how phasors can be drawn. If a line of length V_m rotates in an anti-clockwise direction, as shown in Figure 8.4(a), the vertical displacement of the tip of the line traces out the sine wave in Figure 8.4(b). We can represent this waveform by one of many phasors as illustrated in diagrams (c)–(f).

The length of the phasor is scaled down to represent the r.m.s. value of the waveform, and it is 'frozen' at some particular angle (or, alternatively, at some particular time). In diagram (c) we have 'frozen' the rotating line at angle $\phi = 0$ (or $t = 0$), so that the phasor lies in the horizontal or reference direction. If the rotating line is 'frozen' at angle ϕ_1, we get the phasor in diagram (d); technically, the phasors in (c) and (d) both represent the sine wave, but at different points in time.

If the rotating line is frozen at $\phi = 90°$, we get the phasor in diagram (e). Once again, the length of the phasor is equal to the r.m.s. value of the waveform, i.e. $V_m/\sqrt{2}$. When we allow the rotating line to move to point D in Figure 8.4(a), the resulting phasor is shown in figure 8.4(f).

8.13 Addition of phasors

In a.c. circuits we are often involved in connecting voltage in series which have differing magnitudes and phase angles. In this case **we must add the phasor voltages together**; we *cannot* simply add the magnitudes together and, separately, add the phase angles. Similarly, in parallel circuits, the total current flowing into the circuit is the phasor sum of the current in all the branches. Methods of phasor addition are described below.

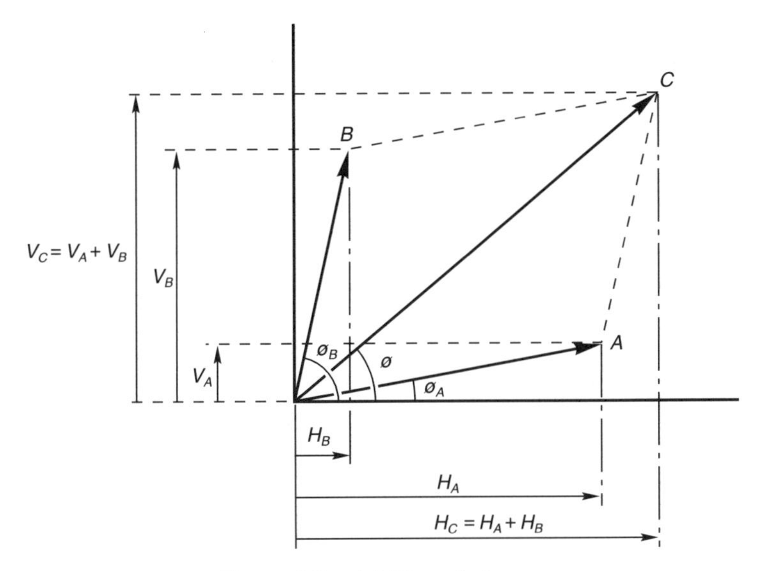

Figure 8.5 Addition of phasors

Figure 8.5 illustrates how phasors A and B are **added graphically** by completing the parallelogram $OACB$. The **magnitude** or **modulus** of the resultant phasor is given by OC, and the **phase angle** or **argument** of OC is equal to ϕ.

The reader should bear in mind that any result obtained graphically is not exact, since the accuracy depends not only on the scale used to draw the phasors, but also on the skill and care employed.

Accurate results are obtained by **calculation** using the length and angle of phasors A and B, as follows: The *resolved components* of phasor A are

$$H_A = A\cos\phi_A$$
$$V_A = A\sin\phi_A$$

and those for phasor B are

$$H_B = B\cos\phi_B$$
$$V_B = B\sin\phi_B$$

The horizontal resolved component of phasor sum C is

$$H_C = H_A + H_B$$

and the vertical resolved component of phasor sum C is

$$V_C = V_A + V_B$$

When we represent a phasor using resolved components in the horizontal and vertical directions, we say that we are using **Cartesian coordinates** or **rectangular components**.

The reader is advised that, when either adding or subtracting phasors using the mathematical method, they *must be* converted into their rectangular components.

The magnitude or modulus of the resultant phasor is

$$C = \sqrt{(H_C^2 + V_C^2)}$$

and its phase angle is

$$\phi = \arctan(V_C/H_C) = \tan^{-1}(V_C/H_C)$$

(**Note**: 'arctan' and '$\tan^{-1}$' mean 'the angle whose tangent is'. The use of arctan is to be preferred because $\tan^{-1}\phi$ could be confused with $1/\tan\phi$ (which are not the same thing)).

As a point of interest, when a phasor is specified in terms of its phase angle and its magnitude, we say that we are giving its **polar components**.

Examples of the addition of phasors are given in Worked Examples 8.16 and 8.17, which follow section 8.14.

8.14 Conversion between rectangular and polar coordinates

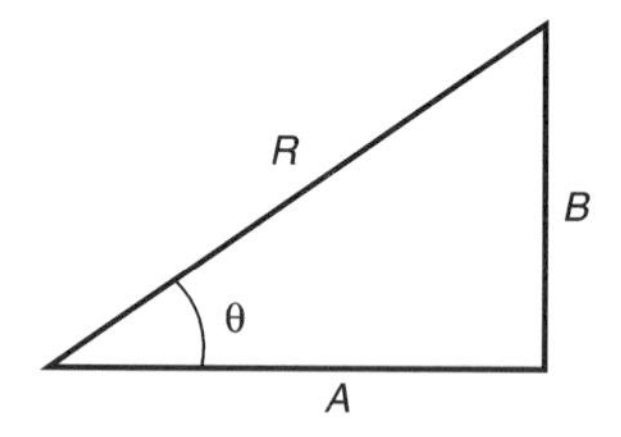

Figure 8.6 Relationship between rectangular and polar coordinates

Information about phasors may be provided either in rectangular or polar coordinates, and we need to be able to convert from one form to the other. Consider the phasor in Figure 8.6; it may be described in its *polar form* as

magnitude $= R$
phase angle $= \theta$

or in its rectangular form as

reference (horizontal) component = A
quadrature (vertical) component = B

Using our knowledge of triangles, we can say that

$$\boldsymbol{R} = \sqrt{(\boldsymbol{A}^2 + \boldsymbol{B}^2)}$$

and

$$\theta = \textbf{arctan}\,(\boldsymbol{B}/\boldsymbol{A})$$

or

$$\boldsymbol{A} = \boldsymbol{R}\cos\theta$$
$$\boldsymbol{B} = \boldsymbol{R}\sin\theta.$$

Worked Example 8.15

(a) What are the rectangular components of a phasor whose magnitude is $V_S = 240$ V, and whose phase angle is 65°?
(b) Determine the polar components of a phasor whose reference component is −6 V, and whose quadrature component is 4 V.

Solution

(a) In this case the component in the reference direction is

$$A = 240\cos 65° = 101.4 \text{ V}$$

and the component in the quadrature direction is

$$B = 240\sin 65° = 217.5 \text{ V}$$

(b) Here the component A (the component in the reference direction) is −6 V, and the component B is 4 V, hence

$$\text{magnitude} = \sqrt{((-6)^2 + 4^2)} = 7.21 \text{ V}$$

and its phase angle is

$$\arctan\theta = B/A = 4/(-6) = -0.6667$$

At this point the reader is cautioned about the use of a calculator in evaluating the result, because a calculator would indicate that $\theta = -33.69°$! However, if we look at the mathematical sign of the rectangular components (A is negative and B is positive), we see that the phasor lies in the *second quadrant*, and not in the third quadrant, as indicated by a calculator. The correct angle is

$$\theta = -33.69° + 180° = 146.31°.$$

Worked Example 8.16

Two alternating voltages are connected in series with one another. Voltage V_1 has an r.m.s. value of 10 V and a phase angle of 80° with respect to the reference direction, and voltage V_2 has a value of 8 V at an angle of 10° to the reference direction.

Draw the corresponding phasor diagram, and determine the magnitude and phase angle of the resultant phasor.

Solution

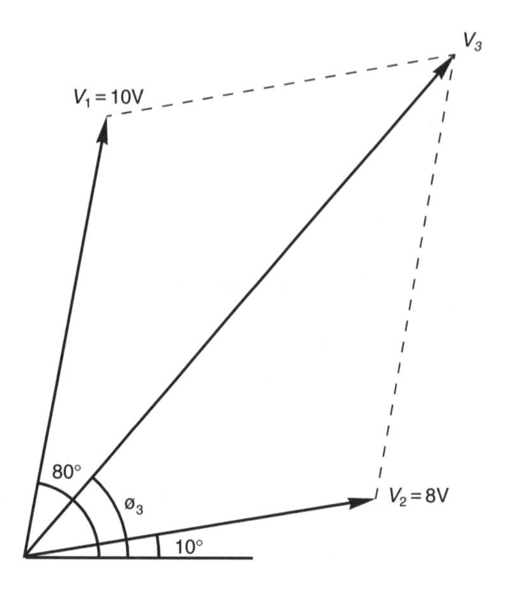

Figure 8.7 Phasor diagram for Worked Example 8.16

The phasor diagram is shown in Figure 8.7, and the reader should draw it to scale to obtain a graphical solution to compare with the mathematical solution obtained below.

For phasor V_1, the resolved components are

$$\text{reference component} = V_1 \cos\phi_1 = 10\cos 80° = 1.736 \text{ V}$$
$$\text{quadrature component} = V_1 \sin\phi_1 = 10\sin 80° = 9.848 \text{ V}$$

For phasor V_2, the resolved components are

$$\text{reference component} = V_2 \cos\phi_2 = 8\cos 10° = 7.878 \text{ V}$$
$$\text{quadrature component} = V_2 \sin\phi_2 = 8\sin 10° = 1.389 \text{ V}$$

The resultant components of the phasor sum are

$$\text{reference component} = 1.736 + 7.878 = 9.614 \text{ V}$$

$$\text{quadrature component} = 9.848 + 1.389 = 11.237 \text{ V}$$

From these figures we can evaluate the magnitude and phase angle of the sum as follows:

$$V_3 = \text{magnitude of the phasor sum}(V_1 + V_2)$$

$$= \sqrt{(9.614^2 + 11.237^2)} = 14.79 \text{ V}$$

and its phase angle is

$$\phi = \arctan(11.237/9.614) = 49.45°$$

(**Note**: The reader should realise that, in practice, we cannot measure either the magnitude or the phase angle to the accuracy given here. All calculations of the type should be tempered by the accuracy not only of the measuring instruments used, but also of the care of the user!)

Worked Example 8.17

The current in one branch of a two-branch parallel circuit is 10 A at a phase angle of −10° with respect to the reference direction, and the resolved components of the current in the other branch are

$$\text{reference component} = -14.1 \text{ A}$$

$$\text{quadrature component} = -5.13 \text{ A}$$

What is the magnitude and phase angle of the total current drawn by the circuit? Draw a phasor diagram showing all the currents in the circuit.

Solution

Branch 1

Before we can add phasors together, we must convert them into their rectangular (resolved) components. In this case, the magnitude of I_1 is 10 A at a phase angle of −10°, so that

$$\text{reference component} = 10\cos(-10°) = 9.848 \text{ A}$$

$$\text{quadrature component} = 10\sin(-10°) = -1.736 \text{ A}$$

Branch 2

The resolved components for the current in this branch are provided in the question and are

$$\text{reference component} = -14.1 \text{ A}$$

$$\text{quadrature component} = -5.13 \text{ A}$$

Complete circuit
The resolved components for the current drawn by the complete circuit are

$$\text{reference current} = 9.848 + (-14.1) = -4.252 \text{ A}$$
$$\text{quadrature current} = -1.736 + (-5.13) = -6.866 \text{ A}$$

Hence

$$\text{magnitude of total current} = \sqrt{((-4.252)^2 + (-6.866)^2)} = 8.076 \text{ A}$$

and

$$\text{phase angle of total current} = \arctan(-6.866/(-4.252)) = \arctan 1.6148.$$

At this point, the reader is cautioned once again about the use of a calculator for the determination of angles. The solution given by the calculator for the above 'arctan' value is +58.23°. However, if we look at the mathematical signs of the reference and quadrature components, we see that they are both negative, and the phase angle must lie in the third quadrant! That is the phase angle is

$$58.23° - 180° = -121.77°$$

The corresponding phasor diagram for the circuit is shown in Figure 8.8.

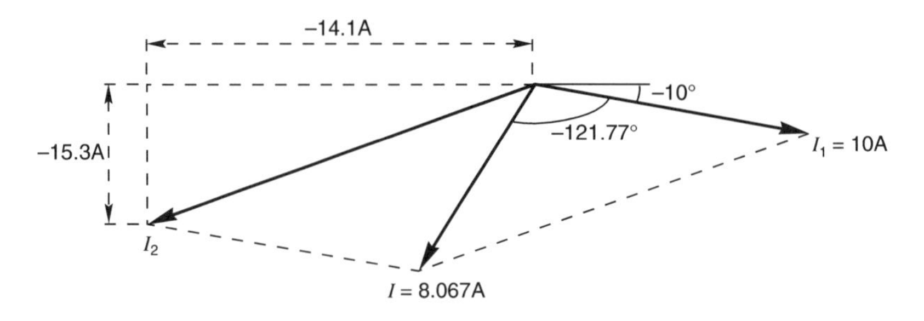

Figure 8.8 Phasor diagram for Worked Example 8.17

8.15 Subtraction of phasors

In the subtraction process (see Figure 8.9)

$$\text{phasor } C = \text{phasor } A - \text{phasor } B$$
$$= \text{phasor } A + (-\text{ phasor } B)$$

phasor A is known as the **minuend**, phasor B as the **subtrahend** and phasor C the **difference**. In effect, phasor subtraction is a process of reversing phasor B and

adding it to phasor A. This is illustrated in Figure 8.9, in which phasor B is reversed (see broken line), and is then added to phasor A to give

$$C = A + (-B) = A - B.$$

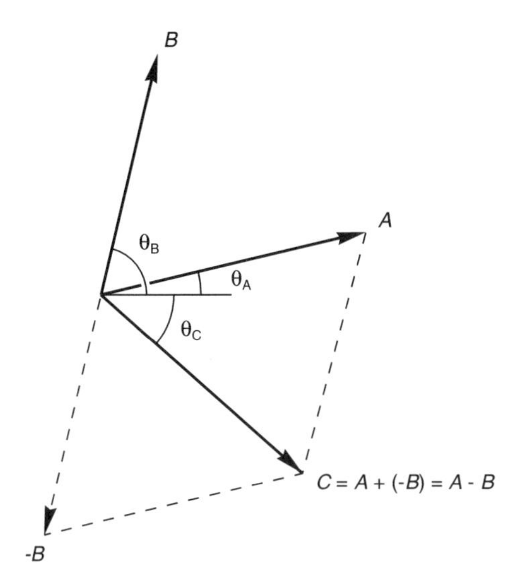

Figure 8.9 Subtraction of phasors

Worked Example 8.18

The total current flowing into a two-branch parallel circuit has a magnitude of 20 A and a phase angle of 10° with respect to the reference direction. The current flowing into branch 1 has a magnitude of 15 A and a phase angle of −40° with respect to the reference direction. Determine the magnitude and phase angle of the current flowing in the second branch. Draw a phasor diagram showing all the currents in the circuit.

Solution

If I_T is the total current flowing into the parallel circuit, then

$$\text{phasor } I_T = \text{phasor sum of}(I_1 + I_2)$$

The phasor for current I_2 is calculated from the following:

$$\begin{aligned} \text{phasor } I_2 &= \text{phasor difference}(I_T - I_1) \\ &= I_T + (-I_1) \end{aligned}$$

Before we can perform the phasor subtraction mathematically, we must first resolve the phasors I_T and I_1 into their reference and quadrature components as follows.

Phasor I_T

$$\text{reference component} = I_T \cos \phi_T = 20 \cos 10^\circ$$
$$= 19.7 \text{ A}$$
$$\text{quadrature component} = I_T \sin \phi_T = 20 \sin 10^\circ$$
$$= 3.47 \text{ A}$$

Phasor I_1

$$\text{reference component} = I_1 \cos \phi_1 = 15 \cos(-40^\circ)$$
$$= 11.5 \text{ A}$$
$$\text{quadrature component} = I_1 \sin \phi_1 = 15 \sin(-40^\circ)$$
$$= -9.64 \text{ A}$$

Calculation of I_2

$$\text{reference component} = I_T \cos \phi_T - I_1 \cos \phi_1$$
$$= 19.7 - 11.5 = 8.2 \text{ A}$$
$$\text{quadrature component} = I_T \sin \phi_T - I_1 \sin \phi_1$$
$$= 3.47 - (-9.64) = 13.11 \text{ A}$$

Hence

$$\text{magnitude of } I_2 = \sqrt{(8.2^2 + 13.11^2)} = 15.46 \text{ A}$$

and the phase angle of I_2 is

$$\phi_2 = \arctan(13.11/8.2) = 58^\circ$$

The phasor diagram for the circuit is shown in Figure 8.10.

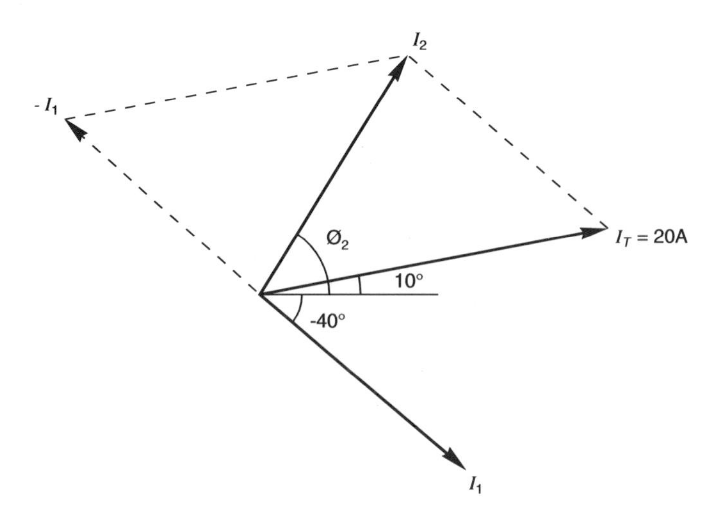

Figure 8.10 Phasor diagram for Worked Example 8.18

8.16 Computer listings

Listing 8.1 Waveform calculations

```
10 CLS '*** CH 8-1 ***
20 PRINT "WAVEFORM CALCULATIONS"
30 DIM D(36): FOR Y = 0 TO 36: D(Y) = 0: NEXT Y
40 PRINT : SUM = 0: SUMSQ = 0: MAX = 0
50 PRINT "Select ONE of the following:": PRINT
60 PRINT "1. Convert degrees to radians."
70 PRINT "2. Convert radians to degrees."
80 PRINT "3. For a SINUSOIDAL WAVE, determine the r.m.s."
90 PRINT "   and average value from the peak value."
100 PRINT "4. Convert Hz to rad/s and wavelength."
110 PRINT "5. For ANY WAVE, determine the average and r.m.s. value."
120 PRINT
130 INPUT "Enter your selection here (1 - 5): ", S
140 IF S < 1 OR S > 5 THEN GOTO 10
150 IF S = 2 THEN GOTO 310
160 IF S = 3 THEN GOTO 430
170 IF S = 4 THEN GOTO 570
180 IF S = 5 THEN GOTO 700
190 CLS : PRINT "CONVERT DEGREES TO RADIANS": PRINT
200 INPUT "Number of values to be converted (max. 5) = ", N
210 IF N < 1 OR N > 5 THEN GOTO 190
220 PRINT
230 FOR Y = 1 TO N
240     INPUT "Angular value in degrees = ", D(Y)
250 NEXT Y
260 PRINT
270 FOR Y = 1 TO N
280     PRINT "Angle = "; D(Y) * 3.14159 / 180; "rad"
290 NEXT Y
300 GOTO 1000
310 CLS : PRINT "CONVERT RADIANS TO DEGREES": PRINT
320 INPUT "Number of values to be converted (max. 5) = ", N
330 IF N < 1 OR N > 5 THEN GOTO 310
340 PRINT
350 FOR Y = 1 TO N
360     INPUT "Angular value in radians = ", D(Y)
370 NEXT Y
380 PRINT
390 FOR Y = 1 TO N
400     PRINT "Angle = "; D(Y) * 180 / 3.14159; "degrees"
410 NEXT Y
420 GOTO 1000
430 CLS : PRINT "R.M.S. AND AVERAGE VALUE FROM PEAK VALUE"
440 PRINT
```

```
450 INPUT "Number of values to be calculated (max. 5) = ", N
460 IF N < 1 OR N > 5 THEN GOTO 430
470 PRINT
480 FOR Y = 1 TO N
490     INPUT "Maximum value = ", D(Y)
500 NEXT Y
510 PRINT
520 FOR Y = 1 TO N
530     PRINT "R.M.S. = "; D(Y) / SQR(2)
540     PRINT "Average = "; D(Y) * 2 / 3.14159
550 NEXT Y
560 GOTO 1000
570 CLS : PRINT "CONVERT Hz TO rad/s AND WAVELENGTH": PRINT
580 INPUT "Number of frequencies to be converted (max. 5) = ", N
590 IF N < 1 OR N > 5 THEN GOTO 570
600 PRINT
610 FOR Y = 1 TO N
620     INPUT "Frequency in Hz = ", D(Y)
630 NEXT Y
640 PRINT
650 FOR Y = 1 TO N
660     PRINT "Frequency = "; D(Y) * 6.28319; "rad/s"
670     PRINT "Wavelength = "; 3E+08 / D(Y); "m"
680 NEXT Y
690 GOTO 1000
700 CLS : PRINT "AVERAGE AND R.M.S. VALUE OF A WAVE": PRINT
710 INPUT "Number of MID-ORDINATES (max. 36) = ", N
720 IF N < 2 OR N > 36 THEN GOTO 700
730 FOR Y = 1 TO N
740     PRINT "For MID-ORDINATE "; Y
750     INPUT "       value = ", D(Y)
760     IF Y = 1 THEN MAX = D(Y)
770     IF Y = 1 THEN MIN = D(Y)
780     IF D(Y) > MAX THEN MAX = D(Y)
790     IF D(Y) < MIN THEN MIN = D(Y)
800     SUM = SUM + D(Y)
810     SUMSQ = SUMSQ + (D(Y)) ^ 2
820 NEXT Y
830 PRINT
840 AV = SUM / N: RMS = SQR(SUMSQ / N)
850 PRINT "Average value = "; AV
860 PRINT "R.M.S. value = "; RMS
870 PRINT "Maximum value = "; MAX
880 PRINT "Minimum value = "; MIN
890 IF MAX < ABS(MIN) THEN MAX = ABS(MIN)
900 PRINT "Magnitude of peak value = "; MAX
910 IF AV = 0 THEN PRINT "Form factor indeterminate": GOTO 930
```

```
920 PRINT "Form factor = "; ABS(RMS / AV)
930 PRINT "Peak factor or crest factor = "; ABS(MAX / RMS)
1000 PRINT
1010 PRINT "Do you wish to:": PRINT
1020 PRINT "1. return to the menu, or"
1030 PRINT "2. leave the program?": PRINT
1040 INPUT "Enter your selection here (1 or 2): ", S
1050 IF S < 1 OR S > 2 THEN CLS : GOTO 1000
1060 IF S = 1 THEN GOTO 10
1070 END
```

There are two Computer listings for this chapter. Listing 8.1 performs five useful calculations as follows:

1. Convert degrees to radians.
2. Convert radians to degrees.
3. For a *sinusoidal wave*, determine the r.m.s. and average value from the peak value.
4. Convert frequency in Hz to angular frequency in rad/s and wavelength (which assumes the wave is in free space).
5. Determine the average and r.m.s. value of a wave using the mid-ordinate rule.

The equations involved in the software have been explained and demonstrated in this chapter.

Lines 700–910 of Listing 8.1 accepts up to 36 mid-ordinate values for a wave and determines

(a) the *mathematical average value* of the wave
(b) the r.m.s. value of the wave
(c) the maximum value
(d) the minimum value
(e) the form factor of the wave
(f) the peak factor or crest factor of the wave

The average value calculated is the mathematical average value, so that if details are provided for a true alternating wave, its average value will be zero, and the form factor is indeterminate (see line 910). If the rectified average value is required, the reader need only provide details for the positive half-cycle of the wave. Line 890 accounts for the case when data is supplied for a waveform which has only negative values.

Listing 8.2 carries out three very important calculations as follows:

A Convert up to five phasors in *polar form* into their equivalent *rectangular form*
B Add up to five phasors together
C Subtract one phasor from another.

Listing 8.2 contains a different approach to its menu. In this case we use the letters *A, B* and *C* to select the options, and it rejects from any other key (see lines 90 and 100).

Listing 8.2 Addition and subtraction of phasors

```
10 CLS '*** CH8-2 ***
20 DIM D(6, 4): PI = 3.14159
30 PRINT "ADDITION AND SUBTRACTION OF PHASORS": PRINT
40 PRINT "Select ONE of the following": PRINT
50 PRINT TAB(4); "A. Convert POLAR to RECTANGULAR coordinates."
60 PRINT TAB(4); "B. ADD up to FIVE phasors together."
70 PRINT TAB(4); "C. SUBTRACT ONE phasor from ANOTHER phasor."
80 PRINT TAB(4); "T. Terminate the program."
90 K$ = INKEY$: IF K$ = "" THEN GOTO 90
100 IF NOT INSTR("AaBbCcTt", K$) > 0 THEN GOTO 90
110 CLS : IF INSTR("Tt", K$) > 0 THEN GOTO 230
120 FOR R = 0 TO 6: FOR C = 0 TO 4: D(R, C) = 0: NEXT C: NEXT R
130 IF INSTR("Aa", K$) > 0 THEN GOSUB 300: GOTO 160 ' * POLAR TO RECT *
140 IF INSTR("Bb", K$) > 0 THEN GOSUB 700: GOTO 160 ' * ADD PHASORS *
150 IF INSTR("Cc", K$) > 0 THEN GOSUB 1100 ' * SUBTACT PHASORS *
160 PRINT
170 PRINT "Do you wish to:": PRINT
180 PRINT "1. return to the main menu, or"
190 PRINT "2. leave the program?": PRINT
200 INPUT "Enter your selection here: ", S
210 IF S < 1 OR S > 2 THEN CLS : GOTO 160
220 IF S = 1 THEN GOTO 10
230 CLS
240 PRINT "Thank you for your company. Have a nice day."
250 END
300 CLS : PRINT "Convert POLAR to RECTANGULAR coordinates."
310 PRINT
320 INPUT "Number of values to be converted (Max. 5) = ", No: PRINT
330 IF No < 1 OR No > 5 THEN GOTO 300
340 FOR N = 1 TO No
350      PRINT TAB(8); "For phasor "; N
360      INPUT "Modulus = ", D(N, 2)
370      INPUT "Angle (deg.) = ", D(N, 3)
380      D(N, 3) = D(N, 3) * PI / 180 ' * Convert deg. to rad. *
390      ' *** CHECK POLAR VALUE ***
400      D(6, 3) = D(N, 3) ' * Move angle to a holding location *
410      IF D(6, 3) > 2 * PI THEN D(6, 3) = D(6, 3) - 2 * PI * INT(D(6, 3) / (2 * PI))
420      IF D(6, 3) > PI THEN D(6, 3) = D(6, 3) - 2 * PI
430      IF D(6, 3) < -2 * PI THEN D(6, 3) = D(6, 3) + 2 * PI * INT(ABS(D(6, 3)) / (2
         * PI))
440      IF D(6, 3) < -PI THEN D(6, 3) = D(6, 3) + 2 * PI
450      ' *** CALCULATE RECTANGULAR COORDINATES ***
460      D(N, 3) = D(6, 3) ' ** REPLACE CORRECTED ANGLE **
470      D(N, 0) = D(N, 2) * COS(D(N, 3)) ' *** Inphase component ***
480      D(N, 1) = D(N, 2) * SIN(D(N, 3)) ' *** Quadrature component ***
490 NEXT N
500 CLS
```

```
510 FOR N = 1 TO No
520     PRINT "For phasor "; N
530     PRINT TAB(3); "Modulus = "; D(N, 2); ": Angle(deg) = "; D(N, 3) * 180 / PI
550     PRINT TAB(3); "Inphase component = "; D(N, 0); ": Quadrature component
        = "; D(N, 1)
580 NEXT N
590 RETURN
700 CLS : PRINT "ADD PHASORS TOGETHER": PRINT
710 INPUT "Number of phasors to be added (max. 5) = ", No: PRINT
720 IF No < 2 OR No > 5 THEN GOTO 700
730 D(6, 0) = 0: D(6, 1) = 0
740 FOR N = 1 TO No
750     PRINT TAB(8); "For phasor "; N
760     INPUT "Inphase component = ", D(N, 0)
770     INPUT "Quadrature component = ", D(N, 1)
780     D(6, 0) = D(6, 0) + D(N, 0)'** total inphase component **
790     D(6, 1) = D(6, 1) + D(N, 1)'** total quadrature component **
800 NEXT N
810 CLS
820 PRINT "Sum of inphase components = "; D(6, 0)
830 PRINT "Sum of quadrature components = "; D(6, 1)
840 PRINT
850 PRINT "Modulus of sum = "; SQR(D(6, 0) ^ 2 + D(6, 1) ^ 2)
860 GOSUB 900 '*** Check special conditions for angle ***
870 PRINT "Phase angle of sum (deg.) = "; D(6, 3) * 180 / PI
880 RETURN
900 '*** SPECIAL CONDITIONS FOR ANGLE CALCULATION ***
910 IF D(6, 0) = 0 AND D(6, 1) > 0 THEN D(6, 3) = PI / 2: RETURN
920 IF D(6, 0) = 0 AND D(6, 1) < 0 THEN D(6, 3) = -PI / 2: RETURN
930 IF D(6, 0) = 0 AND D(6, 1) = 0 THEN D(6, 3) = 0: RETURN
940 IF D(6, 0) < 0 AND D(6, 1) = 0 THEN D(6, 3) = PI: RETURN
950 D(6, 3) = ATN(D(6, 1) / D(6, 0))
960 IF D(6, 0) < 0 AND D(6, 1) > 0 THEN D(6, 3) = D(6, 3) + PI: RETURN
970 IF D(6, 0) < 0 AND D(6, 1) < 0 THEN D(6, 3) = D(6, 3) - PI
980 RETURN
1100 CLS : PRINT "SUBTRACT PHASOR B FROM PHASOR A": PRINT
1110 D(6, 0) = 0: D(6, 1) = 0
1120 INPUT "Inphase component of A = ", D(0, 0)
1130 INPUT "Quadrature component of A = ", D(0, 1): PRINT
1140 INPUT "Inphase component of B = ", D(1, 0)
1150 INPUT "Quadrature component of B = ", D(1, 1): PRINT
1160 D(6, 0) = D(0, 0) - D(1, 0): D(6, 1) = D(0, 1) - D(1, 1)
1170 PRINT "Difference between inphase components = "; D(6, 0)
1180 PRINT "Difference between quadrature components = "; D(6, 1): PRINT
1190 PRINT "Modulus of difference = "; SQR(D(6, 0) ^ 2 + D(6, 1) ^ 2)
1200 GOSUB 900 '*** Check special conditions for angle ***
1210 PRINT "Phase angle of difference (deg.) = "; D(6, 3) * 180 / PI
1220 RETURN
```

Lines 300 to 590 deal with subroutines for converting up to five phasors in polar form into the rectangular equivalents. Details of the phasors are held in array D (a DATA array), which is DIMensioned as a two-dimensional array in line 20. The computer reads the angle in degrees in line 370, and is converted into radians in line 380 (computers always work in radians). Lines 410 to 440 and lines 900 to 980 convert any angle into a 'practical' value in the range 0° to +180° or 0° to −180° before converting it into its rectangular coordinates. For example, it converts 240° into −120°, and 1920° into 120°.

Lines 500–880 deal with the addition of up to five separate phasors. The program is designed to handle phasor values given in *rectangular coordinates*, and it is for this reason that the polar-to-rectangular component conversion routine is included in Listing 8.2. The phasor values and the sum are calculated in both rectangular and polar component versions.

One phasor is subtracted from another using the routine in lines 1100 to 1220. Once again, data must be provided in rectangular coordinate form; the original data and the solution are displayed by the computer in both rectangular and polar coordinate form.

Exercises

8.1 Convert the following: (a) 25° into radians, (b) 125° into radians, (c) −246° into radians, (d) −896° into radians, (e) 0.7 radians into degrees, (f) −4.1 radians into degrees, (g) 9 radians into degrees.

8.2 Determine the angular frequency and periodic time of the following frequencies: (a) 50 Hz, (b) 1 kHz, (c) 10 MHz.

8.3 What is the frequency and periodic time of a wave whose wavelength is 1 cm?

8.4 If a wave has an angular frequency of 600 000 rad/s, what is its frequency and periodic time?

8.5 Data (given in the form of ordinates) for two separate half-waves of voltage are given below. Plot both waves to a base of angle and, by the mid-ordinate method, calculate the mean value for each wave. Assume that the points are linked by straight lines.

θ *(degrees)*	*Wave X*	*Wave Y*
0	4	50
18	10	50
36	37	50
54	50	50
72	37	43
81	10	27.5
90	4	0
99	2	27.5
108	1	43
126	0	50
144	1	50
162	2	50
180	4	50

8.6 For the waves in Problem 8.5, determine the r.m.s. value, the form factor, and the peak factor for each wave (assume that the second half-cycle is a mirror image of the first half-cycle).

8.7 Determine the inductive reactance of (a) a 2 H inductor at a frequency of 50 Hz, (b) a 1 μH inductor at a frequency of 1 MHz.

8.8 A pure inductor is connected to an audio-frequency supply of 25 V, 5 kHz. Calculate the reactance of the inductor, and the r.m.s. value of the current in the circuit.

8.9 A pure inductor of reactance 40 Ω is supplied at 50 Hz. What is the inductance of the inductor?

8.10 At what supply frequency will a 0.08 H inductor have a reactance of 80 Ω?

8.11 Details of two air-cored coils are as follows: the first has an inductance 0.02 H and the inductance of the second is 0.03 H; the supply frequency being 100 Hz. If the effect of mutual inductance can be neglected, what is the effective inductive reactance of the circuit when they are connected (a) in series (b) in parallel?

8.12 A 10 μF capacitor is connected to a variable-frequency supply. Calculate the reactance of the capacitor at 40 Hz, 500 Hz and 1000 Hz.

8.13 When connected to a 500 Hz sinusoidal a.c. supply, a capacitor has a reactance of 50 Ω. What is the capacitance of the capacitor?

8.14 At what frequency will a capacitor of 100 nF have a reactance of 1 kΩ?

8.15 Four 0.1 μF capacitors are connected (a) in parallel, (b) in series to a 10 V, 1 kHz supply. Calculate the current drawn from the supply in each case.

8.16 A low-resistance a.c. ammeter is connected in series to a capacitor, the supply voltage being 240 V, 50 Hz. If the current in the circuit is 4 A, determine (a) the capacitive reactance, (b) the capacitance of the capacitor.

8.17 A capacitor bank of effective capacitance 79.58 μF is connected to an alternator which generates 240 V, 100 Hz when operating at full speed and normal excitation. Calculate the current flowing in the capacitor.

If both the voltage and frequency are proportional to the speed of the alternator rotor, what is the current when the alternator speed is reduced to 50 per cent of normal, the excitation current remaining unchanged?

8.18 A two-branch parallel circuit has a current in one branch of 300 A at $-10°$ to the reference direction, and 400 A at $20°$ to the reference direction in the other branch. Determine the magnitude and phase angle of the resultant current.

8.19 What are the in-phase and quadrature components of *all the currents* in Problem 8.18?

8.20 The following phasor voltages are connected in series with one another in an electronic circuit.

Phasor 1: 10 V at 30° to the reference direction.
Phasor 2: reference component = 8 V, quadrature component = −8 V.
Phasor 3: 15 V at −120° to the reference direction.

Determine (a) the magnitude and phase angle of the resultant voltage, and (b) the reference and quadrature components of the resultant.

8.21 A current of 10 mA at −70° relative to the reference direction flows into a two-branch parallel circuit. If 8 mA at a phase angle of −45° relative to the reference direction flows in one branch, what is the magnitude and phase angle of the current in the other branch?

8.22 Three voltages are corrected in series with one another. The magnitude and phase angle of two of them are

V_1 : 100 V at 30 °to the reference direction

V_2 : 100 V at 90 °to the reference direction

If the resultant voltage has a magnitude of 160 V at an angle of 90° to the reference direction, what is the magnitude and phase angle of the third voltage?

8.23 Using a programming language of your choice, write a program that calculates inductive reactance and capacitive reactance.

Summary of important facts

An **alternating wave** is **periodic**, that is it repeats itself at a regular rate, and the time taken for one cycle to be completed is known as the **periodic time**. The periodic waveform most frequently encountered in electronic and electrical engineering is the *sine wave*.

Each alternating wave has a **positive peak value** and a **negative peak value**, the difference between the two values being known as the **peak-to-peak value** of the wave. In the case of a true alternating wave, the area under the positive half-cycle of a waveform is equal to the area under the negative half-cycle, so that the total area under the curve in one complete cycle is zero.

If the periodic time of the wave is T, then

frequency $= \mathbf{1/T}$ hertz (Hz)

and the **angular frequency**, ω, is

$$\omega = \mathbf{2\pi f} \text{ rad/s}$$

The **wavelength**, λ, of a waveform is given by

$$\lambda = \mathbf{K/f} \text{ metres (m)}$$

where K is a constant which, for free space, is 3×10^8 m/s.

The **average value** or **mean value** of a wave is taken by electronic and electrical engineers to mean the *rectified average value*, or the mean value of the magnitude of the area under one half-cycle. That is

$$\textbf{average value or mean value} = \frac{\textbf{area under the positive half-cycle}}{\textbf{`length' of the base of the wave}}$$

The 'length' of the base may either be measured in time or as an angle. If the mean value is determined by the mid-ordinate rule, and there are n mid-ordinates, then

$$\textbf{mean value} = \frac{\textbf{sum of mid-ordinates}}{n}$$

In the case of a *sine wave*

$$\begin{aligned}\textbf{mean value} &= \frac{2}{\pi} \times \textbf{maximum value}\\ &= \textbf{0.637} \times \textbf{maximum value}\end{aligned}$$

The **root-mean-square** (r.m.s.) of a waveform is the *effective value* of the wave, which has the same heating effect as an equivalent d.c. current.

r.m.s. value = square **root** of the **mean** of the sum of the **squares** of the instantaneous current (or voltage) values of the wave.

In the case of a *sine wave*

$$\begin{aligned}\textbf{r.m.s. value} &= \textbf{maximum value}/\sqrt{2}\\ &= \textbf{0.7071} \times \textbf{maximum value}\end{aligned}$$

The form factor of a wave is

form factor = r.m.s. value/mean value

and the peak factor or crest factor is

peak factor = maximum value/r.m.s. value

In the case of a sine wave

form factor = 1.11
peak factor = 1.414.

The **inductive reactance**, X_L, of a pure inductor is the opposition of the inductor to the flow of alternating current, and is given by

$$X_L = 2\pi f L = \omega L \text{ ohms}$$

and the **capacitive reactance**, X_C, of a pure capacitor is the opposition of the capacitor to the flow of alternating current and is

$$X_C = 1/(2\pi f C) = 1/\omega C \text{ ohms}$$

A **phasor** is a concept introduced by electronic and electrical engineers to 'fix' a sinusoidal quantity in magnitude and phase angle, so that it can easily be visualised. The 'length' of the phasor represents the r.m.s. value of the wave, and its phase angle is its angle with respect to the *reference direction* (which is usually the horizontal direction).

A phasor can either be described in terms of its *magnitude and phase angle*, or in terms of the length of *its horizontal component and its vertical component*. The former is known as the **polar form** of the phasor, and the latter as the **rectangular form** or **Cartesian representation**.

Phasors can be added together or subtracted from one another either *graphically* or *mathematically*, the latter being more accurate. Before we can mathematically either add phasors together or subtract them from one another, we must first convert them into their rectangular component form.

9 Alternating current circuits

9.1 Introduction

In this chapter we look at one of the most useful aspects of electronic and electrical engineering, namely the solution of alternating current circuits. By the end of this chapter, the reader will be able to:

- Solve series resistor-inductor circuits.
- Solve series resistor-capacitor circuits.
- Solve series resistor-inductor-capacitor circuits.
- Solve parallel resistor-inductor circuits.
- Solve parallel resistor-capacitor circuits.
- Solve parallel resistor-inductor-capacitor circuits.
- Write BASIC language programs for the solution of series and parallel a.c. circuits.
- Draw phasor diagrams for a.c. circuits (the author regards this as a vital aspect in the understanding of the solution of a.c. circuits).

9.2 The R-L series circuit

The basic circuit is shown in Figure 9.1(a) in which

V_S = supply voltage

f = supply frequency

X_L = inductive reactance = $2\pi fL$

R = circuit resistance

Z = circuit impedance = $\sqrt{(R^2 + X_L^2)}$

I = magnitude of the current = V_S/Z

ϕ = phase angle of the circuit

= $\arctan(X_L/R) = \arccos(R/Z) = \arcsin(X_L/Z)$

and I lags behind V_S by ϕ (or V_S leads I by ϕ)

V_R = magnitude of the voltage across the resistance = IR

V_L = magnitude of the voltage across the inductance = IX_L

V_S = **phasor sum of V_R and V_L** = $\sqrt{(V_R{}^2 + V_L{}^2)}$.

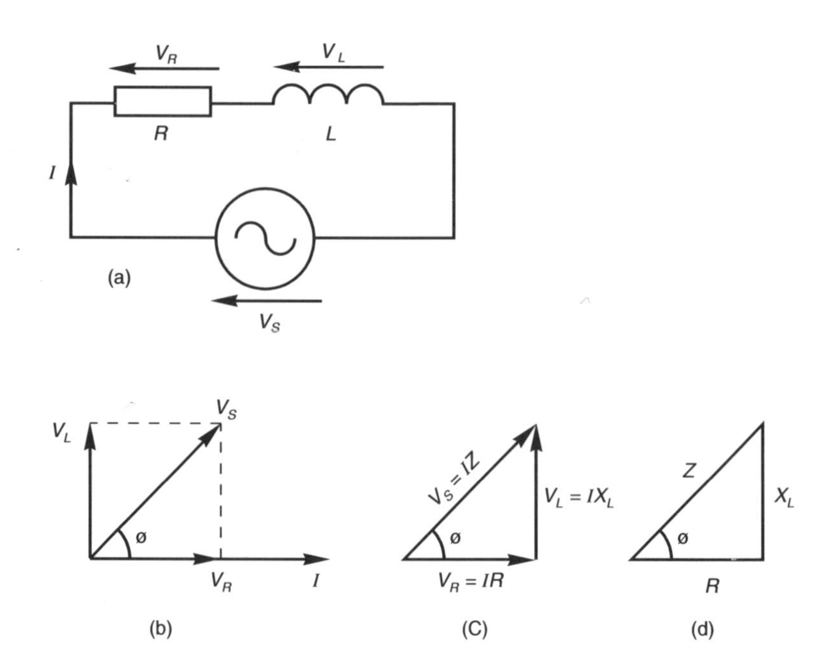

Figure 9.1 The R-L series circuit: (a) the basic circuit; (b) typical phasor diagram; (c) voltage triangle; (d) impedance triangle

In an inductive circuit, the current lags behind the supply voltage by an angle in the range 0° (when $L = 0$) to 90° (when $R = 0$).

In a series circuit, the current flows through every element in the circuit and, by convention, **in series circuit phasor diagrams, the current phasor is drawn in the reference direction**, that is it is shown in the horizontal direction (see Figure 9.1(b)).

As explained in Chapter 8, the voltage drop across the resistor is in phase with the current, so that **the V_R phasor also lies in the reference direction**. Also, the voltage drop across the pure inductor leads the current by 90° (or the current lags behind V_L by 90°), so that **the V_L phasor leads the current phasor by 90°**, i.e. it is vertically upwards with reference to the current phasor (see Figure 9.1(b)).

If the voltage phasors are extracted from the phasor diagram, we are left with the **voltage triangle** of the series *R-L* circuit, as shown in Figure 9.1(c). Furthermore, if we divide each side of the voltage triangle by the magnitude of the current, we have the **impedance triangle** of the circuit in Figure 9.1(d).

Due to the iron loss in an electromagnet, the **effective resistance** of an iron-cored coil may be significantly higher than its 'd.c.' resistance as measured on an ohmmeter. That is, the value of the current may be less than the value which would be calculated if the 'ohmmeter' resistance value was used. See, for example, Worked Example 9.2.

In the following Worked Examples, the circuit and phasor diagrams are generally as shown in Figure 9.1.

Worked Example 9.1

Calculate the impedance of an *R-L* series circuit for which $R = 100\ \omega$, $L = 0.2$ H and the supply frequency is 50 Hz. If the circuit is supplied at 250 V r.m.s., determine (a) the current in the circuit and its phase angle, (b) the magnitude and phase angle of the voltage across each element in the circuit.

Solution

(a) Since the current is given by the equation $I = V/Z$, we must first determine the impedance of the circuit, which is

$$Z = \sqrt{(R^2 + X_L{}^2)}$$

That is, we need to calculate the value of X_L as follows:

$$X_L = 2\pi fL = 2\pi \times 50 \times 0.2 = 62.83\ \Omega$$

hence the magnitude of the impedance is

$$Z = \sqrt{(100^2 + 62.83^2)} = 118.1\ \Omega$$

The magnitude of the current is therefore

$$I = V/Z = 250/118.1 = 2.12\ \text{A}$$

The phase angle of ϕ (see the impedance triangle in Figure 9.1(d)) can be calculated from

$$\phi = \arctan(X_L/R) = \arctan(62.83/100) = 32.14^\circ$$

Since we are dealing with an inductive circuit, the current lags behind thc supply voltage by 32.14°.

(b) The magnitude of the voltage across the resistor is

$$V_R = IR = 2.12 \times 100 = 212\ \text{V}$$

and is in phase with *I*. The magnitude of the voltage across the inductor is

$$V_L = IX_L = 2.12 \times 62.83 = 133.2\ \text{V}$$

and leads *I* by 90°.

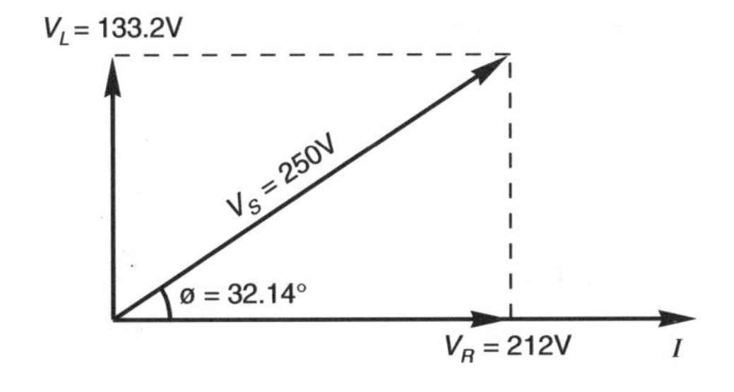

Figure 9.2 Phasor diagram for Worked Example 9.1

The phasor diagram is shown in Figure 9.2. We can check the above results simply by showing that the phasor sum of V_R and V_L is equal to the value of the supply voltage as follows:

$$V_S = \sqrt{(V_R^2 + V_L^2)} = \sqrt{(212^2 + 133.2^2)} = 250.4 \text{ V}$$

The difference between the calculated voltage and the voltage in the problem is only 0.4 V, or 0.16 per cent. This difference is due to 'rounding' errors during the calculation.

Worked Example 9.2

When an iron-cored coil of inductive reactance 15 Ω and 'd.c.' resistance 3 Ω is connected to a 93 V r.m.s. sinusoidal supply, the current in the coil is 6 A. Determine the effective resistance of the coil, and the additional resistance due to the effect of the iron loss.

Solution

The impedance of the coil is

$$Z = V/I = 93/6 = 15.5\ \Omega$$

Applying Pythagoras's theorem to the impedance triangle (Figure 9.1(d)) we see that

$$Z^2 = R_{eff}^2 + X_L^2$$

where R_{eff} is the effective resistance of the coil. Transposing to make R_{eff}^2 the subject of the equation gives

$$R_{eff}^2 = Z^2 - X_L^2$$

or

$$R_{eff} = \sqrt{(Z^2 - X_L^2)} = \sqrt{(15.5^2 - 15^2)} = 3.905\ \Omega$$

If the 'd.c.' resistance of the coil is R, then

$$R_{eff} = R + R_i$$

where R_i is the additional resistance resulting from the iron loss, that is

$$R_i = R_{eff} - R = 3.905 - 3 = 0.905\ \Omega$$

9.3 The R-C series circuit

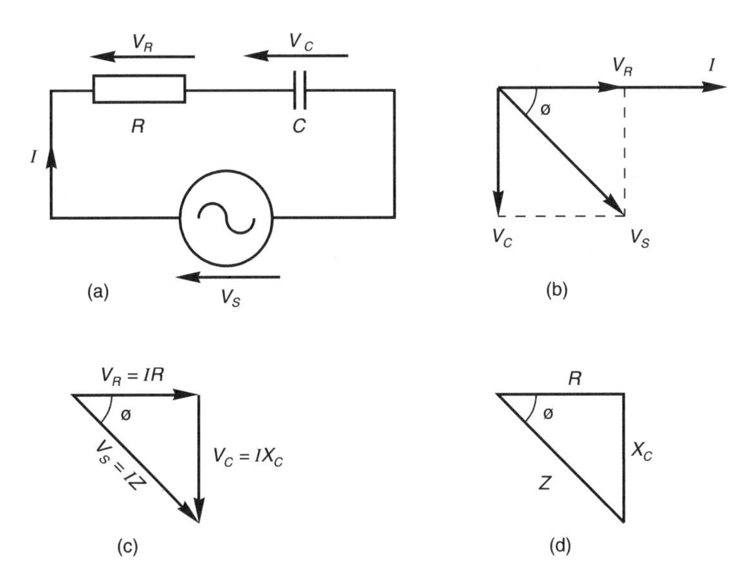

Figure 9.3 The R-C series circuit: (a) the basic circuit; (b) typical phasor diagram; (c) voltage triangle; (d) impedance triangle

An a.c. series R-C circuit is shown in Figure 9.3(a) in which

V_S = supply voltage

X_C = capacitive reactance = $1/(2\pi fC)$

R = circuit resistance

Z = circuit impedance = $\sqrt{(R^2 + X_C{}^2)}$

I = magnitude of the current = V_S/Z

ϕ = phase angle of the circuit = $\arctan(X_C/R)$

= $\arccos(R/Z) = \arcsin(X_C/Z)$

and I leads V_S by ϕ (or V_S lags behind I by ϕ)

V_R = magnitude of voltage across $R = IR$

V_C = magnitude of voltage across $C = IX_C$

V_S = **phasor sum** of V_R and $V_C = \sqrt{(V_R{}^2 + V_C{}^2)}$

As with other series circuits, the current flows through all the elements, and is used as the reference phasor (see Figure 9.3(b)). Moreover, since **V_R is in phase with the current**, it is drawn in the reference direction. Also **V_C lags 90° behind the current through the capacitor** (or I leads V_C by 90°), hence V_C is drawn vertically downwards in Figure 9.3(b).

When the voltage phasors are extracted from the phasor diagram, we have the voltage triangle in Figure 9.3(c). When each side of this triangle is divided by I, we are left with the impedance triangle in Figure 9.3(d).

Worked Example 9.3

A 2 kΩ resistor and a 3.5 nF capacitor are connected in series with one another, and the circuit is energised by a 5 V, 10 kHz sinusoidal supply.

Calculate (a) the capacitive reactance of the capacitor, (b) the impedance of the circuit, (c) the current in the circuit and its phase angle, (d) the magnitude of the voltage across the resistor and across the capacitor, and (e) draw the phasor diagram for the circuit.

Solution

The solution is as follows:

(a) The capacitive reactance is calculated from

$$X_C = 1/(2\pi fC) = 1/(2\pi \times 10\,000 \times 3.5 \times 10^{-9})$$
$$= 4547.3\ \Omega$$

(b) The impedance of the circuit is

$$Z = \sqrt{(R^2 + X_C{}^2)} = \sqrt{(2000^2 + 4547.3^2)} = 4967.7\ \Omega$$

(c) The current is determined by Ohm's law as follows:

$$I = V_S/Z = 5/4967.7 \approx 1 \times 10^{-3} \text{ or } 1\text{ mA}$$

(**Note**: '≈' means 'approximately equal to'.)

At this point, we know the value of all the sides of the impedance triangle, we can calculate the phase angle of the circuit. Using the value of R and Z, we know that

$$\phi = \arccos(R/Z) = \arccos(2000/4967.7) = 66.3°$$

Since this is a capacitive circuit, it follows that I leads V by 66.3°

(d) The magnitude of the voltage across the resistor is

$$V_R = IR = 1 \times 10^{-3} \times 2 \times 10^3 = 2\text{ V}$$

and that across the capacitor is

$$V_C = IX_C = 1 \times 10^{-3} \times 4.55 \times 10^3 = 4.55\text{ V}$$

(**Note**: We can check the calculated voltages as follows:

$$V_S = \sqrt{(V_R{}^2 + V_C{}^2)} = \sqrt{(2^2 + 4.55^2)} = 4.97\text{ V}$$

The difference between the calculated and original value of V_S is due to the 'rounding' process involved in determining I. The difference involved is only 0.6 per cent, which is acceptable.

(e) From the information we now have, the phasor diagram can be drawn in Figure 9.4.

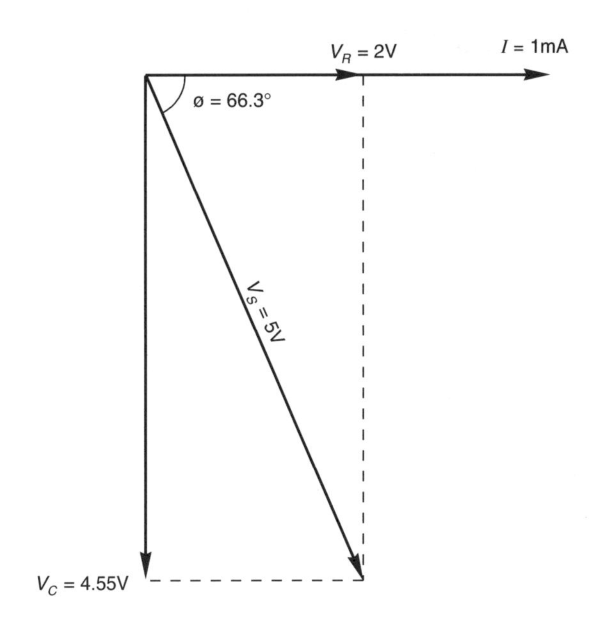

Figure 9.4 Phasor diagram for Worked Example 9.3

Worked Example 9.4

The voltage across the resistor in a series *R-C* circuit is measured to be 34.2 V when the current flowing through it is 1.71 A r.m.s. If the circuit is supplied from a 50 V, 50 Hz source, determine (a) the circuit resistance, (b) the circuit impedance, (c) the reactance of the capacitor, (d) its capacitance. Calculate also (e) the phase angle of the circuit, (f) the voltage across the capacitor. Draw the phasor diagram of the circuit.

Solution

From Ohm's law

$$R = V_R/I = 34.2/1.71 = 20\ \Omega$$

(b) It follows that

$$Z = V_S/I = 50/1.71 = 29.24\ \Omega$$

(c) Now, from the impedance triangle of the circuit we see that

$$Z^2 = R^2 + X_C{}^2$$

or $X_C = \sqrt{(Z^2 - R^2)} = \sqrt{(29.24^2 - 20^2)} = 21.33\ \Omega$

(d) Since $X_C = 1/(2\pi fC)$ then, transposing for C, gives

$$C = 1/(2\pi fX_C) = 1/(2\pi \times 50 \times 21.33)$$
$$= 0.1492 \times 10^{-3} F \text{ or } 149.2\mu F$$

(e) Also, from the impedance triangle, we see that

$$\phi = \arctan(X_C/R) = \arctan(21.33/20) = 46.84°$$

and, since we are dealing with an R-C circuit, I leads V_S by 46.84°.

(f) The voltage across C is

$$V_C = IX_C = 1.71 \times 21.33 = 36.47 \text{ V}$$

We have enough data to draw the phasor diagram of the circuit, which is shown in Figure 9.5.

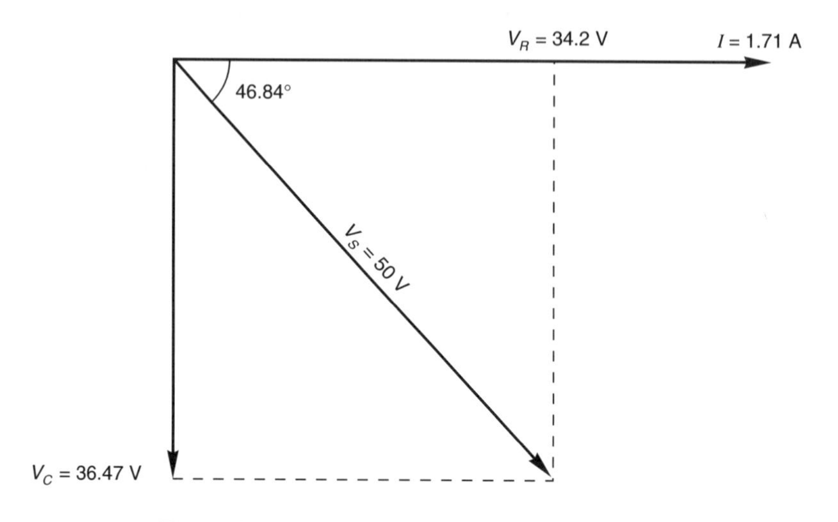

Figure 9.5 Phasor diagram for Worked Example 9.4

9.4 The series R-L-C circuit

The basic R-L-C circuit is shown in Figure 9.6(a). In this case it is important to note that, since X_L and X_C are connected in series, the voltage across them forms a combined reactive voltage. That is, the net reactive voltage across L and C is

$$V_X = V_L - V_C = IX_L - IX_C = I(X_L - X_C)$$

If $(X_L - X_C) > 0$ then the net reactive voltage acts in the direction of V_L, i.e. V_X leads the current by 90°, as shown in figure 9.6(b). The corresponding impedance triangle is shown in Figure 9.6(c). If $(X_L - X_C) < 0$ then the net reactive voltage acts in the direction of V_C, i.e. V_X lags behind the current by 90°, as shown in Figure 9.6(d), and its impedance triangle is in Figure 9.6(e).

The general equations of the circuit are

$$V_S = \sqrt{(V_R^2 + V_X^2)} = \sqrt{((IR)^2 + (I(X_L - X_C)^2))} = IZ$$

where Z is the circuit impedance, and

$$Z = \sqrt{(R^2 + (X_L - X_C)^2)}$$

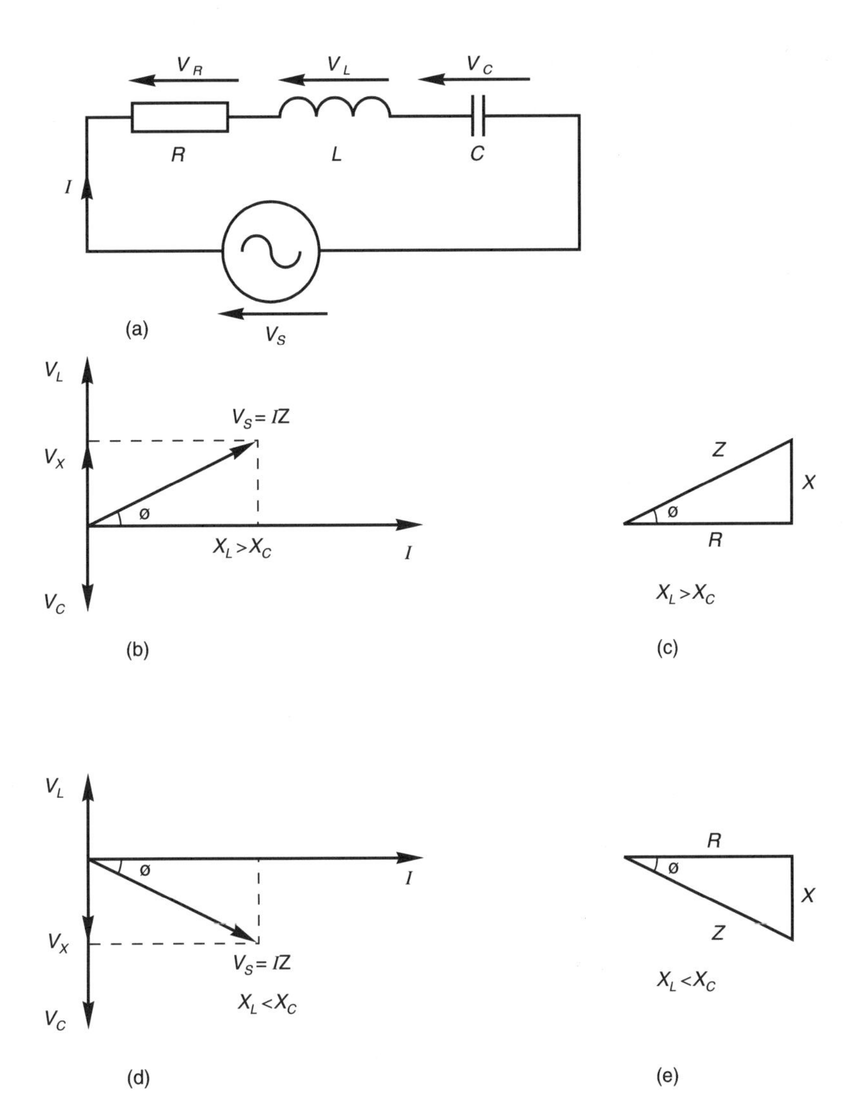

Figure 9.6 The R-L-C series circuit: (a) the basic circuit diagram; (b) phasor diagram for $X_L > X_C$; (c) impedance triangle for $X_L > X_C$; (d) phasor diagram for $X_L < X_C$; (e) impedance triangle for $X_L < X_C$

The phase angle of the circuit depends on whether

(a) X_L is greater than X_C (overall inductive),
(b) X_L is less than X_C (overall capacitive),
(c) X_L is equal to X_C (resistive).

If $X_L > X_C$, the current lags behind V_S (see Figure 9.6(b)), if $X_L < X_C$, the current leads V_S (see Figure 9.6(d)), and if $X_L = X_C$, the current is in phase with V_S. In the latter case, we say that **resonance** has occurred; this condition is so important to electronic and electrical engineers that Chapter 11 is devoted to it.

Worked Example 9.5

A series *R-L-C* circuit in which $R = 11\ \Omega$, $L = 0.07$ H and $C = 290\ \mu$F (see Figure 9.7(a)) carries a current of 5 A at a frequency of 50 Hz. Calculate (a) the impedance and phase angle of the circuit, (b) the voltage across the resistance, the inductance and the capacitance. (c) Draw the phasor diagram of the circuit.

Solution

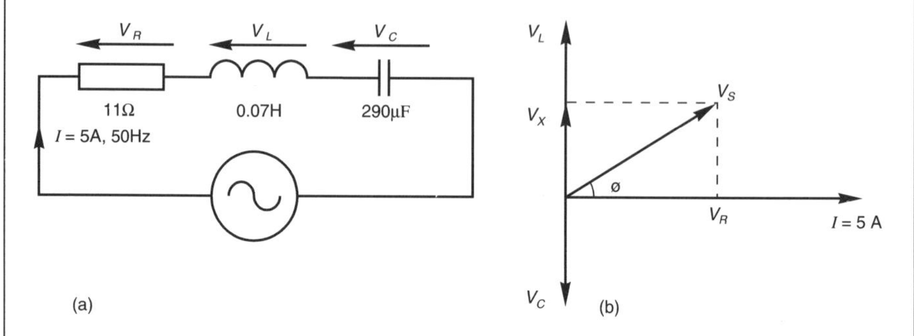

Figure 9.7 (a) Circuit for Worked Example 9.5; (b) its phasor diagram

(a) The reactance of L and C are calculated as follows:

$$X_L = 2\pi fL = 2\pi \times 50 \times 0.07 = 22\ \Omega$$

$$X_C = 1/(2\pi fC) = 1/(2\pi \times 50 \times 290 \times 10^{-6}) = 10.98\ \Omega$$

and the net reactance of the circuit is

$$X = X_L - X_C = 22 - 10.98 = 11.02\ \Omega$$

That is, the circuit has a net inductive reactance at a frequency of 50 Hz. The magnitude of the impedance is, therefore

$$Z = \sqrt{(R^2 + X^2)} = \sqrt{(11^2 + 11.02^2)} = 15.57\ \Omega$$

Since $X_L > X_C$, the phasor diagram is generally similar to that of Figure 9.6(b) (see also Figure 9.7(b), which is the phasor diagram for this problem). From the impedance triangle in Figure 9.6(c), we see that

$$\phi = \arctan(X/R) = \arctan(11.02/11) = 45.05^\circ$$

Also since the circuit has a net inductive reactance, the current lags behind the supply voltage by ϕ.

(b) From Ohm's law, we see that

$$V_R = IR = 5 \times 11 = 55\ \text{V}$$

$$V_L = IXL = 5 \times 22 = 110\ \text{V}$$

$$V_C = IX_C = 5 \times 10.98 = 54.9\ \text{V}$$

That is

$$VX = V_L - V_C = 110 - 54.9 = 55.1 \text{ V}$$

and it follows that

$$V_S = \sqrt{(V_R^2 + VX^2)} = \sqrt{(55^2 + 55.1^2)} = 77.85 \text{ V}$$

These results are interesting because we can see that

1. V_L is greater than the supply voltage, and
2. V_C is practically the same as V_R.

We will not attempt to explain these results here, since we will be going into this type of circuit further in Chapter 11.

(c) Using the values calculated above enable us to draw the phasor diagram in Figure 9.7(b).

Worked Example 9.6

A series *R-L-C* circuit contains a 50 Ω resistor, a 100 Ω inductive reactance and a 16 μF capacitor, the circuit being energised by a 50 V, 50 Hz sinusoidal supply.

Determine (a) the reactance of the capacitor, (b) the impedance of the circuit, (c) the current in the circuit and its phase angle, (d) the magnitude of the voltage across each element in the circuit. Draw the phasor diagram of the circuit.

Solution

(a) Since the supply frequency is 50 Hz, the capacitive reactance is

$$X_C = 1/(2\pi fC) = 1/(2\pi \times 50 \times 16 \times 10^{-6}) = 198.9 \ \Omega$$

(b) The magnitude of the circuit impedance is determined as follows:

$$Z = \sqrt{(R^2 + (X_L - X_C)^2)} = \sqrt{(50^2 + (100 - 198.9)^2)}$$
$$= 110.8 \ \Omega$$

At this point the reader should note that $X_L < X_C$, so that the circuit has a net capacitive reactance, and the current leads the applied voltage by angle ϕ, which is evaluated in part (c).

(c) The magnitude of the current is calculated using Ohm's law, as follows:

$$I = V_S/Z = 50/110.8 = 0.451 \text{ A}$$

The phase angle of the current is given by

$$\phi = \arccos(R/Z) = \arccos(50/110.8) = 63.18°$$

That is, the current leads the applied voltage by 63.18°.

(d) The magnitude of the voltage across each element in the circuit is

$$V_R = IR = 0.451 \times 50 = 22.55 \text{ V (in phase with } I)$$
$$V_L = IXL = 0.451 \times 100 = 45.1 \text{ V (leading } I \text{ by } 90°)$$
$$V_C = IX_C = 0.451 \times 189.9 = 89.7 \text{ V (lagging behind } I \text{ by } 90°)$$

That is

$$V_X = V_L - V_C = 45.1 - 89.7 = -44.6 \text{ V}$$

We can check these values by calculating V_S from the following

$$V_S = \surd(V_R{}^2 + V_X{}^2) = \surd(22.55^2 + (-44.6)^2)$$
$$= 49.98 \approx 50 \text{ V}$$

Using the above values, we can draw the phasor diagram in Figure 9.8.

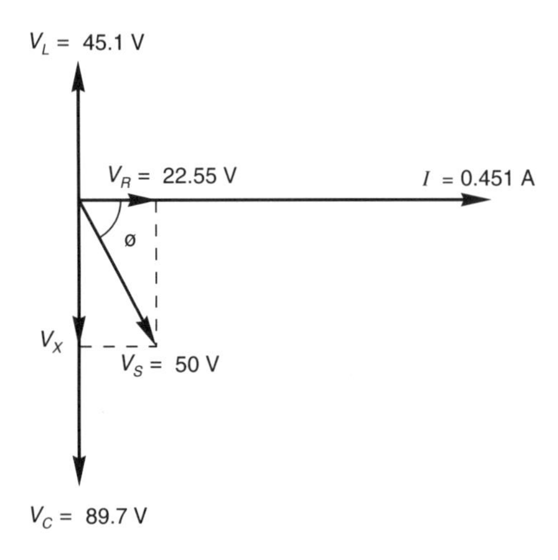

Figure 9.8 Phasor diagram for Worked Example 9.6

9.5 The two-branch R-L parallel circuit

The circuit diagram for a two-branch parallel *R-L* circuit is shown in Figure 9.9(a). Once again we are dealing with ideal elements, so that inductor L is *resistanceless*, and the current in L lags behind the supply voltage by 90°. Moreover, since R is ideal, the current in R is in phase with the supply voltage. For this circuit

V_S = supply voltage
f = frequency
R = resistance of resistor

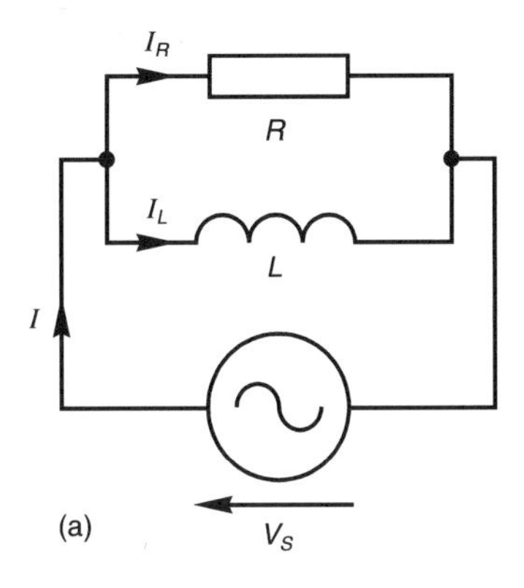

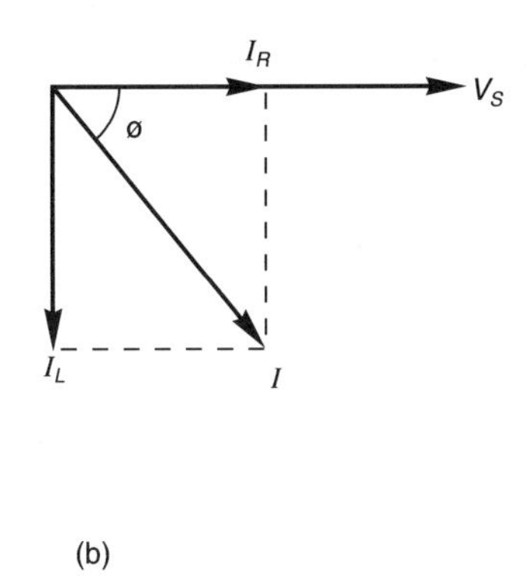

Figure 9.9 (a) Circuit diagram for a parallel R-L a.c. circuit; (b) a typical phasor diagram

$$X_L = \text{inductive reactance} = 2\pi f L$$
$$I_R = \text{current in } R = V_S/R \text{ (in phase with } V_S)$$
$$I_L = \text{current in } L = V_S/X_L \text{ (lagging } V_S \text{ by } 90°)$$
$$I = \text{magnitude of the current drawn by the circuit}$$
$$= \textbf{phasor sum} \text{ of } I_R \text{ and } I_L = \sqrt{(I_R{}^2 + I_L{}^2)}$$
$$\phi = \text{phase angle of the circuit } (I \text{ lags behind } V_S)$$
$$= \arctan(I_L/I_R) = \arccos(I_R/I) = \arcsin(I_L/I).$$

When drawing the phasor diagram of a parallel circuit, **V_S is drawn in the reference (horizontal) direction since it is common to all branches of the parallel circuit**.

Whilst we restrict ourselves to examples of two-branch parallel circuits, there is no limit to the number of branches in the circuit because, to obtain the total current, we simply include as many phasors in the solution as there are branches in the circuit (also see Chapter 8 for the addition of several phasors).

Worked Example 9.7

A parallel circuit contains a 10 Ω resistor in one branch and a pure inductor of 0.05 henrys in the other branch. If the circuit is energised by a 220 V, 50 Hz supply, calculate

(a) the inductive reactance
(b) the current in each branch
(c) the total current drawn by the circuit and its phase angle
(d) the magnitude of the effective impedance of the circuit.

Solution

The circuit diagram is generally as shown in Figure 9.9(a).

(a) The reactance of L is

$$X_L = 2\pi fL = 2\pi \times 50 \times 0.05 = 15.71\ \Omega$$

(b) The current flowing in each branch is determined using Ohm's law as follows: In the resistive branch

$$I_R = V_S/R = 220/10 = 22\ \text{A (in phase with } V_S)$$

and in the inductive branch

$$I_L = V_S/X_L = 220/15.71 = 14\ \text{A (lagging } V_S \text{ by } 90^\circ)$$

and are generally as shown on the phasor diagram in figure 9.9(b)

(c) The total current drawn by the circuit is given by the *phasor sum* of I_R and I_L and, since they are the adjacent and opposite sides, respectively, of a right-angled triangle, then

$$I = \sqrt{(I_R{}^2 + I_L{}^2)} = \sqrt{(22^2 + 14^2)} = 26.08\ \text{A}$$

The phase angle of the circuit is determined from the phasor diagram (Figure 9.9(b)), where we can see that ϕ is given by the expression

$$\phi = \arctan(I_L/I_R) = \arctan(14/22) = 32.47^\circ$$

From the phasor diagram of the circuit, we see that I lags behind V_S by 32.47°.

(d) The magnitude of the circuit impedance is determined from Ohm's law as follows:

$$Z = V_S/I = 220/26.08 = 8.44\ \Omega$$

The reader should take care either when calculating or using the total impedance of the circuit, because the value of the current used above is the *phasor sum* of two currents. That is, each of the currents is at a different angle with respect to the supply voltage. The magnitude impedance is therefore part of a *complex value* which includes both magnitude and phase angle. Complex numbers are discussed in detail in Chapter 19.

9.6 The two-branch R-C parallel circuit

Once again we are dealing with a parallel circuit, so that the current in R is in phase with V_S, and the current in C leads V_S by 90°.

The circuit diagram is shown in Figure 9.10(a) and, as with other parallel circuits, the supply voltage is common to all branches, and is used as the reference (horizontal) phasor. In connection with the circuit

V_S = supply voltage
f = supply frequency
R = resistance of resistor
X_C = reactance of $C = 1/(2\pi fC)$
I_R = current in resistor $= V_S/R$ (in phase with V_S)
I_C = current in capacitor $= V_S/X_C$ (leading V_S by 90°)
I = magnitude of the current drawn by the circuit
= **phasor sum** of I_R and $I_C = \sqrt{(I_R^2 + I_C^2)}$
ϕ = phase angle of the circuit (I leads V_S)
$= \arctan(I_C/I_R) = \arccos(I_R/I) = \arcsin(I_C/I)$.

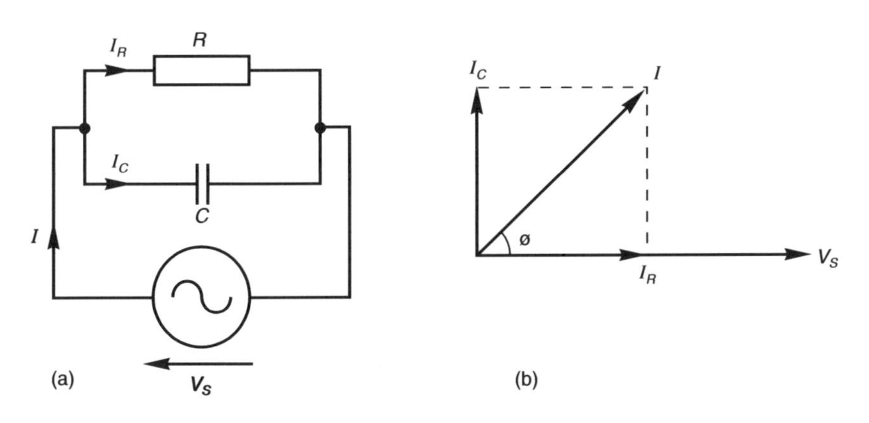

Figure 9.10 (a) Circuit diagram for a parallel R-C a.c. circuit; (b) a typical phasor diagram

Worked Example 9.8

A two-branch parallel R-C circuit has a 20 Ω resistor in one branch and a 10 μF capacitor in the second branch. If the circuit is energised by a 100 V, 1000 rad/s supply, determine
(a) the reactance of the capacitor
(b) the current in each branch
(c) the total current drawn by the circuit and its phase angle
(d) the effective impedance of the circuit.

Solution

Since we are given the angular frequency of the supply in rad/s, we need not calculate the frequency in Hz.
(a) The capacitive reactance of the capacitor is

$$X_C = 1/(\omega C) = 1/(1000 \times 10 \times 10^{-6}) = 100\ \Omega$$

(b) The current in the resistive branch is

$$I_R = V_S/R = 100/20 = 5 \text{ A (in phase with } V_S)$$

and the current in the capacitor is

$$I_C = V_S/X_C = 100/100 = 1 \text{ A (leading } V_S \text{ by } 90°)$$

These phasors are generally as shown in Figure 9.10(b)

(c) The total current drawn by the circuit is

$$I = \sqrt{(I_R^2 + I_C^2)} = \sqrt{(5^2 + 1^2)} = 5.1 \text{ A}$$

From the phasor diagram we see that I leads V_S by

$$\phi = \arctan(I_C/I_R) = \arctan(1/5) = 11.31°$$

(d) The magnitude of the impedance of the parallel circuit is

$$Z = V_S/I = 100/5.1 = 19.61 \ \Omega$$

Once again, the reader is reminded that the magnitude of the impedance must be calculated from the Ohm's law expression $Z = V_S/I$, where I is the magnitude of the current determined from the *phasor sum* of I_R and I_C.

9.7 A practical parallel circuit (R-L in parallel with C)

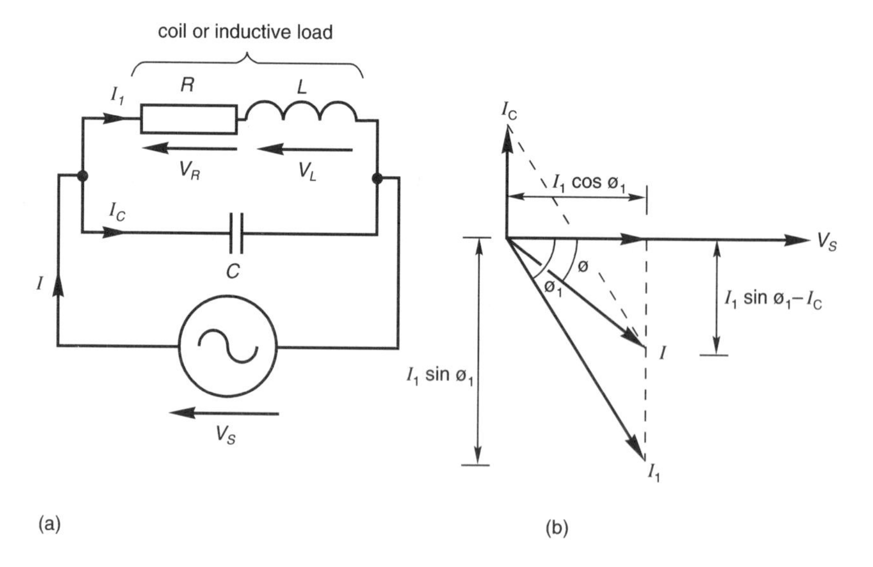

Figure 9.11 (a) Circuit diagram for a practical parallel circuit; (b) a typical phasor diagram

Many practical parallel circuits comprise a coil or an inductive circuit in parallel with a capacitor (see Figure 9.11). Moreover, a practical coil has resistance as well as inductance, and we look at an example of this kind here.

A practical application of this kind occurs in what is known as *power factor correction* in power engineering, which is dealt with in more detail in Chapter 10.

The *R-L* section would correspond either to a coil or a motor which, generally, has a large lagging phase angle. In many cases, the purpose of capacitor *C* is to reduce the lagging phase angle of the current drawn from the supply to a smaller phase angle than that of the inductive circuit itself. The reason why we need to do this is explained in Chapter 10.

In the case of Figure 9.11(a)

V_S = supply voltage
f = supply frequency
R = resistance of *L-R* branch
L = inductance of *L-R* branch
C = capacitance
I_1 = current drawn by the inductive branch
I_C = current drawn by the capacitor $= V_S/X_C$
I = total current drawn by the circuit
= **phasor sum** of I_1 and I_C
ϕ_1 = lagging phase angle of *L-R* branch
ϕ = phase angle of the complete circuit.

The reader should note that ϕ could either be a lagging or a leading phase angle, dependent on the relative value of I_1 and I_C.

Worked Example 9.9

An industrial circuit comprises a two-branch parallel circuit energised by a 500 V, 50 Hz single-phase supply. One branch of the circuit contains a 10.63 Ω resistor in series with a 0.021 H inductor, and the other branch contains a 80 μF capacitor. Determine

(a) the impedance and phase angle of the *L-R* circuit,
(b) the current drawn by the *L-R* circuit
(c) the voltage across *R*
(d) the voltage across *L*
(e) the reactance of the capacitor
(f) the current drawn by the capacitor
(g) the total current drawn by the circuit and its phase angle
(h) the magnitude of the impedance of the complete circuit.

Solution

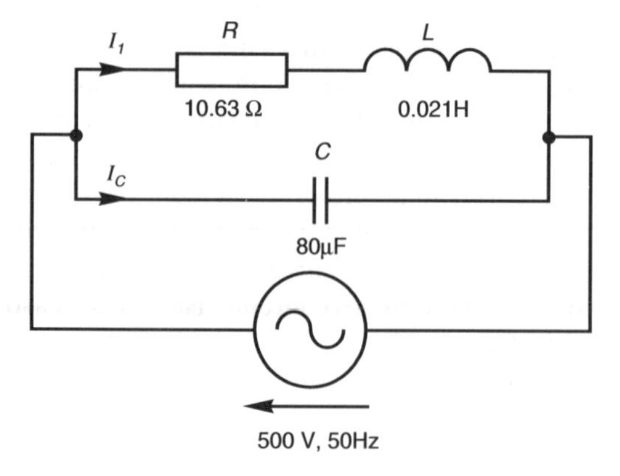

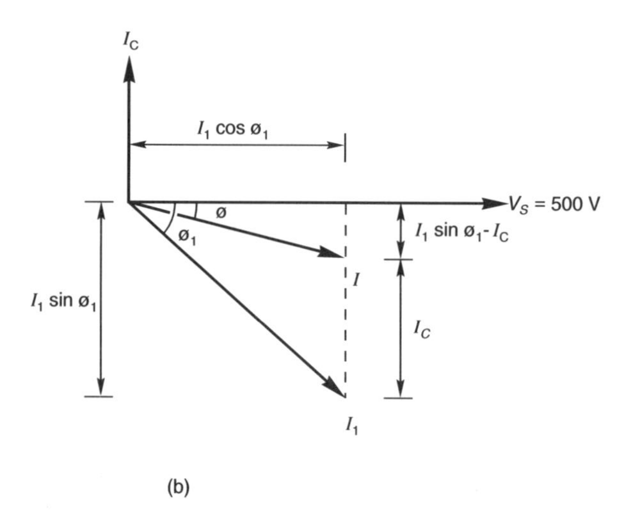

Figure 9.12 (a) Circuit diagram for Worked Example 9.9; (b) its phasor diagram

This could be a typical example using practical values.

(a) The inductive reactance of L is

$$X_L = 2\pi fL = 2\pi \times 50 \times 0.021 = 6.6\ \Omega$$

and the magnitude of the impedance of the L–R branch is

$$Z_1 = \sqrt{(R^2 + X_L{}^2)} = \sqrt{(10.63^2 + 6.6^2)} = 12.51\ \Omega$$

and its phase angle is

$$\begin{aligned}\phi_1 &= \arctan(X_L/R) = \arctan(6.6/10.63)\\ &= 31.84°(I_1 \text{ lagging behind } V_S)\end{aligned}$$

(b) The magnitude of the current drawn by the L-R circuit is determined by Ohm's law as follows:

$$I_1 = V_S/Z_1 = 500/12.51$$
$$= 40 \text{ A(lagging behind } V_S \text{ by } 31.84^\circ)$$

(c) The magnitude of the voltage across R is given by

$$V_R = I_1 R = 40 \times 10.63 = 425.2 \text{ V}$$

(d) The voltage across L is

$$V_L = I_1 X_L = 40 \times 6.6 = 264 \text{ V}$$

(**Note**: the voltage across the L-R circuit is

$$\sqrt{(V_R{}^2 + V_L{}^2)} = \sqrt{(425.2^2 + 264^2)} \approx 500 \text{ V}.$$

(e) The reactance of the capacitor is calculated from

$$X_C = 1/(2\pi f C) = 1/(2\pi \times 50 \times 80 \times 10^{-6})$$
$$= 39.8\ \Omega$$

(f) The capacitor current is

$$I_C = V_S/X_C = 500/39.8 = 12.56 \text{ A(leading } V_S \text{ by } 90^\circ)$$

(g) To determine the total current drawn by the circuit we need to look carefully at the phasor diagram of the circuit, which is drawn in Figure 9.12(b).

One method of solution is to *draw the phasor diagram very accurately and to a large scale*. This method is reliable, but is not very accurate because of errors which can arise during the drawing process (the reader should, in any event, draw the phasor diagram to give some idea of the values involved).

The alternative, and more accurate solution, is to calculate the answer as follows: The *lagging component* of I_1 is

$$I_1 \sin \phi_1 = 40 \sin 31.84^\circ = 21.1 \text{ A}$$

Consequently, the *lagging component* of I is

$$I_1 \sin \phi_1 - I_C = 21.1 - 12.56 = 8.54 \text{ A}$$

The *in-phase component* of I is

$$I_1 \cos \phi_1 = 40 \cos 31.84^\circ = 34 \text{ A}$$

Hence the magnitude of the current drawn by the complete circuit is

$$I = \sqrt{(34^2 + 8.54^2)} = 35.06 \text{ A}$$

and the phase angle of the complete circuit is

$$\phi = \arctan\left[\frac{\text{lagging component of } I}{\text{in-phase component of } I}\right]$$
$$= \arctan(8.54/34) = 14.1^\circ \text{ (I lagging } V_S)$$

(h) The magnitude of the impedance of the complete circuit is

$$Z = V_S/I = 500/35.06 = 14.26\ \Omega.$$

9.8 BASIC language programs for the solution of series and parallel circuits

Three BASIC language programs are listed here for the solution of circuits of the type described in this chapter. In fact, the solution of the majority of the Worked Examples and Exercises in this chapter have been checked using these programs.

The programs in this section of the book can either be used for the direct solution either of a series or a parallel circuit or (and equally importantly) they can be used to verify the solution of a circuit you have obtained by hand calculation. For example, in Worked Example 9.4, we needed to calculate the resistance and capacitance of an *R-C* series circuit. Having calculated these values, they can be insert into the computer program to verify the original data in the problem, i.e. we can check that the specified value of the current and the voltage across *R* are those given in the problem.

The programs adopt the convention that the supply voltage is the reference phasor, and the phase angle of the current(s) is evaluated with reference to the supply voltage.

All the programs accept data either in 'normal' format or in 'scientific' format. That is, the value 1057 can be supplied as 1057, or as 1.057E3, or as 0.001057E6, etc, and 0.0058 can be given as 0.0058 or as 5.8E-3 or as 5800E-6, etc.

Computer listing 9.1 provides a solution for a series *R-L-C* a.c. circuit, and Computer listing 9.2 gives the solution for a two branch *R-L-C* parallel circuit. In both cases, the parameters used in the program are all are set to zero in line 30. This is good practice in any program, because it is possible for the computer memory to contain information from a previous calculation, which can produce errors in the present calculation.

In the case of the series circuit, you are asked for the value of the supply voltage and frequency, followed by the value of R, L and C in the circuit. If the circuit does not include one (or more) of the elements, you either give the value zero or, alternatively, press the ENTER key. In the case of a capacitor this may appear to present a problem, because a capacitor of zero value has infinite reactance! This is overcome in line 110 of listing 9.1, where a zero-value capacitor is given a reactance of zero, thereby replacing it by a short-circuit.

Lines 180 and 120 overcome a 'divide by zero' error when we attempt to calculate the phase angle of a circuit if $R = 0$. Since all computers used angles in radians, line 300 changes the calculated angle in radians into degrees.

Lines 330 to 400 refer to the work in Chapters 10 and 11, where we shall be performing further calculations using this software. Line 500 refers to the work in Chapter 11.

Listing 9.1 R-L-C a.c. series circuit analysis

```
10 CLS '******** CH9-1.BAS *******
20 PRINT "R-L-C A.C. Series Circuit Analysis": PRINT
30 V = 0: F = 0: R = 0: L = 0: C = 0
40 INPUT "Supply Voltage (volts) = ", V
50 INPUT "Supply Frequency (Hz) = ", F: PRINT
60 INPUT "Circuit Resistance (ohms) = ", R
70 INPUT "Inductance (henrys) = ", L
80 INPUT "Capacitance (farads) = ", C
90 PI = 3.141593
100 XL = 2 * PI * F * L
110 IF C = 0 THEN XC = 0: GOTO 130
120 XC = 1 / (2 * PI * F * C)
130 X = XC - XL
140 Z = SQR((R ^ 2) + (X ^ 2))
150 IF Z = 0 THEN PRINT "Fatal error: Z = 0": END
160 I = V / Z
170 VL = I * XL
180 VC = I * XC
190 IF R = 0 AND X > 0 THEN PHI = PI / 2: GOTO 220
200 IF R = 0 AND X < 0 THEN PHI = -PI / 2: GOTO 220
210 PHI = ATN(X / R)
220 PRINT
230 PRINT "Inductive reactance = "; XL; " ohms"
240 PRINT "Capacitive reactance = "; XC; " ohms"
250 PRINT "Impedance = "; Z; " ohms"
260 PRINT "Current = "; I; " A"
270 PRINT "Voltage across R = "; I * R; " V"
280 PRINT "Voltage across L = "; I * XL; " V"
290 PRINT "Voltage across C = "; I * XC; " V"
300 PRINT "Phase angle = "; PHI * 180 / PI; " degrees"
310 IF X = 0 THEN PRINT TAB(10); "I IN PHASE WITH V": GOTO 350
320 IF X > 0 THEN PRINT TAB(10); "I LEADING V" ELSE PRINT TAB(10); "I LAGGING V"
330 '**** Lines 340 - 380 refer to Chapter 10 ****
340 IF (PHI = PI / 2 OR PHI = -PI / 2) THEN PRINT "Power factor = 0": GOTO 350
350 PRINT "Power factor = "; COS(PHI)
360 PRINT "Power consumed = "; I ^ 2 * R; " watts"
370 PRINT "VA consumed = "; V * I; " VA"
380 PRINT "VAr consumed = "; V * I * ABS(SIN(PHI)); " VAr"
390 '**** Lines 400, 500 and 510 refer to Chapter 11 ****
400 IF VL > .99 * VC AND VL < 1.01 * VC AND R > 0 THEN GOSUB 500
410 END
500 PRINT "Q-factor = "; VC / V
510 RETURN
```

The series circuit program, Computer listing 9.1, determines the inductive and capacitive reactance of the circuit, the magnitude of the impedance of the circuit and the current in it, the voltage across R, L and C, the phase angle of the circuit, and it determines whether the current lags or leads the supply voltage (the reader will find it an interesting exercise to follow the operation of these programs in detail).

The two-branch parallel circuit program, Computer listing 9.2, determines the magnitude of the impedance of both branches and of the complete circuit, together with the phase angle of each branch and the current in each branch. Finally, it calculates the current drawn from the supply and its phase angle, and states whether the current lags behind or leads the supply voltage.

Lines 500 to 580 refer to work in Chapter 10, and line 700 refers to Chapter 11.

Listing 9.2 R-L-C two-branch a.c. circuit analysis

```
10 CLS '***** CH9-2.BAS *****
20 PRINT "R-L-C 2-Branch Parallel A.C. Circuit Analysis": PRINT
30 V = 0: F = 0: R1 = 0: L1 = 0: C1 = 0: R2 = 0: L2 = 0: C2 = 0
40 INPUT "Supply Voltage (volts) = ", V
50 INPUT "Supply Frequency (Hz) = ", F
60 INPUT "Branch 1 resistance (ohms) = ", R1
70 INPUT "Branch 1 inductance (henrys) = ", L1
80 INPUT "Branch 1 capacitance (farads) = ", C1
90 INPUT "Branch 2 resistance (ohms) = ", R2
100 INPUT "Branch 2 inductance (henrys) = ", L2
110 INPUT "Branch 2 capacitance (farads) = ", C2
120 PI = 3.141593
130 REM **** Branch 1 ****
140 XL1 = 2 * PI * F * L1
150 IF C1 = 0 THEN XC1 = 0: GOTO 170
160 XC1 = 1 / (2 * PI * F * C1)
170 X1 = XC1 - XL1
180 Z1 = SQR((R1 ^ 2) + (X1 ^ 2))
190 IF Z1 = 0 THEN PRINT : PRINT "Fatal error: Z1 = 0": END
200 I1 = V / Z1
210 IF R1 = 0 AND X1 > 0 THEN PHI1 = PI / 2: GOTO 240
220 IF R1 = 0 AND X1 < 0 THEN PHI1 = -PI / 2: GOTO 240
230 PHI1 = ATN(X1 / R1)
240 REM **** Branch 2 ****
250 XL2 = 2 * PI * F * L2
260 IF C2 = 0 THEN XC2 = 0: GOTO 280
270 XC2 = 1 / (2 * PI * F * C2)
280 X2 = XC2 - XL2
290 Z2 = SQR((R2 ^ 2) + (X2 ^ 2))
300 IF Z2 = 0 THEN PRINT : PRINT "Fatal error: Z2 = 0": END
```

```
310 I2 = V / Z2
320 IF R2 = 0 AND X2 > 0 THEN PHI2 = PI / 2: GOTO 350
330 IF R2 = 0 AND X2 < 0 THEN PHI2 = -PI / 2: GOTO 350
340 PHI2 = ATN(X2 / R2)
350 REM **** Complete circuit ****
360 IH = I1 * COS(PHI1) + I2 * COS(PHI2)
370 IV = I1 * SIN(PHI1) + I2 * SIN(PHI2)
380 I = SQR((IH ^ 2) + (IV ^ 2))
390 PHI = ATN(IV / IH)
400 PRINT
410 PRINT "Magnitude of Z1 = "; V / I1; " ohms"
420 PRINT "I1 = "; I1; " at angle "; PHI1 * 180 / PI; " degrees"
430 PRINT "Magnitude of Z2 = "; V / I2; " ohms"
440 PRINT "I2 = "; I2; " at angle "; PHI2 * 180 / PI; " degrees"
450 PRINT "Magnitude of total Z = "; V / I; " ohms"
460 PRINT "Total current = "; I; " A"
470 PRINT "Overall phase angle = "; PHI * 180 / PI; " degrees"
480 IF IV = 0 THEN PRINT TAB(10); "I IN PHASE WITH V": GOTO 510
490 IF IV > 0 THEN PRINT TAB(10); "I LEADING V" ELSE PRINT TAB(10); "I LAGGING V"
500 '**** Lines 510 - 550 refer to Chapter 10 ****
510 IF (PHI = PI / 2 OR PHI = -PI / 2) THEN PRINT "Power factor = 0": GOTO 530
520 PRINT "Overall power factor = "; COS(PHI)
530 PRINT "Total power consumed = "; (I1 ^ 2 * R1) + (I2 ^ 2 * R2); " watts"
540 PRINT "Total VA consumed = "; V * I; " VA"
550 PRINT "Total VAr consumed = "; V * I * ABS(SIN(PHI)); " VAr"
560 '**** Lines 570 and 700 refer to Chapter 11 ****
570 IF I1 > .99 * I2 AND I1 < 1.01 * I2 THEN GOSUB 700
580 END
700 PRINT "Q-factor = "; I1 / I
710 RETURN
```

Although Computer listing 9.3 is nominally for the solution of a right-angled triangle, it has several purposes. It can, for example, be used to solve the voltage or the impedance triangle of a series circuit. Also readers who are studying more advanced courses can use it when dealing with complex numbers. It provides a 'rectangular' to 'polar' number complex number convertor (see Chapter 19 for details of 'complex' numbers).

Computer listing 9.3 is a simple menu-driven program having two options, namely solving a triangle given the 'opposite' and 'adjacent' sides, and solving a triangle given the hypotenuse and the angle.

Since computers, like calculators, prefer to deal with angles in the range $-90°$ to $+90°$, lines 190 and 200 deal with the case where the angle is in the range $90°$ to $180°$, and $-90°$ to $-180°$. The reader should try to understand the manipulations involved.

Listing 9.3 Solution of a right-angled triangle

```
10 CLS '**** CH9-3 ****
20 PRINT "SOLUTION OF A RIGHT-ANGLED TRIANGLE": PRINT
30 PI = 3.141593: A = 0: V = 0: H = 0: ANG = 0: S = 0
40 PRINT "Do you wish to:": PRINT
50 PRINT "1. provide horizontal and vertical data to"
60 PRINT "   solve for the hypotenuse and the angle, or"
70 PRINT "2. provide hypotenuse and angle data to"
80 PRINT "   solve for the vertical and horizontal?": PRINT
90 INPUT "Enter your selection here (1 or 2): ", S: PRINT
100 IF S < 1 OR S > 2 THEN GOTO 10 ELSE IF S = 2 THEN GOTO 230
110 INPUT "Horizontal length (reference side) = ", A
120 INPUT "Vertical length (opposite side) = ", V
130 PRINT
140 PRINT "Length of hypotenuse = "; SQR((A ^ 2) + (V ^ 2))
150 IF A = 0 AND V > 0 THEN ANG = PI / 2: GOTO 210
160 IF A = 0 AND V < 0 THEN ANG = -PI / 2: GOTO 210
170 IF A < 0 AND V = 0 THEN ANG = PI: GOTO 210
180 ANG = ATN(V / A)
190 IF A < 0 AND V > 0 THEN ANG = ANG + PI
200 IF A < 0 AND V < 0 THEN ANG = ANG - PI
210 PRINT "Angle = "; ANG * 180 / PI; " degrees"
220 END
230 PRINT : INPUT "Length of hypotenuse = ", H
240 INPUT "Angle (degrees) = ", ANG
250 ANG = ANG * PI / 180: PRINT
260 PRINT "Length of horizontal (reference side) = "; H * COS(ANG)
270 PRINT "Length of vertical (opposite side) = "; H * SIN(ANG)
280 END
```

Exercises

The reader should draw the circuit and the phasor diagram for each of the following exercises.

9.1 A coil has a resistance of 10 Ω and inductive reactance of 7.5 Ω. Calculate the magnitude of the impedance of the coil and its phase angle.

9.2 A coil of resistance 15 Ω and inductive reactance 20 Ω is connected in series with the coil in Problem 9.1. What is the magnitude of the total impedance of the circuit and the overall phase angle?

9.3 A coil of resistance 7.32 Ω and inductive reactance 10 Ω is connected in series with a non-inductive resistor. If the impedance of the circuit is 20 Ω, determine the resistance of the resistor. What is the phase angle of the complete circuit?

9.4 For the following R-L circuits, determine the value of the specified quantities:

(a) $R = 10\ \Omega, L = 0.01$ H, $f = 100$ Hz; determine the magnitude of Z and ϕ
(b) $Z = 6.4\ \Omega, L = 0.01$ H, $f = 60$ Hz; calculate the resistance
(c) $R = 10\ \Omega, Z = 15.8\ \Omega, f = 50$ Hz; calculate the inductive reactance and the inductance
(d) $R = 30\ \Omega, Z = 40\ \Omega, L = 0.021$ H; determine the inductive reactance and the supply frequency.

9.5 A coil of resistance of 10 Ω and inductance of 0.05 H is supplied by a 50 V, 50 Hz sinusoidal supply. Calculate the current in the circuit and its phase angle. What value of voltage appears across the resistance and across the inductance?

9.6 A current of 2 A (d.c.) flows in a coil when 100 V (d.c.) is applied to it. When a voltage of 260 V, 50 Hz is applied to the coil, the r.m.s. value of current in it is 2 A. Determine the inductance of the coil.

9.7 A 5 Ω resistor is connected in series with a 6 Ω capacitive reactance. Evaluate the magnitude of the circuit impedance and its phase angle.

9.8 The following particulars refer to a series R–C circuit.

(a) $R = 60\ \Omega$, $C = 50\ \mu$F, $f = 50$ Hz. Determine Z and ϕ
(b) $Z = 30\ \Omega$, $R = 20\ \Omega$, $f = 50$ Hz. Calculate the value of C
(c) $Z = 60\ \Omega$, $C = 70\mu F$, $f = 50$ Hz. Estimate the value of R and ϕ
(d) $Z = 60\ \Omega$, $R = 30\ \Omega$, $C = 30.6\ \mu$F. Calculate the value of f.

9.9 A capacitor is connected in series with a pure resistor of 100 Ω, the circuit being supplied by a 250 V, 50 Hz supply. If the current in the circuit is 2 A calculate (a) the impedance of the circuit, (b) the reactance of the capacitor, (c) the capacitance of the capacitor, (d) the phase angle of the circuit, (e) the voltage across the resistor, (f) the voltage across the capacitor.

9.10 A 400 Ω resistor is connected in series with two series-connected capacitors of $C_1 = 0.6\ \mu$F and $C_2 = 0.4\ \mu$F. If the supply voltage is 10 V, 1 kHz, calculate (a) the effective capacitance of the two series-connected capacitors, (b) the impedance of the circuit, (c) the current flowing in the circuit and its phase angle, (d) the voltage across each element in the circuit.

9.11 A 15 V metal-filament lamp which requires a current of 1.0 A, is to be operated from a 20 V, 100 Hz source. The excess voltage is to be 'dropped' by connecting a capacitor in series with the lamp. Determine the capacitance of the capacitor. What voltage appears between the capacitor terminals?

9.12 A coil of resistance 6 Ω and inductance 100 mH is connected in series with a 80 μF capacitor to a 100 V, 50 Hz supply. Determine (a) the reactance of the inductor and of the capacitor, (b) the impedance of the complete circuit, (c) the current drawn by the circuit and its phase angle, (d) the voltage across the coil and across the capacitor.

9.13 An a.c. series circuit carries a current of 1 mA. The circuit components comprise a 1 kΩ pure resistor, a 2 kΩ inductive reactance and a 3 kΩ capacitive reactance. Determine:

(a) the impedance of the circuit
(b) the value of the applied voltage
(c) the phase angle of the circuit
(d) the voltage drop across each circuit element.

9.14 A series *R-L-C* circuit contains a resistance of 11 Ω, an inductance of 0.07 H, and a 290 μF capacitance. If the current flowing is 5 A, 50 Hz, determine (a) the impedance of the circuit, (b) the voltage across each element in the circuit, (c) the supply voltage, (d) the phase angle of the circuit.

9.15 An inductor of 0.08 H is connected in parallel with a 33 Ω resistor to a 100 V, 50 Hz supply. Calculate (a) the current in each branch and its phase angle, (b) the total current drawn from the supply and its phase angle, (c) the impedance of the circuit.

9.16 A 250 Ω resistor is connected in parallel with a 0.5 μF capacitor, the combination being connected to a 10 V, 1 kHz supply. Determine (a) the current in each branch and its phase angle, (b) the total current drawn from the supply and its phase angle, (c) the impedance of the circuit.

9.17 A coil of inductance 0.1 H and resistance 1 kΩ is connected in parallel with a 0.01 μ F capacitor, the combination being supplied by a 10 V supply at a frequency of (a) 2.5 kHz, (b) 10 kHz. Determine for each frequency (i) the current in each branch and its phase angle, (ii) the current drawn from the supply and its phase angle, (iii) the impedance of the circuit.

9.18 A 3-branch a.c. parallel circuit carries the following currents:

branch 1: 1.8 A at an angle of 20°
branch 2: 5 A at an angle of 80°
branch 3: 3 A at an angle of −60°

the phase angles being relative to the supply voltage. Determine the total current drawn from the supply and its phase angle.

9.19 A feeder cable carries a current to a distribution board, from which the following currents are taken.

Load 1: 100 A at an angle of 30° lagging
Load 2: 150 A at an angle of 60° leading
Load 3: 80 A at an angle of 0°
Load 4: 50 A at an angle of 90° leading

all angles being relative to the supply voltage. Calculate the magnitude and phase angle of the current in the feeder cable.

9.20 A two-branch parallel circuit has a coil of resistance 10 Ω and inductive reactance 8 Ω in one branch, and a resistance of 8 Ω and capacitive reactance of 10 Ω in the other branch. If the supply voltage is 10 V r.m.s., calculate (a) the current in each branch and its phase angle, (b) the total current drawn from the supply and its phase angle, and (c) the voltage across each element in the circuit.

9.21 Modify the three BASIC program Computer listings in this chapter into a single menu-driven program so that you are given the option either of solving a series circuit, or a parallel circuit, or a triangle in two ways.

Summary of important facts

The **impedance** of a circuit is the total opposition of the circuit to current flow. For *any circuit*, the magnitude of the impedance, Z, can be calculated from Ohm's law expression

$$Z = V_S/I$$

where V_S is the *magnitude of the supply voltage*, and I is the *magnitude of the current* in the circuit. The reader should note that the impedance, the voltage and the current are *complex quantities* (for details, see Chapter 14), that is they all have magnitude and phase angle (usually specified with respect to the reference or horizontal direction).

For a **series *R-L* circuit**, the magnitude of the impedance is given by

$$\begin{aligned} Z &= \sqrt{(R^2 + X_L^2)} = \sqrt{(R^2 + (\omega L)^2)} \\ &= \sqrt{(R^2 + (2\pi f L)^2)} \end{aligned}$$

where R is the *circuit resistance*, X_L is the *inductive reactance* of the circuit, ω is the *angular frequency* (rad/s) of the supply, f is the *supply frequency* (Hz) of the supply, and L is the *circuit inductance.*

In an R-L series circuit, the **current lags behind the supply voltage** by angle ϕ, where

$$\phi = \arctan(X_L/R) = \arccos(R/Z) = \arcsin(X_L/Z)$$

The voltage across the circuit elements is

$$\begin{aligned} V_R &= IR \text{ (in phase with } I) \\ V_L &= IX_L = 2\pi f L I \text{ (leading } I \text{ by } 90^\circ) \end{aligned}$$

and

$$V_S = \sqrt{(V_R^2 + V_L^2)}$$

In a **series *R-C*** circuit

$$\begin{aligned} Z &= \sqrt{(R^2 + X_C^2)} = \sqrt{(R^2 + (1/\omega C)^2)} \\ &= \sqrt{(R^2 + (1/2\pi f C)^2)} \end{aligned}$$

where X_C is the *capacitive reactance of the circuit* and C is the *capacitance*. **The current leads the supply voltage by ϕ**, where

$$\phi = \arctan(X_C/R) = \arccos(R/Z) = \arcsin(X_C/Z)$$

The voltage across the circuit elements is

$$V_R = IR \text{ (in phase with } I)$$
$$V_C = IX_C = I/(2\pi fC) \text{ (lagging } I \text{ by } 90°)$$

and

$$V_S = \sqrt{(V_R^2 + V_C^2)}$$

In a **series *R-L-C*** circuit

$$Z = \sqrt{(R^2 + (X_L - X_C)^2)} = \sqrt{(R^2 + X^2)}$$
$$= \sqrt{(R^2 + (\omega L - 1/\omega C)^2)}$$
$$= \sqrt{(R^2 + (2\pi fL - 1/2\pi fC)^2)}$$

Depending on the relative value of X_L and X_C, the net reactance, $X(= X_L - X_C)$, may either be positive or negative. If $X_L > X_C$, the net reactance is inductive and I lags behind V_S. If $X_L < X_C$, the net reactance is capacitive and I leads V_S. The phase angle of the circuit is given by

$$\phi = \arctan(X/R) = \arccos(R/Z) = \arcsin(X/Z)$$

and the voltage across the circuit elements is

$$V_R = IR \text{ (in phase with } I)$$
$$V_L = IX_L = 2\pi fLI \text{ (leading } I \text{ by } 90°)$$
$$V_C = IX_C = I/(2\pi fC) \text{ (lagging } I \text{ by } 90°)$$

and

$$V_S = \sqrt{(V_R^2 + (V_L - V_C)^2)}$$

In the special case where $X_L = X_C$, the net reactance is zero and the current is in phase with the applied voltage. This condition is known as **series resonance** and is described in Chapter 11.

Where many elements are connected in series, the equivalent circuit can be reduced to a series *R-L-C* circuit as follows: The total equivalent resistance is given by

$$R_E = R_1 + R_2 + R_3 + \ldots$$

The total equivalent inductance is

$$L_E = L_1 + L_2 + L_3 + \ldots$$

and the total inductive reactance is

$$X_{LE} = X_{L1} + X_{L2} + X_{L3} + \ldots$$

The reciprocal of the total equivalent capacitance of n series-connected capacitors is

$$\frac{1}{C_E} = \frac{1}{C_1} + \frac{1}{C_2} + \frac{1}{C_3} + \ldots + \frac{1}{C_n}$$

whose capacitive reactance is

$$X_{CE} = 1/(\omega C_E)$$

The analysis of the circuit then follows that of a series R-L-C circuit.

In the case of a **two-branch parallel R-L circuit**, the current in the resistive branch is

$$\boldsymbol{I_R = V_S/R} \text{ (in phase with } V_S)$$

and in the inductive circuit is

$$\boldsymbol{I_L = V_S/X_L} \text{ (lagging behind } V_S \text{ by } 90^\circ)$$

The current drawn by the circuit is the phasor sum of I_R and I_L, and its *magnitude* is given by

$$\boldsymbol{I = \sqrt{(I_R^2 + I_L^2)}} \text{ (lagging behind } V_S \text{ by } \phi)$$

The phase angle ϕ of the circuit is

$$\boldsymbol{\phi = \arctan (I_L/I_R) = \arccos (I_R/I) = \arcsin (I_L/I)}$$

In the case of a **two-branch parallel R–C circuit**, the current in the resistive branch is

$$\boldsymbol{I_R = V_S/R} \text{ (in phase with } V_S)$$

and in the capacitive circuit is

$$\boldsymbol{I_C = V_S/X_C} \text{ (leading } V_S \text{ by } 90^\circ)$$

The current drawn by the circuit is the phasor sum of I_R and I_C, and its *magnitude* is given by

$$\boldsymbol{I = \sqrt{(I_R^2 + I_C^2)}} \text{ (leading } V_S \text{ by } \phi)$$

The phase angle ϕ of the circuit is

$$\boldsymbol{\phi = \arctan (I_C/I_R) = \arccos (I_R/I) = \arcsin (I_C/I)}$$

In a circuit in which a **R-L circuit** (i.e. a coil) **is connected in parallel with a capacitor**, the magnitude of the current in the R-L circuit is

$$\boldsymbol{I_1 = V_S/Z_1 = V_S/\sqrt{(R^2 + X_L^2)}}$$

where Z_1 is the *impedance* of the *R-L* circuit. The current in the *R-L* branch lags behind V_S by angle ϕ_1, where

$$\boldsymbol{\phi_1 = \arctan(X_L/R)}$$

The in-phase (reference) component of the total current is $I_1 \cos \phi_1$, and the reactive (quadrature) component of the total current is the difference between I_C and $I_1 \sin \phi_1$. The total current drawn from the supply is

$$\boldsymbol{I = \sqrt{((I_1 \cos \phi_1)^2 + (I_1 \sin \phi_1 - I_C)^2)}}$$

and the phase angle is

$$\boldsymbol{\phi = \arctan((I_1 \sin \phi_1 - I_C)/(I_1 \cos \phi_1))}$$

In the special case where $I_1 \sin \phi_1 = I_C$, the current is in phase with the supply voltage, and is known as **parallel resonance**. This is described in Chapter 11.

Where ***n* circuits are connected in parallel**, the total current is the *phasor sum of the currents* in the *n* branches. Perhaps the simplest method of doing this is by drawing the phasor diagram, but the most accurate method is by calculation.

Volt-amperes, power, reactive VA and power factor consumption

10.1 Introduction

In this chapter we look at the components of power in a.c. circuits. By the end of this chapter, the reader will be able to

- Understand the power triangle.
- Be able to calculate the components of the power triangle.
- Understand and calculate the value of real power or active power, reactive power or quadrature power, and apparent power.
- Appreciate the meaning of power factor and perform power factor correction calculations.
- Use software to solve the power triangle and power factor calculations.

10.2 The power triangle

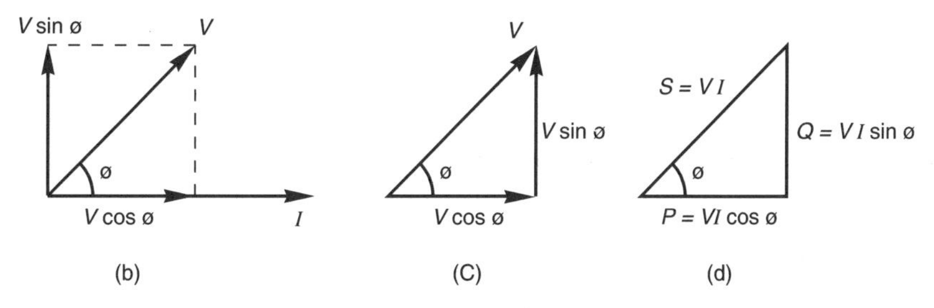

Figure 10.1 (a) Phasor diagram for a series R-L circuit; (b) its voltage; (c) its power triangle

The phasor diagram for a series *R-L* circuit is shown in Figure 10.1(a); since the diagram is for a series circuit, the current is shown in the reference direction. When the components of voltage for the circuit elements are extracted, we are left with the **voltage triangle** in figure 10.1(b). If each side of the voltage triangle is multiplied by the magnitude of the current, we have the **power triangle** of the circuit in Figure 10.1(c), in which

P = **active power** or **real power** = $VI\cos\phi$

Q = **reactive power** or **quadrature power** = $VI\sin\phi$

S = **apparent power** = VI

ϕ = **phase angle** between V and I

If V is in volts and I in amperes, then

P is in watts (W)

Q is in volt-amperes reactive (VAr)

S is in volt-amperes (VA)

also

$$S = \sqrt{(P^2 + Q^2)} \text{ and } \phi = \arctan(Q/P)$$

It can also be shown that

$$P = (\textit{magnitude} \text{ of the current})^2 \times \text{resistance}$$
$$= I^2 R.$$

Worked Example 10.1

The power consumed by a single-phase circuit is 4.2 kW. If the phase angle between the current and the voltage is 53°, determine the apparent power and the reactive power consumed.

Solution

From the power triangle in Figure 10.1(c) we see that $\tan\phi = Q/P$, or

$$Q = P\tan\phi = 4.2 \times \tan 53^\circ = 5.57 \text{ kVAr}$$

and since $\cos\phi = P/S$, then

$$S = P/\cos\phi = 4.2/\cos 53^\circ = 7 \text{ kVA}$$

(**Note**: we can verify the result of the calculation as follows:

$$\sqrt{(P^2 + Q^2)} = \sqrt{(4.2^2 + 5.57^2)} = 7 \text{ kVA} = S.)$$

Worked Example 10.2

An inductive circuit has an active component of current of 7.5 A, and a lagging reactive component of current of 5.5 A, the *voltage* being used as the reference phasor. If the supply voltage is 240 V, determine (a) the magnitude of the current, (b) the phase angle of the circuit, (c) the power consumed, (d) the VAr consumed, (e) the VA consumed.

Solution

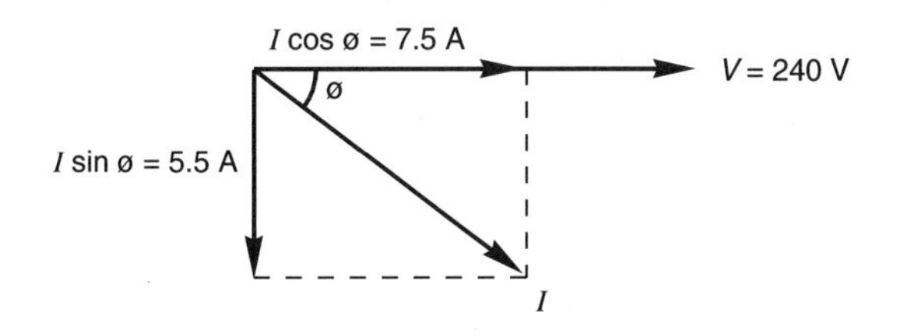

Figure 10.2 Phasor diagram for Worked Example 10.2

The phasor diagram for the circuit is shown in Figure 10.2, in which the voltage is drawn in the reference direction
(**Note**: If we wish, we can put the current in the reference direction simply by rotating the *whole of the diagram* in an anti-clockwise direction by ϕ.)

(a) The magnitude of the current is

$$I = \surd((I\cos\phi)^2 + (I\sin\phi)^2)$$
$$= \surd(7.52 + 5.52) = 9.3 \text{ A}$$

(b) Since we know the resolved components of the current, then

$$\tan\phi = \frac{I\sin\phi}{I\cos\phi} = \frac{5.5}{7.5} = 0.7333$$

hence

$$\phi = \arctan 0.7333 = 36.25^\circ$$

(c) The power consumed is

$$P = VI\cos\phi = 240 \times \text{in-phase component of current}$$
$$= 240 \times 7.5 = 1800 \text{ W}$$

(d) The VAr is

$$Q = VI\sin\phi = 240 \times \text{quadrature component of current}$$
$$= 240 \times 5.5 = 1320 \text{ VAr}$$

(e) The apparent power may be calculated from

$$S = VI = 240 \times 9.3 = 2232 \text{ VA}$$

or

$$S = \surd(P^2 + Q^2) = \surd(1800^2 + 1320^2) = 2232 \text{ VA}.$$

10.3 Power factor

Engineers are concerned about the amount of power a circuit consumes, and the **power factor** of a circuit is defined as

$$\textbf{power factor} = \frac{\textbf{active power consumed}}{\textbf{apparent power consumed}}$$

$$= \frac{VI\cos\phi}{VI} = \cos\phi$$

Since $\cos\phi$ is positive both for leading and lagging phase angles, it is *always necessary* to state whether the power factor is leading or is lagging, i.e. whether the current leads the voltage or lags behind it.

Electrical power circuits usually involve inductive elements such as motors, transformers, etc. and in these cases the power factor is usually lagging.

Worked Example 10.3

What is the power factor of the circuit in Worked Example 10.2?

Solution

It is usual to assume, when dealing with power factor calculations, that the voltage lies in the reference direction, and it advisable to sketch the phasor diagram of the circuit. Fortunately it has been drawn for Worked Example 10.2 (see Figure 10.2), and we see that the *current lags behind the voltage* by ϕ, hence

$$\text{power factor} = \cos\phi = \cos 36.25° = 0.806 \text{ } \textit{lagging}.$$

Worked Example 10.4

A single-phase 440 V induction motor consumes 3.5 kW at a power factor of 0.8 lagging. Determine (a) the current drawn from the supply, (b) the output power if the overall efficiency of the motor is 85 per cent, (c) the active and reactive components of the input current, (d) the apparent power and reactive power consumed.

Solution

(a) The power consumed is given by $P = VI\cos\phi$, hence

$$I = P/(V\cos\phi) = 3.5 \times 10^3/(440 \times 0.8)$$
$$= 9.94 \text{ A}$$

Since we are told that the motor has a lagging power factor, the current lags behind the supply voltage by $\arccos 0.8 = 36.9°$

(b) The output power from the motor is

$$\text{output power} = \text{input power} \times \text{per-unit efficiency}$$
$$= 3.5 \times 0.85 = 2.975 \text{ kW}$$

(c) The phasor diagram for the current is shown in Figure 10.3(a). the active component of the current is

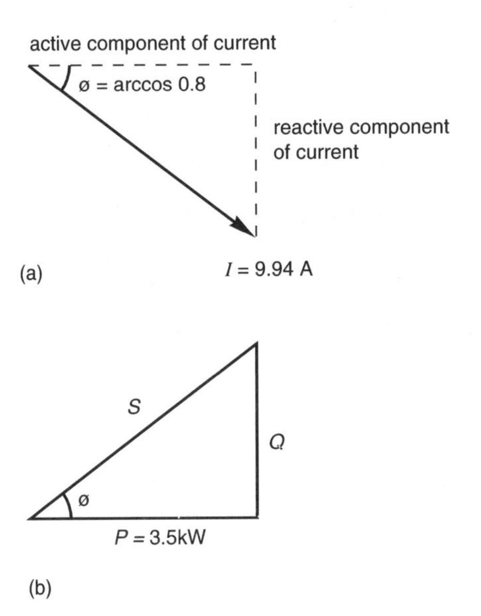

Figure 10.3 (a) Phasor diagram for Worked Example 10.4(c); (b) power triangle for Worked Example 10.4(d)

$$I \cos \phi = 9.94 \times 0.8 = 7.952 \text{ A}$$

and the reactive component of the current is

$$I \sin \phi = 9.94 \times \sin(\arccos 0.8)$$
$$= 9.94 \times \sin 36.9^\circ = 5.97 \text{ A}$$

(d) The power triangle for this case is shown in Figure 10.3(b). The reader is asked to note that, by international convention, a lagging reactive power is assumed to have a positive sign. From the power triangle, $\cos \phi = P/S$, hence the apparent power consumed is

$$S = P/\cos \phi = 3.5/0.8 = 4.375 \text{ kW}$$

Since $\tan \phi = Q/P$, the reactive power consumed is

$$Q = P \tan \phi = 3.5 \times \tan 36.9^\circ = 2.63 \text{ kVAr}$$

10.4 Power factor of combined loads

Power system engineers are often concerned with the overall power factor imposed on a substation or a generating station. This type of problem can often be dealt with by using the method of phasor addition described in Chapter 8, and is outlined in Worked Example 10.5 (appropriate Computer listings have been given at the end of Chapter 8).

Worked Example 10.5

A substation has the following loads connected to it:

load A: 500 kVA at a lagging power factor of 0.9.
load B: 200 kW at a lagging power factor of 0.4.
load C: 300 kVA at a leading power factor of 0.6.

What is the overall power, reactive power, apparent power, and power factor connected to the substation?

Solution

The solution can either be obtained graphically or by calculation. We will do both.

Graphical solution
There are two things we need to do with the data. First, we must make sure that all the magnitude values are given in the same units, *which must be in kVA* and, secondly, we must convert each power factor into its associated angle. The results are given below.

load A: 500 kVA at 25.84° lagging
load B: 500 kVA at 66.42° lagging
load C: 300 kVA at 53.13° leading

The graphical solution is shown in Figure 10.4. The accuracy of the result depends on the care taken in drawing the phasors, and it is estimated that the overall load is about

940 kVA at 30° lagging (or a power factor of 0.866 lagging)

The reader should draw the phasor diagram and see what his/her result is.

Solution by calculation
In this case we must convert the values for each load into its resolved components, which are

load A: 450 kW, 218 kVAr lagging
load B: 200 kW, 458.3 kVAr lagging
load C: 180 kW, 240 kVAr leading

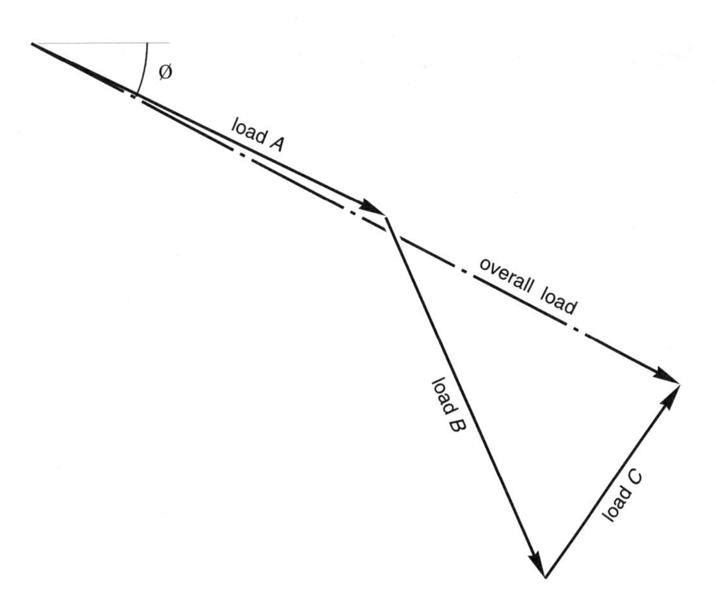

Figure 10.4 Graphical solution of Worked Example 10.5

Hence

$$\text{total power} = 450 + 200 + 180 = 830 \text{ kW}$$
$$\text{total kVAr} = 218 + 458.3 - 240 \text{ kVAr lagging}$$
$$= 436.3 \text{ kVAr lagging}$$

Therefore

$$\text{total kVA} = \surd((\text{kW})^2 + (\text{kVAr})^2)$$
$$= \surd(830^2 + 436.3^2) = 937.7 \text{ kVA}$$

and the overall power factor is

$$\frac{\text{total kW}}{\text{total kVA}} = \frac{830}{937.7} = 0.885 \text{ lagging}$$

10.5 Power factor correction

Industrial installations have many inductive loads, with the consequence that the power factor is lagging, and its value is usually fairly low. Industry is penalised for having a low power factor by means of special clauses in the supply tariff, which encourages them to improve the power factor to a value which is closer to unity (the ideal value!). The simplest, and often the most economic, method is to connect a capacitor across the terminals of any device which has a low power factor, i.e. a motor or other inductive load.

Industrial capacitors are usually rated in kVAr at the rated supply voltage (rather than in microfarads); physically, they often look like industrial transformers rather than capacitors!

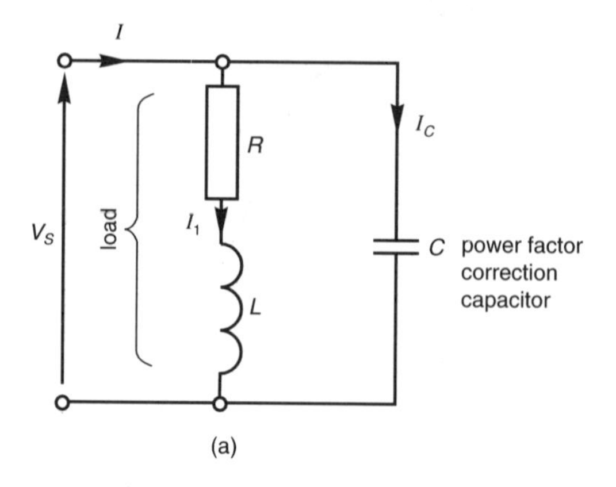

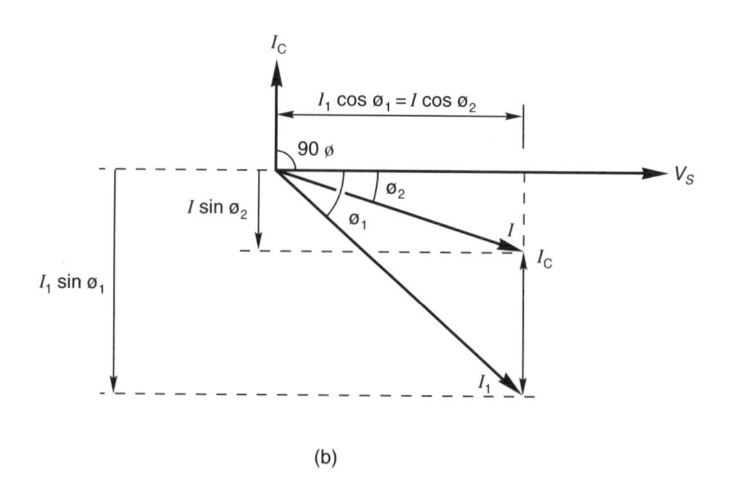

Figure 10.5 (a) Circuit diagram showing power factor correction by capacitor and (b) its phasor diagram

A circuit diagram for a typical single-phase power-factor correction problem is shown in Figure 10.5(a). Elements R and L represent the inductive load, and C is the power factor correcting capacitor. Current I_1 at lagging angle ϕ_1 flows into the load (see Figure 10.5(b)); since the power loss in the capacitor is negligible, it is reasonable to assume that the capacitor current, I_C, leads the supply voltage by 90°.

The net effect of the capacitor current is shown on the phasor diagram, in which the overall phase angle is improved from ϕ_1 to ϕ_2. Since the power loss in the capacitor is negligible, the in-phase component of the supply current is unaltered when the capacitor is connected to the load, i.e.

$$I_1 \cos \phi_1 = I \cos \phi_2$$

It is rather unusual to be able to correct the overall power factor to unity, because the cost of the capacitor usually does not justify the saving achieved in the electricity tariff.

The capacitor current is determined from the relationship

$$I_C + I \sin \phi_2 = I_1 \sin \phi_1$$

or

$$I_C = I_1 \sin \phi_1 - I \sin \phi_2$$

where I is the current drawn from the supply *after* the capacitor is connected.

From Ohm's law for a.c. circuits

$$I_C = V_S / X_C = V_S / (1/2\pi f C) = 2\pi f C V_S$$

and the capacitance of the capacitor is given by

$$C = \frac{I_C}{2\pi f V_S} = \frac{I_1 \sin \phi_1 - I \sin \phi_2}{2\pi f V_S} \; F$$

$$= \frac{(I_1 \sin \phi_1 - I \sin \phi_2) \times 10^6}{2\pi f V_S} \; \mu\text{F}$$

The reactive power consumed by the capacitor is

$$Q_C = V_S I_C \text{ VAr} = \frac{V_S I_C}{1000} = \text{kVAr}$$

$$= P(\tan \phi_1 - \tan \phi_2) \text{ kVAr}$$

where P is the input power in kW.

Worked Example 10.6

A single-phase load consumes a current of 24 A at a power factor of 0.6 lagging. If the overall power factor of the load is improved to 0.9 lagging when a capacitor is connected in parallel with it, determine (a) the total current drawn from the supply when the capacitor is connected, (b) the value of the capacitor current.

Solution

The phasor diagram for this example is shown in Figure 10.6, in which

$$I_1 = 24 \text{ A}$$

$$\phi_1 = \arccos 0.6 = 53.13^\circ \text{ lagging}$$

$$\phi_2 = \arccos 0.9 = 25.84^\circ \text{ lagging}$$

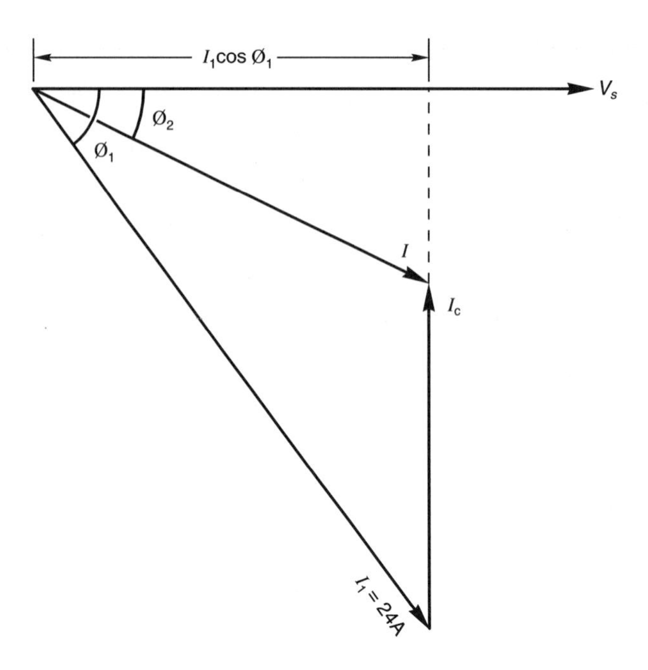

Figure 10.6 Phasor diagram for Worked Example 10.6

(a) The in-phase component of the load current is

$$I_1 \cos \phi_1 = 24 \times 0.6 = 14.4 \text{ A}$$

hence

$$14.4 = I \cos \phi_2 = 0.9I$$

That is, the current drawn from the supply after the capacitor is connected is

$$I = 14.4/0.9 = 16 \text{ A}$$

That is, one effect of connecting the capacitor to the system is to reduce the current drawn by the load from 24 A to 16 A! Two side effects of this are (i) the copper loss in the cable is reduced, and (ii) the voltage drop in the cable is reduced.

(b) The capacitor current is given by

$$\begin{aligned} I_C &= I_1 \sin \phi_1 - I \sin \phi_2 \\ &= 24 \sin 53.13^\circ - 16 \sin 25.84^\circ \\ &= 19.2 - 6.97 = 12.23 \text{ A.} \end{aligned}$$

Worked Example 10.7

Determine the capacitance and the kVAr rating of the capacitor in Worked Example 10.6 if the supply voltage is 440 V and the frequency is 50 Hz.

Solution

The equation for the capacitance of the power-factor correcting capacitor is

$$C = I_C/(2\pi f V_S) = 12.23/(2\pi \times 50 \times 440)$$
$$= 88.5 \times 10^{-6} \text{ F or } 88.5\ \mu\text{F}$$

and its kVAr rating is

$$Q_C = V_S I_C/1000 = 440 \times 12.23/1000 = 5.4 \text{ kVAr}$$

10.6 Computer listings

There are two Computer listings in this section. Listing 10.1 solves the power triangle given two sides of the triangle. That is we can

1. solve for VA and angle using power and VAr, or
2. solve for VAr and angle using power and VA, or
3. solve for power and angle using VA and VAr.

Since most versions of the BASIC language can only solve inverse trigonometric ratios for tangents, we can only use the ATN instruction and Pythagoras's theorem.

It is left as an exercise for the reader to extend the program to deal with the other cases where we know the angle and ONE of the sides.

Listing 10.2 deals with power factor calculations. The options offered are

1. convert power factor into a phase angle, or
2. convert a phase angle into a power factor, or
3. deal with a power factor correction problem.

The first of these options uses the trigonometric identity

$$1 + \tan^2 \phi = 1/\cos^2 \phi$$

to determine the tangent of the power factor you supply to it (see line 230), from which the phase angle in degrees is calculated. This identity is used twice in the third option of Listing 10.2 in lines 470 and 490.

The third option deals with a power factor correction problem of a type similar to that in Worked Examples 10.6 and 10.7. The software asks for the magnitude of the load current and power factor of the inductive load, and for the required value of the power factor for the final system. From this it determines the magnitude of the current drawn from the supply after the power factor has been corrected, together with the capacitor current. If you can give the supply voltage and frequency, it also calculates the capacitance of the capacitor in microfarads, the reactance of the capacitor, and its kVA rating.

Listing 10.1 Solution of a power triangle

```
10 CLS '*** CH10-1 ***
20 PRINT "SOLUTION OF A POWER TRIANGLE": PRINT
30 PRINT "Do you wish to: ": PRINT
40 PRINT "1. Solve for VA and ANGLE using Power and VAr?"
50 PRINT "2. Solve for VAr and ANGLE using Power and VA?"
60 PRINT "3. Solve for Power and ANGLE using VA and VAr?": PRINT
70 INPUT "Enter your selection here (1, 2 or 3): ", Sel: PRINT
80 IF Sel < 1 OR Sel > 3 THEN GOTO 10
90 PI = 3.141593: P = 0: S = 0: Q = 0: PHI = 0
100 IF Sel = 2 THEN CLS : GOTO 300
110 IF Sel = 3 THEN CLS : GOTO 400
200 CLS : PRINT "Solve for VA and ANGLE using Power and VAr": PRINT
210 INPUT "Power consumed (W) = ", P
220 INPUT "Reactive power consumed (VAr) = ", Q: PRINT
230 PRINT "Apparent power (VA) = "; SQR(P ^ 2 + Q ^ 2)
240 PHI = ATN(Q / P)
250 PRINT "Phase angle (deg) = "; PHI * 180 / PI
260 END
300 PRINT "Solve for VAr and ANGLE using Power and VA": PRINT
310 INPUT "Apparent power consumed (VA) = ", S
320 INPUT "Power consumed (W) = ", P: PRINT
330 PRINT "Reactive power (VAr) = "; SQR(S ^ 2 - P ^ 2)
340 PHI = ATN(SQR(S ^ 2 - P ^ 2) / P)
350 PRINT "Phase angle (deg) ="; PHI * 180 / PI
360 END
400 PRINT "Solve for Power and ANGLE using VA and VAr": PRINT
410 INPUT "Apparent power consumed (VA) = ", S
420 INPUT "Volt-amperes reactive consumed (VAr) = ", Q: PRINT
430 PRINT "Power (W) = "; SQR(S ^ 2 - Q ^ 2)
440 PHI = ATN(Q / SQR(S ^ 2 - Q ^ 2))
450 PRINT "Phase angle (deg) ="; PHI * 180 / PI
460 END
```

Listing 10.2 Power factor calculations

```
10 CLS ' *** CH 10-2 ***
20 PRINT "POWER FACTOR CALCULATIONS": PRINT
30 PRINT "Do you wish to:": PRINT
40 PRINT "1. Convert POWER FACTOR into a PHASE ANGLE, or"
50 PRINT "2. Convert a PHASE ANGLE into a POWER FACTOR, or"
60 PRINT "3. Solve a power factor correction problem?": PRINT
70 INPUT "Enter you selection here (1, 2 or 3): ", Sel: PRINT
```

```
80 IF Sel < 1 OR Sel > 3 THEN GOTO 10
90 PI = 3.141593
100 IF Sel = 2 THEN GOTO 300
110 IF Sel = 3 THEN GOTO 400
200 CLS : PRINT "Convert a POWER FACTOR into a PHASE ANGLE": PRINT
210 INPUT "Power factor = ", PF
220 IF PF <= 0 OR PF > 1 THEN GOTO 200
230 T = SQR((1 - PF ^ 2) / PF ^ 2) '** Tangent of phase angle **
240 PHI = ATN(T) * 180 / PI
250 PRINT "Phase angle (deg) = "; PHI
260 END
300 CLS : PRINT "Convert a PHASE ANGLE into a POWER FACTOR": PRINT
310 INPUT "Phase angle (deg) = ", PHI: PRINT
320 IF PHI > 90 OR PHI < -90 THEN GOTO 300
330 PHI = PHI * PI / 180
340 PRINT "Power factor = "; COS(PHI)
350 END
400 CLS : PRINT "Power factor correction calculation": PRINT
410 INPUT "Magnitude of load current = ", I1
420 INPUT "Load power factor = ", PF1
430 INPUT "Required value of power factor = ", PF
440 IF (PF1 OR PF) > 1 OR (PF1 OR PF) < 0 THEN GOTO 400
450 IF PF <= PF1 THEN GOTO 400
460 I = I1 * PF1 / PF '** Calculate final current **
470 T1 = SQR((1 - PF1 ^ 2) / PF1 ^ 2) '** tangent of PF1 **
480 PHI1 = ATN(T1) '** Initial angle (rad) **
490 T = SQR((1 - PF ^ 2) / PF ^ 2) '** tangent of final power factor **
500 PHI = ATN(T) '** final phase angle **
510 Q1 = I1 * SIN(PHI1) '** Quadrature component of I1 **
520 Q2 = I * SIN(PHI) '** Quadrature component of final current **
530 Ic = ABS(Q1 - Q2) '** Capacitor current **
540 PRINT "Do you know the supply voltage (Y/N) "
550 K$ = INKEY$: IF K$ = "" THEN GOTO 550
560 IF NOT INSTR("YyNn", K$) > 0 THEN GOTO 550
570 IF K$ = "N" OR K$ = "n" THEN GOTO 610
580 INPUT "Supply voltage (V) = ", Vs
590 INPUT "Frequency (Hz) = ", f
600 C = Ic / (2 * PI * f * Vs)
610 PRINT : PRINT "Magnitude of overall current (A) = "; I
620 PRINT "Magnitude of capacitor current (A) = "; Ic
630 IF K$ = "Y" OR K$ = "y" THEN GOSUB 800
640 END
800 PRINT "Capacitance (microfarads) = "; C * 10 ^ 6
810 PRINT "Reactance of capacitor (ohms) = "; 1 / (2 * PI * f * C)
820 PRINT "Capacitor rating (kVAr) = "; Vs * Ic / 1000
830 RETURN
```

If the problem posed is not of the type in Worked Examples 10.6 and 10.7, then the software can be used to verify the solutions you have obtained by hand calculation.

The reader is reminded here that Listings 9.1 and 9.2, which refer to *R-L-C* series and parallel circuits respectively, also contain extensions which determine the power, reactive power, apparent power and power factor of *R-L-C* circuits.

Exercises

10.1 A circuit consumes a power of 10 MW at a phase angle of (a) 60°, (b) −35°. What VA and VAr is consumed in each case?

10.2 If the quadrature power consumption in an electronic circuit is 100 mVAr, what power and VA are consumed, given that the phase angle of the circuit is 50°?

10.3 An inductive load consumes an apparent power of 100 MVA at a power factor 0.8. What power and reactive power are consumed?

10.4 If the power and apparent power consumed by a circuit are 10 W and 15 VA, determine the reactive power consumed and power factor of the circuit.

10.5 What is the power consumption and phase angle of a circuit which consumes 300 kVA and 200 kVAr?

10.6 A single-phase load draws a current of 40 A at a power factor of 0.65 lagging. If a capacitor which takes a current of 20 A is connected in parallel with the load, determine the new value of the supply current taken by the complete system, and the overall power factor.

10.7 A capacitor is connected to the terminals of a single-phase motor, and the overall current drawn is 9.33 A. If the current taken by the capacitor is 4.5 A, determine (a) the current drawn by the motor itself and (b) the phase angle and power factor of the motor. (c) What is the power factor of the motor and capacitor combination?

10.8 A capacitor of 24.4 μF is connected to the terminals of an inductive load. The load is supplied at 440 V, 50 Hz, and the inductive load takes a current of 10 A at a power factor of 0.8 lagging. Determine (a) the value of the capacitor current, (b) the total current drawn from the supply when the capacitor is connected, (c) the power factor of the combination when the capacitor is connected and (d) the kVAr rating of the capacitor.

10.9 A substation supplies the following single-phase loads:

200 kw at a power factor of 0.8 leading
600 kW at a power factor of 0.8 lagging
400 kW at unity power factor
500 kVA at a power factor of 0.9 lagging

What is the overall kVA consumption of the loads and the effective power factor?

10.10 What reactive power must be connected to the substation load in Problem 10.9 to improve the overall power factor to unity?

10.11 A 9 kVA load at unity power factor is connected in parallel with a 15 kVA load at 0.6 power factor lagging. What is the overall kVA and power factor of the load? If a capacitor is connected in parallel with this combination to give an overall power factor of 0.95, determine
(a) the reactive power consumption of the capacitor
(b) the total apparent power consumed when the capacitor is connected
(c) the power factor of the circuit if the inductive load is switched off.

10.12 An inductive load draws a current of 200 A at 150 V, the power factor of the load being 0.5 lagging. A cable is available to supply the load with a rating of 170 A. If this cable is used to supply the load, determine
(a) the current rating of the capacitor which must be connected to the terminals of the load to limit the cable current to 170 A, and
(b) the overall power factor when the capacitor is connected.

10.13 A 500 W, 240 V, 50 Hz single-phase induction motor has a power factor of 0.7 lagging. The power factor of the system is to be improved by connecting a capacitor across the terminals of the motor. Determine
(a) the active and reactive components of the original current
(b) the total current supplied when the capacitor is connected
(c) the reactance of the capacitor and its current
(d) the capacitance of the capacitor.

Summary of important facts

The power consumed by an a.c. circuit can be divided into one of three types, namely

1. **active power** or **real power**, $P = VI\cos\phi$
2. **reactive power** or **quadrature power**, $Q = VI\sin\phi$
3. **apparent power** or **volt-amperes**, $S = VI$

in which V is the *magnitude of the voltage* across the element or circuit, I is the *magnitude of the current* in the circuit, and ϕ is the *phase angle* between V and I. Also

$$S = \sqrt{(P^2 + Q^2)}$$

The **power factor** of an element or circuit is given by

$$\textbf{power factor} = \frac{\textbf{active power}}{\textbf{apparent power}} = \frac{VI\cos\phi}{VI} = \cos\phi$$

The active power, reactive power and apparent power are related to one another by the **power triangle** (see Figure 10.10(c)).

In an industrial situation, a low power factor is penalised by a high tariff. If an item of equipment has a low power factor, it can be corrected to a higher value by connecting a capacitor to its terminals.

11 Resonant circuits

11.1 Introduction

In this chapter we look at special cases of series and parallel *R-L-C* a.c. circuits, namely those in which the current drawn from the supply is in phase with the supply voltage. This condition is known as **resonance**.

Resonance is a natural condition which occurs when an object (in our case, the components in a circuit) have a natural frequency which vibrates in sympathy with the forcing signal (in our case, the supply frequency). In terms of a mechanical system, we have the case where an opera singer can cause a glass to shatter when a very high note is produced.

By the end of this chapter, the reader will be able to

- Appreciate the significance of resonance in series and parallel circuits.
- Calculate the current in, and the voltage across elements in resonant circuits.
- Determine the power consumed by resonant circuits.
- Calculate the *Q*-factor of resonant circuits.
- Evaluate the dynamic resistance or dynamic impedance of a parallel resonant circuit.
- Use computer software associated with resonant circuits.

11.2 The series resonant circuit

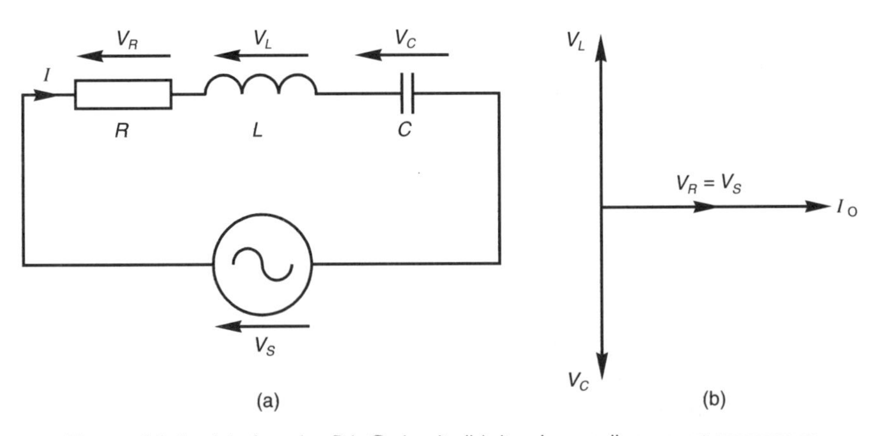

Figure 11.1 (a) A series R-L-C circuit; (b) its phasor diagram at resonance

The circuit diagram of a series R-L-C circuit is shown in Figure 11.1(a), and its phasor diagram at resonance is in diagram (b). When resonance occurs in a series circuit

$$V_L = V_C$$

or

$$I_0\omega_0 L = I_0/(\omega_0 C)$$

where ω_0 is the **resonant frequency** in rad/s, and I_0 is the current at resonance. That is

$$\boldsymbol{\omega_0 = 1/\sqrt{(LC)}}\ \text{rad/s}$$

or

$$\boldsymbol{f_0 = 1/(2\pi\sqrt{(LC)})}\ \text{Hz}$$

Since the reactive components of voltage cancel each other out, the total current in the circuit is restricted only by the resistance of the circuit, hence the current at resonance is

$$\boldsymbol{I_0 = V_S/R}$$

Since the resistance of an electrical power circuit has a low value, series resonance in a power circuit can be very dangerous because of the high current involved. Associated with this is the fact that a dangerously high voltage appears across L and C.

The voltage across the resistance at resonance is

$$V_R = I_0 R = V_S$$

That is, a voltage equal to the supply voltage appears across R. This *does not* imply that the voltage across L and C at resonance is zero! The voltage across these elements is (see also section 11.3)

$$V_L = I_0 X_L = I_0\omega_0 L$$
$$V_C = I_0 X_C = I_0/(\omega_0 C)$$

The power consumed by the circuit at resonance is

$$P = I_0^2 R.$$

Worked Example 11.1

A series circuit contains a 10 μF capacitor, a pure inductor of 100 mH, and a resistance of 10 Ω. Determine (a) the resonant frequency of the circuit, (b) the current at resonance when the supply voltage is 10 V, (c) the voltage across the capacitance and the inductance at resonance, (d) the power consumed.

Solution

(a) The resonant frequency of the circuit is

$$\omega_0 = 1/\sqrt{(LC)} = 1/\sqrt{(100 \times 10^{-3} \times 10 \times 10^{-6})}$$
$$= 1000 \text{ rad/s or } 159.2 \text{ Hz}$$

(b) The impedance of the circuit at resonance is equal to R, hence

$$I_0 = V_S/R = 10/10 = 1 \text{ A}$$

(c) The capacitive reactance at resonance is

$$X_C = 1/\omega_0 C = 1/(1000 \times 10 \times 10^{-6}) = 100 \ \Omega$$

and the inductive reactance at resonance is

$$X_L = \omega_0 L = 1000 \times 100 \times 10^{-3} = 100 \ \Omega$$

The voltage across C and L are therefore

$$V_C = I_0 X_C = 1 \times 100 = 100 \text{ V}$$
$$V_L = I_0 X_L = 1 \times 100 = 100 \text{ V}$$

(**Note**: the magnitude of V_C and V_L are both ten times greater than the supply voltage! (see also Worked Example 11.2).)

(d) The power consumed by the circuit at resonance is

$$P = I_0^2 R = 1^2 \times 10 = 10 \text{ W.}$$

11.3 Q-factor of a series resonant circuit

The ***Q*-factor** or **quality factor** of a series resonant circuit can be thought of as *the voltage magnification across either the capacitor or the inductor when compared with the supply voltage*. That is

$$\textbf{\textit{Q}-factor} = \frac{V_L}{V_S} = \frac{I_0 \omega_0 L}{I_0 R} = \frac{\omega_0 L}{R} = \frac{X_L}{R}$$

or

$$\textbf{\textit{Q}-factor} = \frac{V_C}{V_S} = \frac{I_0/\omega_0 C}{I_0 R} = \frac{X_C}{R} = \frac{1}{\omega_0 CR}$$

or

$$\textbf{\textit{Q}-factor} = \frac{1}{R}\sqrt{\left[\frac{L}{C}\right]}.$$

Worked Example 11.2

Determine the Q-factor at resonance of the series circuit in Worked Example 11.1.

Solution

We can use any of the above equations as follows:

$$Q\text{-factor} = X_L/R = 100/10 = 10$$
$$Q\text{-factor} = X_C/R = 100/10 = 10$$
$$Q\text{-factor} = \frac{1}{R}\sqrt{\left[\frac{L}{C}\right]} = \frac{1}{10}\sqrt{\left[\frac{100 \times 10^{-3}}{10 \times 10^{-6}}\right]} = 10.$$

11.4 Pure L in parallel with pure C at resonance

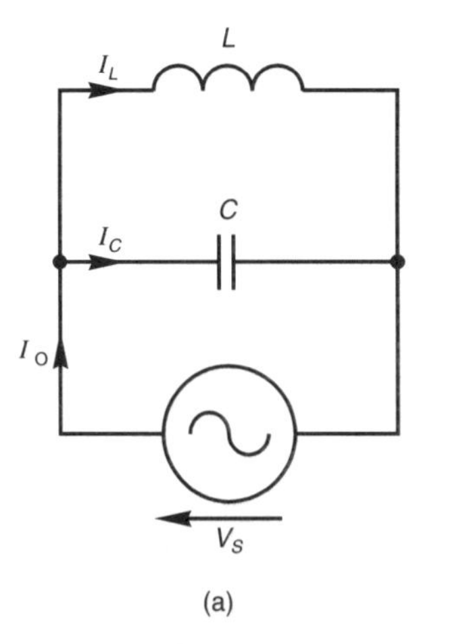

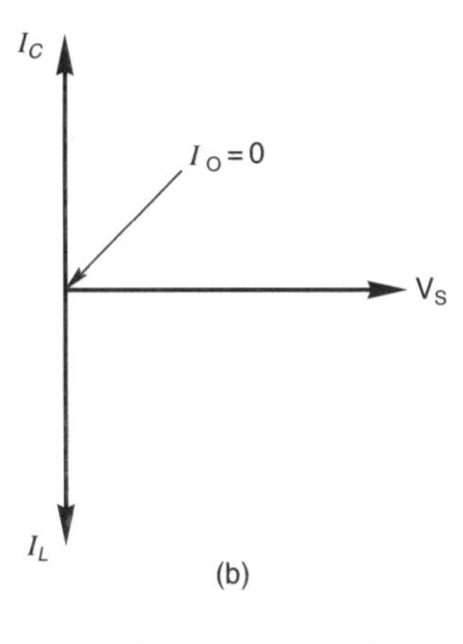

Figure 11.2 (a) Circuit diagram showing pure L in parallel with pure C; (b) phasor diagram at resonance

As with the series circuit, resonance occurs when the current drawn from the supply is in phase with the supply voltage. This occurs in Figure 11.2 when $I_L = I_C$, or

$$\frac{V_S}{X_L} = \frac{V_S}{X_C}$$

that is when

$$X_L = X_C$$

or

$$\omega_0 L = 1/\omega_0 C$$

where ω_0 is the **resonant frequency** of the circuit. Hence

$$\boldsymbol{\omega_0 = 1/\sqrt{(LC)}} \text{ rad/s}$$

or

$$\boldsymbol{f_0 = 1/(2\pi\sqrt{(LC)})} \text{ Hz.}$$

Since both L and C are assumed to be 'pure' (which, in practice, is not possible), then the two currents cancel out, *and the current drawn from the supply is zero*, i.e. $I_0 = 0$. However, in practice (see section 11.5), the current at resonance can have a very low value. In the ideal situation posed here, no current is drawn from the supply, and *no power is consumed by the circuit*.

Worked Example 11.3

Calculate the resonant frequency of a two-branch parallel circuit which has a pure inductor of 1.0 mH in one branch, and a 0.1 μF capacitor in the other. If 10 V at the resonant frequency is applied to the circuit, determine the current in each branch of the circuit.

Solution

The circuit and phasor diagrams are generally as shown in Figure 11.2. The resonant frequency of the circuit is

$$\begin{aligned}\omega_0 &= 1/\sqrt{(LC)} = 1/\sqrt{(1 \times 10^{-3} \times 0.1 \times 10^{-6})} \\ &= 100\,000 \text{ rad/s or } 15915 \text{ Hz or } 15.915 \text{ kHz}\end{aligned}$$

The reactance of the inductive branch is

$$X_L = \omega_0 L = 100\,000 \times 1 \times 10^{-3} = 100\ \Omega$$

and the current in that branch is

$$I_L = V_S/X_L = 10/100 = 0.1 \text{ A lagging } V_S \text{ by } 90°.$$

The reactance of the capacitive branch is

$$X_C = 1/(\omega_0 C) = 1/(100\,000 \times 0.1 \times 10^{-6}) = 100\ \Omega$$

and the current in the capacitor is

$$I_C = V_S/X_C = 10/100 = 0.1 \text{ A leading } V_S \text{ by } 90°.$$

Since $I_L = I_C$, and the two are in phase opposition to one another, no current is drawn from the supply.

11.5 L-R in parallel with pure C at resonance

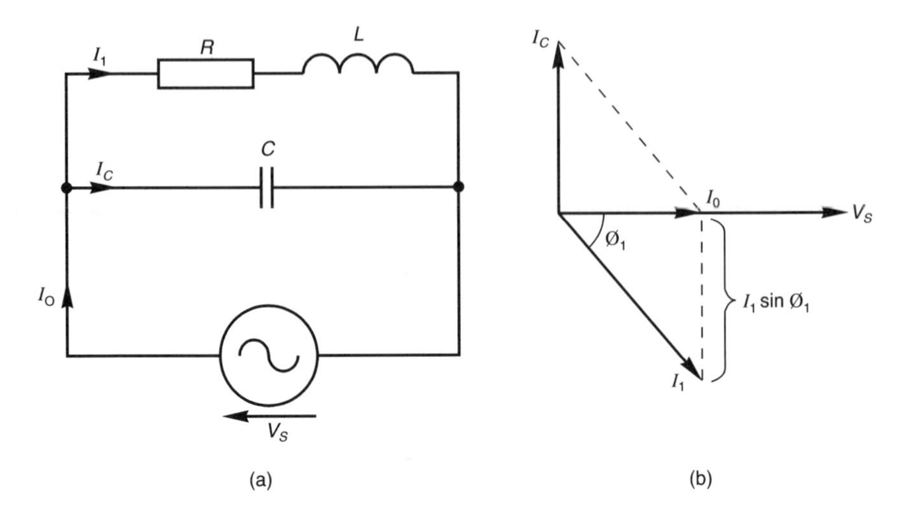

Figure 11.3 (a) Circuit diagram of R-L in parallel with C; (b) its phasor diagram at resonance

The circuit is shown in Figure 11.3(a), and a typical phasor diagram in diagram (b); this is a more practical form of circuit, since the inductive branch has some resistance.

Once again, resonance occurs when the total quadrature current drawn from the supply is zero, that is when the current drawn from the supply is in phase with the supply voltage, or when

$$I_C = I_1 \sin \phi_1$$

It can be shown that this occurs at the **resonant frequency** of the circuit, which is

$$\omega_0 = \sqrt{\left[\frac{1}{LC} - \frac{R^2}{L^2}\right]} \text{ rad/s}$$

and $f_0 = \omega_0 / 2\pi$ Hz

In many practical circuits, the value (R^2/L^2) is much less than $(1/LC)$, in which case the resonant frequency is approximately

$$\omega_0 = \sqrt{(1/LC)} \text{ rad/s}$$

The reader will note from the phasor diagram in Figure 11.3(b) that, when the coil has some resistance, the circuit draws a current from the supply at the resonant frequency. In fact, the greater the value of resistance, the greater the current drawn at resonance!

The increase in current with resistance in the coil is accounted for by the fact that the coil consumes power, and current must flow in order to supply the power.

Since current flows into the circuit at resonance, it is said to have a **dynamic resistance** (resistance at resonance), R_D, or **dynamic impedance**, Z_D. Its value is calculated from the equation

$$R_D = \frac{L}{CR}$$

The current drawn from the supply at resonance is

$$I_0 = V_S/R_D$$

and the power consumed is I_1^2R, where I_1 is the current in the coil (see Figure 11.3(a)).

11.6 Q-factor of a parallel circuit at resonance

The ***Q*-factor** of a parallel circuit is *the current magnification between the reactive (quadrature) current in either branch of the circuit when compared with the current drawn from the supply.*

In a parallel circuit containing *pure L and pure C* (see Figure 11.2(a)), the Q-factor is

$$\textbf{\textit{Q}-factor} = \frac{\text{reactive current}}{\text{supply current}} = \frac{I_C}{I_0} = \frac{I_L}{I_0}$$

In this case, since $I_0 = 0$, the Q-factor is infinity!

In a practical case of the type in Figure 11.3(a), the Q-factor is

$$\textbf{\textit{Q}-factor} = \frac{I_C}{I_0} = \frac{I_1 \sin \phi_1}{I_0}.$$

Worked Example 11.4

A coil of resistance 12 Ω and inductance 0.12 H is connected in parallel with a 60 μF capacitor to a 100 V supply at the resonant frequency. Calculate the value of the resonant frequency. Determine (a) the dynamic impedance of the circuit at this frequency, (b) the current taken from the supply, (c) the magnitude of the current in each branch, (d) the Q-factor of the circuit at resonance, (e) the total power consumed. Draw the phasor diagram of the circuit at resonance.

Solution

The resonant frequency is calculated from

$$\omega_0 = \sqrt{\left[\frac{1}{LC} - \frac{R^2}{L^2}\right]}$$

$$= \sqrt{\left[\frac{1}{0.12 \times 60 \times 10^{-6}} - \frac{12^2}{0.12^2}\right]}$$

$$= \sqrt{(138\,889 - 10\,000)} = 359 \text{ rad/s}$$

and

$$f_0 = \omega_0/2\pi = 57.1 \text{ Hz}$$

(**Note**: had we used the approximate equation, we would have obtained

$$\omega_0 = 1/\sqrt{(LC)} = 372.7 \text{ rad/s or } 59.3 \text{ Hz.)}$$

(a) The dynamic impedance at resonance is

$$Z_D = L/CR = 0.12/(60 \times 10^{-6} \times 12) = 166.7\ \Omega$$

(b) The current drawn at resonance is

$$I_0 = V_S/Z_D = 100/166.7 = 0.6 \text{ A}$$

(c) The impedance of the coil at resonance is

$$Z_1 = \sqrt{(R^2 + \omega_0{}^2L^2)} = \sqrt{(12^2 + (359 \times 0.12)^2)}$$
$$= 44.72\ \Omega$$

The magnitude of the current in the coil is

$$I_1 = V_S/Z_1 = 100/44.72 = 2.24 \text{ A lagging}$$

The reactance of the capacitor is

$$X_C = 1/\omega_0 C = 1/(359 \times 60 \times 10^{-6}) = 46.43\ \Omega$$

and the current in the capacitor is

$$I_C = V_S/X_C = 100/46.43 = 2.15 \text{ A leading } V_S \text{ by } 90^\circ.$$

(**Note**: $I_1 > I_C$, but the quadrature component of I_1 is equal to I_C.)

(d) The Q-factor of the circuit is calculated from

$$Q = 1/\omega_0 CR = 1/(359 \times 60 \times 10^{-6} \times 12) = 3.86$$

(e) The total power consumed is

$$P = I_1{}^2R = 2.24^2 \times 12 = 60.2 \text{ W}$$

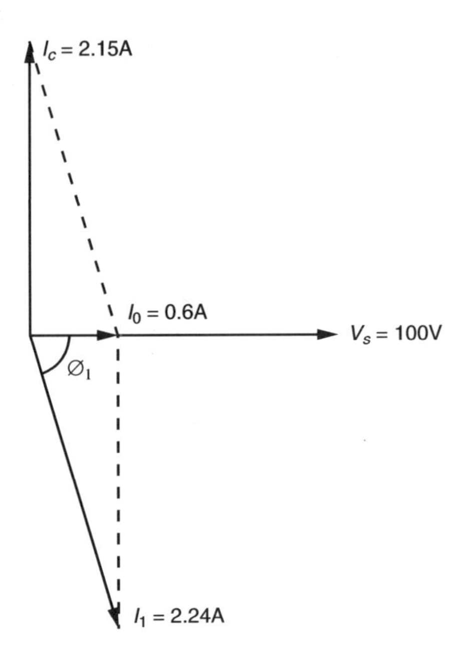

Figure 11.4 Phasor diagram for Worked Example 11.4

We now have all the information to draw the phasor diagram (see Figure 11.4), except the angle ϕ_1 of the coil. This is $-74.4°$; it is left as an exercise for the reader to verify this value.

11.7 Computer listing

The Computer listing associated with this chapter is given in listing 11.1. This allows the user to evaluate the resonant frequency of

1. a series *R-L-C* circuit
2. a parallel circuit with *R-L* in parallel with *C*.

In both cases the resonant frequency is calculated both in rad/s and in hertz, and both determine the impedance at resonance. Provided that the circuit resistance has a positive value, the *Q*-factor of the circuit is also evaluated.

The reader is reminded that Listings 9.1 and 9.2 deal, respectively, with the analysis of series and parallel a.c. circuits (including the resonant condition). The final few lines of each of the latter Listings calculate, for a resonant or near-resonant circuit, the *Q*-factor of the circuit.

Listing 11.1 Resonant frequency and impedance at resonance

```
10 CLS ' *** Ch11-1 ***
20 PRINT "RESONANT FREQUENCY AND IMPEDANCE AT RESONANCE": PRINT
30 PRINT "Select ONE of the following": PRINT
40 PRINT "1. Resonant frequency of a SERIES RLC circuit."
50 PRINT "2. Resonant frequency of R-L in PARALLEL with C.": PRINT
60 INPUT "Enter your selection here (1 or 2): ", Sel
70 IF Sel < 1 OR Sel > 2 THEN GOTO 10
80 PI = 3.141593
90 IF Sel = 2 THEN GOTO 400
200 CLS
210 PRINT "Resonant frequency of a SERIES circuit": PRINT
220 INPUT "Resistance (ohms) = ", R
230 IF R < 0 THEN GOTO 200
240 INPUT "Inductance (H) = ", L
250 INPUT "Capacitance (F) = ", C: PRINT
260 w0 = 1 / SQR(L * C)
270 PRINT "Resonant frequency = "; w0; "rad/s"
280 PRINT "                   = "; w0 / (2 * PI); "Hz"
290 PRINT "Impedance at resonance = "; R; "ohms"
300 IF R > 0 THEN PRINT "Q-factor = "; SQR(L / C) / R
310 END
400 CLS
410 PRINT "Resonant frequency of R-L in PARALLEL with C": PRINT
420 PRINT "For the R-L branch"
430 INPUT "    Resistance (ohms) = ", R
440 IF R < 0 THEN GOTO 400
450 INPUT "    Inductance (H) = ", L
460 INPUT "Capacitance (F) of the capacitive branch = ", C: PRINT
470 w0 = SQR((1 / (L * C)) - (R ^ 2 / L ^ 2))
480 PRINT "Resonant frequency = "; w0; "rad/s"
490 PRINT "                   = "; w0 / (2 * PI); "Hz"
500 IF R > 0 THEN GOSUB 600
510 END
600 PRINT "Dynamic impedance (ohms) = "; L / (C * R)
610 PRINT "Q-factor = "; 1 / (w0 * C * R)
620 RETURN
```

Exercises

11.1 A series circuit contains a 1.5 μF capacitor in series with an inductor of 50 mH. If the circuit has a resistance of 100 Ω, determine the resonant frequency of the circuit, and the current in the circuit if 15 V at the resonant frequency is applied.

11.2 What is the Q-factor of the circuit in Exercise 11.1?

11.3 An iron-cored coil is connected in series with a variable capacitor. If resonance occurs at a frequency of 1 kHz when the capacitance is 0.1 μF, what is the inductance of the inductor?

11.4 An a.c. circuit consists of a coil of resistance 10 Ω and inductance 0.2 H in series with a 60 μF capacitor. If the circuit is energised by a 100 V supply at the resonant frequency, calculate

(a) the resonant frequency of the circuit
(b) the current at resonance
(c) the power consumed at resonance
(d) the Q-factor of the circuit at resonance.

11.5 A resistor, a capacitor and an inductor are connected in series with a 10 V variable-frequency supply. The maximum current in the circuit is measured as 1 A at a frequency of 1 kHz, at which time the voltage across the capacitor is 62.9 V. Determine the resistance, capacitance and inductance in the circuit.

11.6 If the Q-factor of a series R-L-C circuit is 12, its resistance is 5 Ω and its inductance is 90 mH, calculate the capacitance in the circuit and the resonant frequency of the circuit.

11.7 A series R-L-C circuit contains a coil of resistance 10 Ω and inductance 10 mH, together with a capacitance of 0.1 μF. If the circuit is supplied by a 100 mV source at the resonant frequency, calculate

(a) the resonant frequency of the circuit
(b) the current in the circuit at resonance
(c) the voltage across the coil and across the capacitor
(d) the Q-factor of the circuit.

11.8 A series circuit comprises a coil of inductance 5 mH and resistance 10 Ω, in series with a variable capacitor. If the supply voltage is 100 mV at a frequency of 100 000 rad/s, what is the capacitance of the capacitor to give resonance at this frequency?

11.9 If a variable resistor is connected in series with the circuit in Exercise 11.8, what will be the voltage across the capacitor when the additional resistance is (a) 0 Ω, (b) 20 Ω, (c) 40 Ω?

11.10 A 10 mH pure inductor is connected in parallel with a 0.2 μF capacitor. What is the resonant frequency of the circuit?

11.11 A parallel circuit has a resonant frequency of 562.7 kHz, and it contains a 10 nF capacitor in one branch. What is the value of the ideal inductance in the other branch?

11.12 A coil of inductance 31.8 mH and resistance 10 Ω is connected in parallel with a capacitor to a 50 Hz supply. What value of capacitance must be connected in the other branch to give resonance?

11.13 If the supply voltage in Exercise 11.12 is 20 V at the resonant frequency, what current is drawn from the supply, and what is the dynamic impedance of the circuit?

11.14 A coil of inductance 10 mH and resistance 20 Ω is connected in parallel with a variable capacitance, the circuit being supplied by a 100 mV, 11.25 kHz supply. What value of capacitance causes the circuit to be resonant?

What current is drawn by the circuit at resonance, and what power is consumed by the circuit? What is the *Q*-factor of the circuit at resonance?

11.15 A circuit which is resonant at 50 Hz comprises a coil of impedance 100 Ω and resistance 60 Ω in parallel with a capacitor. The circuit is supplied by a 50 V, 50 Hz source. Determine

(a) the inductance of the coil
(b) the capacitance required to give resonance
(c) the current drawn from the supply
(d) the power consumed at resonance
(e) the Q-factor and dynamic impedance at resonance.

Summary of important facts

A circuit containing an inductor and a capacitor is **resonant** when the current drawn from the supply is in phase with the supply voltage.

In the case of a **series circuit** and a **parallel circuit** containing a *pure* L and a *pure* C, the **resonant frequency** is given by

$$\omega_0 = 1/\sqrt{(LC)} \text{ rad/s}$$

or

$$f_0 = 1/(2\pi\sqrt{(LC)}) \text{ Hz}$$

In the case of a *series circuit at resonance*, the current in the circuit at resonance is

$$I_0 = V_S/R$$

and the power consumed at resonance is

$$P = I_0^2 R$$

The ***Q*-factor** or **quality factor** of a series resonant circuit is given by the *voltage magnification produced by the circuit at resonance*, that is

$$Q = \frac{X_L}{R} = \frac{X_C}{R} = \frac{1}{R}\sqrt{\left[\frac{L}{C}\right]}$$

Resonance occurs in a parallel circuit containing L-R in one branch, i.e. a coil, and C in the other branch at a frequency of

$$\omega_0 = \sqrt{\left[\frac{1}{LC} - \frac{R^2}{L^2}\right]} \text{ rad/s}$$

and

$$f_0 = \omega_0/2\pi \text{ Hz}$$

The **dynamic impedance** or **resistance at resonance** of *L-R* in parallel with *C* is

$$Z_D = \frac{L}{CR}\ \Omega$$

and the ***Q*-factor of a parallel circuit** at resonance is

$$Q\text{-factor} = \frac{\text{reactive current in one branch}}{\text{current drawn from the supply}}$$

In the case of *pure L in parallel with pure C*, no current is drawn from the supply, and the Q-factor is infinite.

Where the parallel circuit consists of a coil in parallel with pure C

$$Q\text{-factor} = \frac{I_C}{I_0} = \frac{I_1 \sin \phi_1}{I_0}$$

where I_1 is the current in the coil, and ϕ_1 is the phase angle of the coil.

12 Transformers

12.1 Introduction

A knowledge of transformers is vital to both power and electronics engineers.

Broadly speaking, transformers can be classified into either **multi-winding transformers** and **auto-transformers**; the most popular type of multi-winding transformer is the *two-winding transformer*, which is the subject of the majority of this chapter. The auto-transformer is a *single-winding transformer*, and is described in section 12.10.

Transformers can be used either to 'step up' a voltage (or 'step down' a current), or to 'step down' a voltage (or 'step up' a current). The supply voltage is connected to the **primary winding** of the transformer, and power is extracted from the **secondary winding**. The two windings have a *common magnetic circuit* which, in the case of a power transformer, is an iron circuit.

In some electronic circuits, the transformer is used for **impedance matching**, which enables *maximum power* to be extracted from the secondary winding.

By the end of this chapter, the reader will be able to

- Understand and use the e.m.f. equation of the transformer.
- Calculate the voltage, the current and the turns ratio of a two-winding and an auto-transformer.
- Determine the efficiency of a transformer.
- Draw the phasor diagram for a transformer.
- Determine the condition for maximum power transfer into a load.
- Use computer software for transformer calculations.

12.2 Basic principles

The circuit diagram of an **ideal power transformer** is shown in Figure 12.1. We shall assume that the windings are resistanceless.

When V_1 is applied to the primary winding, current I_1 flows in it and produces a magnetic flux in the iron core. In power transformers, we assume that there is no magnetic loss, so that all the flux links with the secondary winding.

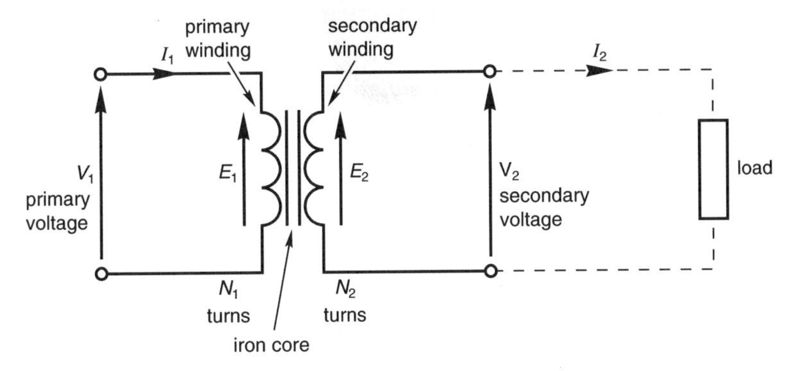

Figure 12.1 Ideal two-winding transformer

We also assume for the moment that there is no power loss in the transformer itself, so that the 'back e.m.f.' induced in the primary winding is equal to the supply voltage. That is $V_1 = E_1$.

The e.m.f., E_2, induced in the secondary winding causes current I_2 to flow in the load. The value of E_2 depends on the primary voltage, and on the number of turns on the primary and secondary windings.

12.3 e.m.f. equation

The equation for the e.m.f. induced in any winding of a transformer which is excited by a *sinusoidal supply* is

$$E = 4.44\,Nf\Phi_m$$

where N is the number of turns on the winding
f is the supply frequency, and
Φ_m is the maximum flux density in the core.

In the case of a two-winding transformer, we have

$$E_1 = 4.44N_1f\Phi_m = 4.44N_1fB_ma$$
$$E_2 = 4.44N_2f\Phi_m = 4.44N_2fB_ma$$

where B_m is the maximum value of flux density in the core, and a is the cross-sectional area of the core.

Worked Example 12.1

An ideal power transformer is energized by a 240 V, 50 Hz supply. The transformer core has a cross-sectional area of 8 cm^2, and the maximum flux density in the core is 0.4 T. If the secondary induced voltage is 50 V, calculate the number of turns on both windings.

Solution

The applied voltage is given by the equation

$$V_1 = 4.44N_1 f\Phi_m = 4.44N_1 fB_m a$$

hence

$$\begin{aligned} N_1 &= V_1/(4.44fB_m a) \\ &= 240/(4.44 \times 50 \times 0.4 \times 8 \times 10^{-4}) \\ &= 3378 \text{ turns} \end{aligned}$$

and

$$\begin{aligned} N_2 &= V_2/(4.44fB_m a) \\ &= 50/(4.44 \times 50 \times 0.4 \times 8 \times 10^{-4}) \\ &= 704 \text{ turns} \end{aligned}$$

If the reader carries out the above calculations, he/she will find that each winding appears to have an additional fractional part of a turn, i.e. the result is not exact. In practice, a winding can only have an integral number of turns.

12.4 Voltage, current and turns ratio

In an ideal transformer, two relationships hold good, and are:

each winding supports the same number of volts per turn

that is

$$\frac{V_1}{N_1} = \frac{V_2}{N_2}$$

and

there is ampere-turn balance between the windings

or

$$I_1 N_1 = I_2 N_2$$

These two relationships can be combined into the following

$$\frac{V_2}{V_1} = \frac{I_1}{I_2} = \frac{N_2}{N_1}$$

where N_2/N_1 is known as the **turns ratio** of the transformer. If $V_2 > V_1$, we say that the transformer has a **step-up voltage ratio** or **step-up turns ratio**. If $V_2 < V_1$, we say that the transformer has a **step-down voltage ratio**.

Also, if the transformer has a **step-up voltage ratio, it has a step-down current ratio**, and vice versa.

If we cross-multiply the terms in the first two parts of the equation, we get

$$V_1 I_1 = V_2 I_2$$

That is, each winding **supports the same number of volt-amperes**.

The reader is asked to note that transformers are rated in volt-amperes, or in multiples of VA, i.e. kVA, MVA, etc., and *not in watts*. The reason is that the transformer rating is limited by the current that a winding can carry rather than the in-phase component of the current (which is the part of the current which produces power).

It is true to say that the majority of transformers used in electronic circuits are not ideal so that, when measurements are made on them, the above equations do not always necessarily agree with the measured results.

Worked Example 12.2

Calculate the full-load primary and secondary current of a 25 kVA, 440/240 V transformer.

Solution

Since we are dealing with an ideal transformer, each winding will handle 25 kVA. That is

$$V_1 I_1 = 25\,000 \text{ VA}$$

or $I_1 = 25\,000/440 = 56.82$ A

and $I_2 = 25\,000/240 = 104.2$ A.

Worked Example 12.3

A 440/6600 V single-phase transformer has a rating of 50 kVA, and has 1500 turns on the secondary (h.v.) winding. Calculate the number of turns on the primary winding, and the full-load current in each winding.

Solution

Since the secondary (h.v.) winding has 1500 turns, each winding supports

$$6600/1500 = 4.4 \text{ volts/turn}$$

Hence, for the primary winding

$$\text{number of turns} = \text{voltage/(volts/turn)}$$
$$= 440/4.4 = 100 \text{ turns}$$

and for the primary winding

$$\text{voltage} \times \text{full-load current} = 50\,000 \text{ VA}$$

or

$$I_{1(\text{full-load})} = 50\,000/440 = 113.6 \text{ A}$$

and for the secondary

$$I_{2(\text{full-load})} = 50\,000/6600 = 106.1 \text{ A}.$$

12.5 Transformer efficiency

Each winding of a *practical transformer* has some resistance and, when current flows in the transformer, each winding produces an I^2R loss. The sum of the I^2R losses in the transformer is known as the **copper loss**. The copper loss is a *variable loss* because it varies with the current flowing through the transformer at that particular time.

If the transformer has a primary winding resistance of R_P and a secondary winding resistance of R_S, and the respective primary and secondary currents are I_P and I_S, then the *total copper loss* in the transformer is

$$I_P{}^2 R_P = I_S{}^2 R_S$$

Similarly, there are losses in the magnetic circuit, i.e. hysteresis loss and eddy current loss. The sum of these is known as the **iron loss** of the transformer. The iron loss is a *fixed loss* because it does not vary with the load supplied by the transformer.

The **per-unit efficiency** of the transformer is given by

$$\textbf{efficiency} = \frac{\textbf{output power}}{\textbf{input power}} \text{ per unit (p.u.)}$$

$$= \frac{\textbf{output power}}{\textbf{output power + total loss}}$$

$$= \frac{\textbf{input power} - \textbf{total loss}}{\textbf{input power}}$$

The **per-cent efficiency** is

$$\textbf{efficiency} = \textbf{per unit efficiency} \times \textbf{100\%}$$

The **total power loss** in the transformer is given by

$$\textbf{total loss} = \textbf{copper loss} + \textbf{iron loss}$$

The efficiency of a transformer varies with the load current, as shown in Figure 12.2. It can be shown that the maximum efficiency of a transformer occurs when

$$\textbf{iron loss} = \textbf{copper loss}.$$

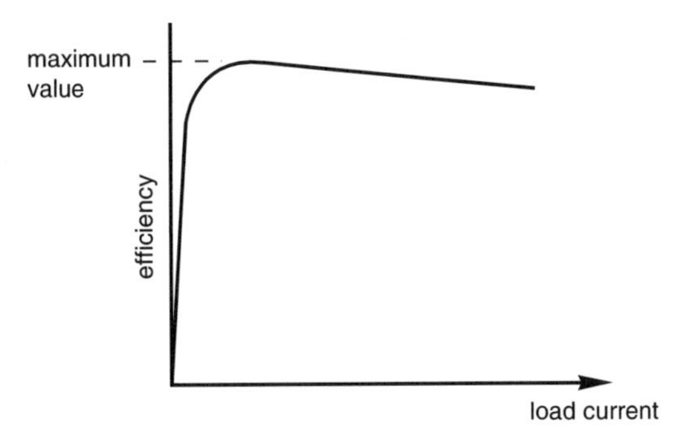

Figure 12.2 Graph showing how the efficiency of a transformer varies with load current

Worked Example 12.4

The output power from a single-phase transformer is 200 kW and, at this load, the total losses in the transformer are 8 kW. Determine the efficiency of the transformer.

Solution

The input power to the transformer is

$$\text{input power} = \text{output power} + \text{total loss} = 200 + 8 = 208\ \text{kW}$$

hence

$$\text{per-cent efficiency} = \frac{200}{208} \times 100 = 96.15\%.$$

Worked Example 12.5

A single-phase 2.5 kVA, 240/24 V transformer has a primary winding resistance of 2.4 Ω, and a secondary winding resistance of 0.02 Ω. Determine the full-load copper loss in the transformer.

Solution

The full-load primary winding current is

$$I_{1FL} = 2500/240 = 10.42\ \text{A}$$

hence

$$\text{primary full-load copper loss} = I_P{}^2 R_P = 10.42^2 \times 2.4 = 260.6\ \text{W}$$

The full-load current in the secondary winding is

$$I_{2FL} = 2500/24 = 104.2 \text{ A}$$

and the corresponding copper loss in the secondary is

$$I_S{}^2 R_S = 104.2^2 \times 0.02 = 217.2 \text{ W}$$

The total copper loss is therefore

$$260.6 + 217.2 = 477.8 \text{ W}.$$

12.6 Relationship between the copper loss and the load kVA

The copper loss (*CuLoss*) is proportional to $(\text{current})^2$, so that the full-load copper loss is proportional to $(\text{full-load current})^2$, and the copper loss at some other current, say I_2, we shall call $CuLoss_2$. That is

$$CuLoss_2 \propto I_2{}^2$$

and at full-load

$$CuLoss_{FL} \propto I_{FL}{}^2$$

That is

$$\frac{CuLoss_2}{CuLoss_{FL}} = \left[\frac{I_2}{I_{FL}}\right]^2$$

hence

$$CuLoss_2 = CuLoss_{FL} \times (I_2/I_{FL})^2$$

Also, assuming that the load voltage is constant, then

$$\text{load kVA} \propto \text{current}$$

therefore

$$CuLoss_2 = CuLoss_{FL} \times (kVA_2/kVA_{FL})^2$$

Also, if we need to determine the copper loss at a particular *power value*, then we can use the relationship

$$\frac{\text{power}}{\text{volt amperes}} = \text{power factor}$$

hence

$$\text{volt amperes} = \text{power/power factor}.$$

Worked Example 12.6

A 500 kVA transformer has a full-load copper loss of 5 kW. Determine the copper loss when the transformer delivers a load of (a) 250 kVA, (b) 400 kVA, (c) 250 kW at 0.8 power factor, (d) 200 kW at 0.5 power factor.

Solution

(a) From the preceding, we see that

$$\begin{aligned}\text{copper loss at 250 kVA} &= CuLoss_{FL} \times (kVA_2/kVA_{FL})^2 \\ &= 5 \times (250/500)^2 = 1.25\ \text{kW}\end{aligned}$$

(b) The copper loss at a load of 400 kVA is

$$\text{copper loss} = 5 \times (400/500)^2 = 3.2\ \text{kW}$$

(c) When the load is 250 kW at 0.8 power factor, the kVA consumed is

$$\frac{\text{power}}{\text{power factor}} = \frac{250}{0.8} = 312.5\ \text{kVA}$$

hence

$$\text{copper loss} = 5 \times (312.5/500)^2 = 1.953\ \text{kW}$$

(d) At a load of 300 kW and a power factor of 0.5, the kVA consumption is

$$\frac{\text{power}}{\text{power factor}} = \frac{200}{0.5} = 400\ \text{kVA}$$

and

$$\text{copper loss} = 5 \times (400/500)^2 = 3.2\ \text{kW}.$$

12.7 No-load phasor diagram

The no-load phasor diagram for an ideal transformer is shown in Figure 12.3; the phasors neglect the effect of the voltage drop in the primary winding produced by the no-load current, I_0. In the diagram

$$\begin{aligned}I_0 &= \text{no-load current} = \sqrt{(I_C{}^2 + I_{\text{mag}}{}^2)} \\ I_C &= \text{core-loss component of no-load current} \\ &= I_0 \cos \phi_0 \\ I_{\text{mag}} &= \text{magnetising component of no-load current} \\ &= I_0 \sin \phi_0\end{aligned}$$

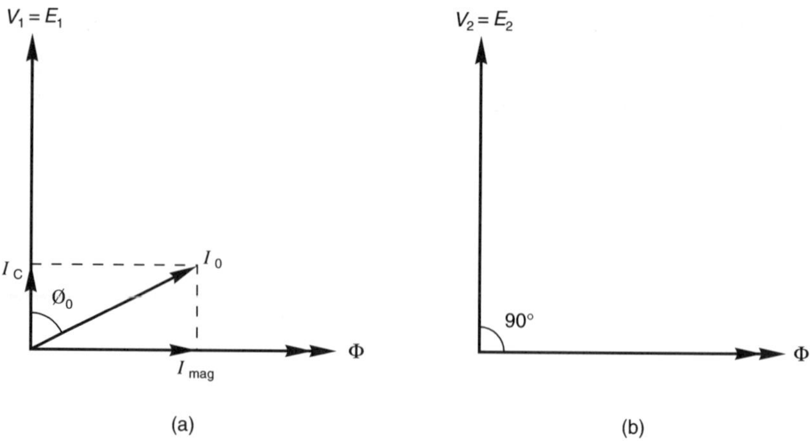

Figure 12.3 No-load phasor diagram for a simple transformer: (a) primary winding; (b) secondary winding

$$\begin{aligned} \phi_0 &= \text{no-load phase angle} \\ &= \arctan(I_{mag}/I_C) = \arccos(I_C/I_0) \\ \Phi &= \text{magnetic flux, lagging } 90^\circ \text{behind } V_1 \end{aligned}$$

and

$$\begin{aligned} P_0 &= \text{core loss or iron loss or no-load power} \\ &= V_1 I_0 \cos\phi_0. \end{aligned}$$

In common with other phasor diagrams, the quantity which is common to all parts of the circuit is drawn in the reference direction. In a transformer, this is the magnetic flux, so that it is shown in the reference (horizontal) direction.

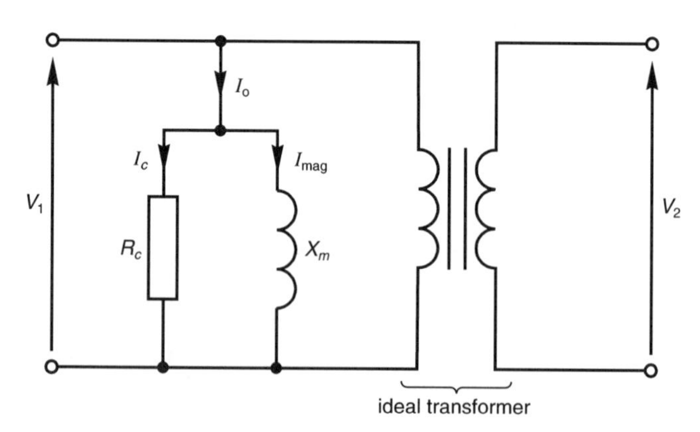

Figure 12.4 No-load equivalent circuit for the simple transformer

Only if the transformer has no copper loss or iron loss can it be thought of as an ideal transformer. However, even a **simplified transformer** has some losses in the iron circuit, and a simplified **equivalent circuit** of a transformer under no-load conditions is shown in Figure 12.4.

The no-load current of the transformer is considered to be due to two elements, namely a pure resistor R_C, in which the *core loss component* of the no-load current flows, and X_m (a pure inductor) in which the *magnetising component* of the no-load current flows. It follows that

$$R_C = \frac{V_1}{I_C} = \frac{V_1}{I_0 \cos \phi_0}$$

$$X_m = \frac{V_1}{I_{mag}} = \frac{V_1}{I_0 \sin \phi_0}$$

The value of R_C and X_m are determined from the results of the **no-load test** or **open-circuit test**, in which the supply voltage, V_1, the magnitude of the no-load current, I_0, and the no-load power, P_0, consumed by the transformer are measured.

Worked Example 12.7

An open-circuit test on a 4 kVA transformer gave the following results:

power consumption = 70 W
primary voltage = 200 V
no-load current = 0.7 A

Determine (a) the no-load power factor and phase angle, (b) the core loss and magnetising component of current, (c) the value of R_C and X_m (see Figure 12.4).

Solution

(a) The no-load power factor is calculated as follows:

$$\text{power factor} = \frac{\text{no-load power}}{\text{no-load VA}} = \frac{70}{200 \times 0.7} = 0.5$$

hence

$$\phi_0 = \arccos 0.5 = 60^\circ$$

That is, I_0 lags behind V_1 by 60°.

(b) The core loss component of the no-load current is

$$I_C = I_0 \cos \phi_0 = 0.7 \times 0.5 = 0.35 \text{ A}$$

and the magnetising component of the no-load current is

$$I_{mag} = I_0 \sin \phi_0 = 0.7 \times \sin 60^\circ = 0.606 \text{ A}$$

(**Note**: $\sqrt{(I_C{}^2 + I_{mag}{}^2)} = \sqrt{(0.35^2 + 0.606^2.)} = 0.7 \text{ A} = I_0$)

(c) The value of R_C is calculated from

$$R_C = V_1/I_C = 200/0.35 = 571.4\ \Omega$$

and

$$X_m = V_1/I_{mag} = 200/0.606 = 330\ \Omega.$$

12.8 Phasor diagram for a loaded transformer

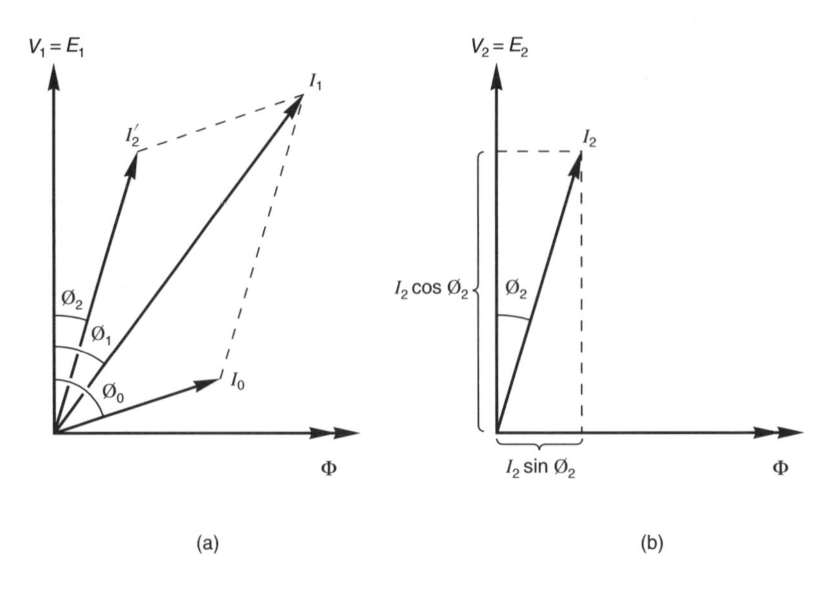

Figure 12.5 Phasor diagram for a transformer with an inductive load (neglecting the voltage drop in the windings): (a) primary winding; (b) secondary winding

When an inductive load is connected to the secondary winding, the phasor diagram for the transformer is shown in Figure 12.5. It should be noted that the phasor diagram neglects the effect of the voltage drop in the windings.

Referring to the phasor diagram for the secondary winding in Figure 12.5(b) (remember that the magnetic flux is the reference phasor), we see that the secondary current, I_2, lags behind the secondary voltage E_2 by angle ϕ_2, which is the phase angle of the load.

A corresponding current I'_2 flows in the primary winding so as to maintain the ampere-turn balance between the windings. That is

$$I'_2 N_1 = I_2 N_2$$

or

$$I'_2 = I_2 N_2 / N_1$$

and the current I'_2 lags behind E_1 by angle ϕ_2.

The total current, I_1, drawn by the primary winding is the *phasor sum* of I'_2 and the no-load current I_0, as shown in Figure 12.5(a).

The *simplified equivalent circuit* of the transformer corresponding to the phasor diagram in Figure 12.5 is shown in Figure 12.6.

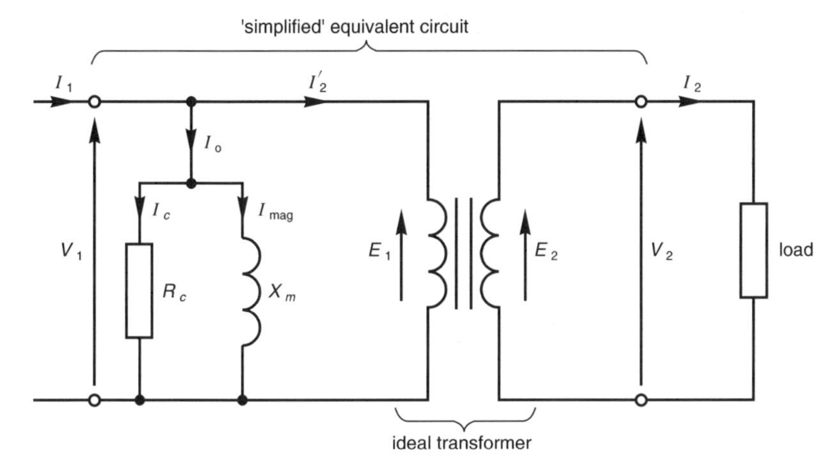

Figure 12.6 Simplified equivalent circuit of a transformer

Worked Example 12.8

A single-phase transformer with a step-down voltage ratio of 10:1 draws a primary current of 5.5 A at a power factor of 0.9 lagging when the secondary current is 50 A at a power factor of 0.95 lagging. Determine the magnitude and power factor of the no-load current; calculate also the core loss component and the magnetising component of the no-load current. Neglect the effect of the voltage drop in the windings.

Solution

Since the transformer has a step-down voltage ratio of 10:1, it has a step-up current ratio of 10:1. That is (neglecting the no-load current) the primary current is 0.1 of the corresponding secondary current, or

$$I'_2 = I_2/10 = 50/10 = 5 \text{ A}$$

This current lags behind the primary voltage by an angle corresponding to the load power factor of 0.95, i.e. an angle of

$$\phi_2 = \arccos 0.95 = 18.19^\circ$$

Next we determine the horizontal and vertical components of I_1 and I'_2 (see Figure 12.5(a)) as follows:

$$\begin{aligned}
\text{Horizontal component of } I'_2 &= I'_2 \sin \phi_2 \\
&= 5 \sin 18.19^\circ = 1.56 \text{ A} \\
\text{Horizontal component of } I_1 &= I_1 \sin(\arccos 0.9) \\
&= 5.5 \sin 25.84^\circ = 2.4 \text{ A} \\
\text{Vertical component of } I'_2 &= I'_2 \cos \phi_2 \\
&= 5 \times 0.95 = 4.75 \text{ A} \\
\text{Vertical component of } I_1 = I_1 \cos \phi_1 &= 5.5 \times 0.9 = 4.95 \text{ A}
\end{aligned}$$

Since I_1 = phasor sum $(I'_2 + I_0)$, then

I_0 = phasor difference $(I_1 - I'_2)$

Hence

horizontal component of $I_0 = 2.4 - 1.56 = 0.84$ A

and

vertical component of $I_0 = 4.95 - 4.75 = 0.2$ A

Therefore

magnitude of $I_0 = \sqrt{(0.84^2 + 0.2^2)} = 0.863$ A

and its phase angle is

$$\phi_0 = \arctan\frac{\text{horizontal component of } I_0}{\text{vertical component of } I_0}$$

$$= \arctan = \frac{0.84}{0.2} = 76.6^\circ$$

That is I_0 lags behind V_1 by 76.6°. The no-load power factor is

$\cos 76.6^\circ = 0.232$

The core-loss component of the no-load current is

$I_C = I_0 \cos\phi_0 = 0.863 \times 0.232 = 0.2$ A

and the magnetising component is

$I_{mag} = I_0 \sin\phi_0 = 0.863 \times \sin 76.6^\circ = 0.84$ A

(**Note:** $\sqrt{(I_C{}^2 + I_{mag}{}^2)} = \sqrt{(0.2^2 + 0.84^2)} = 0.863 \text{ A} = I_0$.)

12.9 Maximum power transfer to a resistive load

If the load is a pure resistance, and we are using an ideal loss-free transformer, then **maximum power is transferred** from the supply or source into the load **when the connected load appears to have the same resistance as that of the power source.**

If the source resistance is R_1, and that of the load is R_L, then maximum power transfer occurs when

$$R_1 = \left[\frac{N_1}{N_2}\right]^2 \times R_L$$

where N_1 and N_2 are the respective number of turns on the primary and secondary winding.

That is, a transformer can be used as an *impedance-level convertor*. This, of course, is only applicable to electronic circuits, where the resistance levels involved are fairly high.

When conditions for maximum power transfer occurs, the load is said to be **matched** to the source, and is referred to in electronic engineering as **impedance matching**.

Worked Example 12.9

An electronic circuit has an output resistance of 375 ohms. Determine the turns ratio of a transformer to give maximum power transfer to a 15 Ω load.

Solution

From the above equation, we see that

$$\left[\frac{N_1}{N_2}\right]^2 = \frac{R_1}{R_L} = \frac{375}{15} = 25$$

or

$$\frac{N_1}{N_2} = \sqrt{25} = 5$$

That is, a step-down voltage ratio of 5:1.

12.10 The auto-transformer

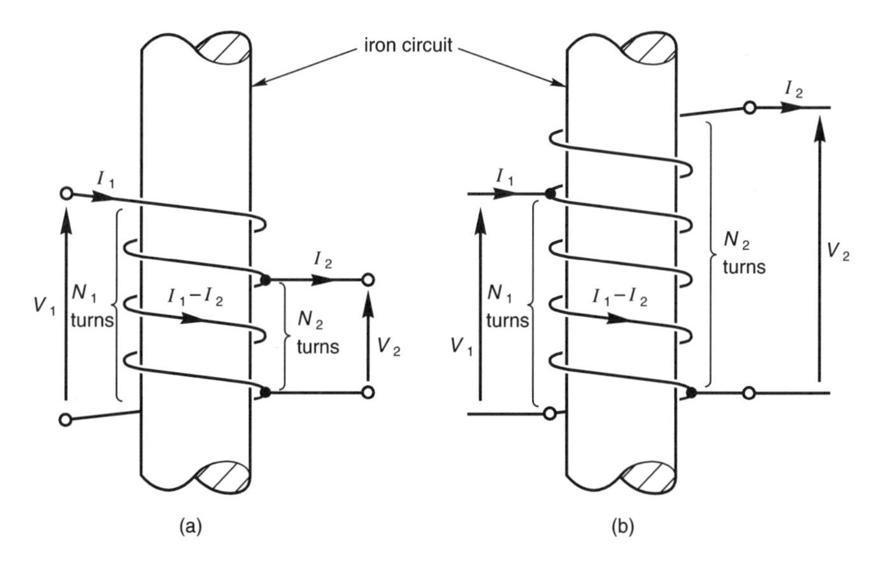

Figure 12.7 The auto-transformer with: (a) a step-down voltage ratio; (b) a step-up voltage ratio

An **auto-transformer** has a single winding, part being common to both the primary and secondary windings (see Figure 12.7). In common with all power transformers, ampere-turn balance is maintained between the windings, and the number of volts per turn is the same on both the secondary and primary sections of the winding. The general equation for the transformer is therefore

$$\frac{V_1}{V_2} = \frac{N_1}{N_2} = \frac{I_2}{I_1}$$

The winding arrangement for a *step-down voltage ratio* is shown in Figure 12.7(a), and for a *step-up voltage ratio* in Figure 12.7(b).

Auto-transformers are particularly useful where the primary and secondary voltage are similar (but not equal) to one another.

Worked Example 12.10

An auto-transformer is used to step up a voltage from 240 to 260 V. If the transformer has 2080 turns in total, calculate the number of turns on the primary section of the transformer.

If the transformer supplies a load of 8 kVA, determine the current in each part of the winding.

Solution

Since the maximum voltage (the secondary voltage) is 260 V, then the number of volts per turn on the transformer is

$$\frac{V_2}{N_2} = \frac{260}{2080} = 0.125$$

hence $V_1/N_1 = 0.125$, or

$$N_1 = \frac{V_1}{0.125} = \frac{240}{0.125} = 1920 \text{ turns}$$

The VA supplied by the secondary winding is

$$8000 = 260I_2$$

hence

$$I_2 = 8000/260 = 30.77 \text{ A}$$

and since $8000 = 240I_1$, then

$$I_1 = 8000/240 = 33.33 \text{ A}$$

That is to say $I_2 = 30.77$ A flows in the upper part of the winding (see Figure 12.7(b)) and $(I_1 - I_2) = 2.56$ A flows in the lower part of the winding.

12.11 Computer listing

Listing 12.1 Transformer calculations

```
10 CLS ' **** CH 12-1 ***
20 PRINT "TRANSFORMER CALCULATIONS": PRINT
30 PRINT TAB(6); "Select ONE of the following:": PRINT
40 PRINT TAB(4); "1. Voltage induced in a winding.": PRINT
50 PRINT TAB(4); "2. Primary and secondary voltage.": PRINT
60 PRINT TAB(4); "3. Primary and secondary current.": PRINT
70 PRINT TAB(4); "4. Maximum power transfer.": PRINT
80 INPUT "Enter your selection here (1 - 4): ", S
90 IF S < 1 OR S > 4 THEN GOTO 10
100 IF S = 2 THEN GOTO 260
120 IF S = 3 THEN GOTO 440
130 IF S = 4 THEN GOTO 620
140 CLS : PRINT "VOLTAGE INDUCED IN A WINDING": PRINT
150 PRINT "The formulae for the r.m.s. voltage is:": PRINT
160 PRINT "          E = 4.44 * N * f x PHIm": PRINT
170 PRINT " where N is the number of turns on the winding"
180 PRINT "        f is the supply frequency (Hz)"
190 PRINT "        PHIm is the maximum flux (Wb) in the core"
200 PRINT
210 INPUT "N = ", N
220 INPUT "f (Hz) = ", f
230 INPUT "Maximum flux (Wb) = ", PHIm: PRINT
240 PRINT "E = "; 4.44 * N * f * PHIm; "V (r.m.s.)"
250 END
260 CLS : PRINT "PRIMARY AND SECONDARY VOLTAGE": PRINT
270 PRINT TAB(6); "Select ONE of the following:": PRINT
280 PRINT TAB(4); "1. Calculate the primary voltage.": PRINT
290 PRINT TAB(4); "2. Calculate the secondary voltage.": PRINT
300 INPUT "Enter your selection here (1 or 2): ", S
310 PRINT
320 IF S < 1 OR S > 2 THEN GOTO 260
330 IF S = 2 THEN GOTO 390
340 PRINT "          The equation is E1 = E2 * N1/N2": PRINT
350 INPUT "E2 = ", E2
360 INPUT "N1 = ", N1
370 INPUT "N2 = ", N2: PRINT
380 PRINT "E1 = "; E2 * N1 / N2; " V (r.m.s.)": END
390 PRINT "          The equation is E2 = E1 * N2/N1": PRINT
400 INPUT "E1 = ", E1
410 INPUT "N1 = ", N1
420 INPUT "N2 = ", N2: PRINT
430 PRINT "E2 = "; E1 * N2 / N1; " V (r.m.s.)": END
440 CLS : PRINT "PRIMARY AND SECONDARY CURRENT": PRINT
450 PRINT TAB(6); "Select ONE of the following:": PRINT
460 PRINT TAB(4); "1. Calculate the primary current.": PRINT
```

```
470 PRINT TAB(4); "2. Calculate the secondary current.": PRINT
480 INPUT "Enter your selection here (1 or 2): ", S
490 PRINT
500 IF S < 1 OR S > 2 THEN GOTO 440
510 IF S = 2 THEN GOTO 570
520 PRINT "          The equation is I1 = I2 * N2/N1": PRINT
530 INPUT "I2 = ", I2
540 INPUT "N1 = ", N1
550 INPUT "N2 = ", N2: PRINT
560 PRINT "I1 = "; I2 * N2 / N1; " A": END
570 PRINT "          The equation is I2 = I1 * N1/N2": PRINT
580 INPUT "I1 = ", I1
590 INPUT "N1 = ", N1
600 INPUT "N2 = ", N2: PRINT
610 PRINT "I2 = "; I1 * N1 / N2; " A": END
620 CLS : PRINT "MAXIMUM POWER TRANSFER": PRINT
630 PRINT TAB(6); "Select ONE of the following:": PRINT
640 PRINT TAB(4); "1. Input resistance.": PRINT
650 PRINT TAB(4); "2. Load resistance.": PRINT
660 PRINT TAB(4); "3. Turns ratio.": PRINT
670 INPUT "Enter your selection here (1 - 3): ", S
680 PRINT
690 IF S < 1 OR S > 3 THEN GOTO 620
700 IF S = 2 THEN GOTO 800
710 IF S = 3 THEN GOTO 880
720 CLS : PRINT "The equation for maximum power transfer is": PRINT
730 PRINT "     R1 = RL * (N1/N2)^2": PRINT
740 PRINT "where R1 = input resistance"
750 PRINT "      RL = load resistance"
760 PRINT "      N1/N2 = turns ratio": PRINT
770 INPUT "RL (ohms) = ", RL
780 INPUT "Turns ratio (N1/N2) = ", TR: PRINT
790 PRINT "R1 = "; RL * (TR) ^ 2; " ohms": END
800 CLS : PRINT "The equation for maximum power transfer is": PRINT
810 PRINT "      RL = R1/(N1/N2)^2": PRINT
820 PRINT "where R1 = input resistance"
830 PRINT "      RL = load resistance"
840 PRINT "      N1/N2 = turns ratio": PRINT
850 INPUT "R1 (ohms) = ", R1
860 INPUT "Turns ratio (N1/N2) = ", TR: PRINT
870 PRINT "RL = "; R1 / (TR) ^ 2; " ohms": END
880 CLS : PRINT "The equation for maximum power transfer is": PRINT
890 PRINT "     (N1/N2) = SQR(R1/RL)": PRINT
900 PRINT "where R1 = input resistance"
910 PRINT "      RL = load resistance"
920 PRINT "      N1/N2 = turns ratio": PRINT
930 INPUT "R1 (ohms) = ", R1
940 INPUT "RL (ohms) = ", RL: PRINT
950 PRINT "(N1 / N2) = "; SQR(R1 / RL): END
```

Computer listing 12.1 offers four options, namely

1. the solution of the e.m.f. equation of a transformer
2. calculation of primary and secondary voltage
3. calculation of primary and secondary current
4. determination of input resistance, load resistance or turns ratio to give maximum power transfer to a resistive load.

The calculations follow the equations introduced in this chapter.

For problems involving phasor calculations, such as the determination of the load current drawn by a transformer which carries a no-load current, the reader should use the program in Listing 8.2 of Chapter 8.

Calculations involving the efficiency of a transformer can be handled by the program in Listing 3.1 of Chapter 3, which deals with power and efficiency.

Exercises

12.1 Determine the full-load current flowing in the primary and secondary windings of a 3300/240 V, 20 kVA single-phase transformer.

12.2 If the full-load secondary current of a 10 kVA single-phase transformer is 50 A, what is the rated secondary voltage?

12.3 A 440/110 V single-phase transformer draws 10 A from the 440 V supply when fully loaded. Neglecting losses within the transformer, determine (a) the kVA rating of the transformer, (b) the full-load secondary current.

12.4 A single-phase transformer has voltage ratio of 440/250 V. If the 440 V winding has 308 turns on it, calculate the number of turns on the low-voltage winding.

12.5 A single-phase transformer supplies a load current of 20 A. If the primary to secondary winding voltage ratio is 225:30, what is the value of the value of the secondary current?

12.6 A 30 kVA single-phase transformer has a primary to secondary winding turns ratio of 2100:70, the primary winding voltage being 3300 V. For full-load conditions, determine (a) the full-load primary and secondary winding current and (b) the secondary voltage.

12.7 A single-phase transformer has a primary to secondary winding turns ratio of 10:1. If 3300 V a.c. is applied to the primary winding, and a pure resistive load of 100 Ω is connected to the secondary terminals, determine (a) the secondary voltage, (b) the secondary current, (c) the primary current. Neglect the effect of losses.

12.8 A 3300/240 V, 2 kVA single-phase transformer supplies a number of 240 V, 60 W tungsten-filament parallel-connected lamps. Calculate (a) the maximum number of lamps the transformer can supply without becoming overloaded, (b) the full-load primary current under this condition. Neglect the effect of losses.

12.9 A single-phase transformer has a primary to secondary winding voltage ratio of 10:1, and provides an output power of 10 kW at 0.8 power factor. If the low-voltage winding supplies its output at 110 V determine, neglecting the effect of losses, the current in each winding and the rating of the transformer.

12.10 The primary winding of a single-phase transformer is rated at 440 V, and has 308 turns on it. The transformer has two secondary windings, X and Y. The open-circuit voltage of winding X is 25 V, and winding Y has 7 turns on it. Neglecting the effect of losses, calculate (a) the number of turns on winding X and (b) the open-circuit voltage between the terminals of winding Y.

12.11 A 50 Hz single-phase transformer having 2080 turns on one winding and 104 turns on the other winding, has a sinusoidal core flux of peak value 1.4 mWb. Determine the r.m.s. voltage induced in each winding.

12.12 The core of a single-phase transformer has a net cross-sectional area of 250 cm^2, the number of turns on the primary and secondary winding being 200 and 18, respectively. If the maximum value of the flux density in the core is 1.0 T, and the supply frequency is 50 Hz, calculate the value of the e.m.f. induced in each winding.

12.13 A 3.3 kV/440 V, 50 Hz single-phase transformer has an induced voltage per turn of 2 V. If the maximum value of the flux density in the core is 1.2 T, determine (a) the cross-sectional area of the core, (b) the number of turns on each winding.

12.14 The 50-turn primary winding of a transformer is energised by a 240 V sinusoidal supply, the peak value of flux in the core being 25 mWb. Determine the supply frequency.

12.15 A two-winding, single-phase transformer has a step-down voltage ratio of 1.9 kV/240 V, the supply frequency being 50 Hz and the induced voltage per turn being 1.5. If the peak value of flux density in the core is 1.2 T, determine (a) the cross-sectional area of the core, (b) the number of turns on each winding.

If the rating of the transformer is 15 kVA, calculate the full-load current in each winding.

12.16 A transformer supplies a power of 160 kW to a load, the total internal loss in the transformer being 5 kW. Estimate the percentage efficiency of the transformer.

12.17 A 5 kVA transformer has an iron loss of 75 W and a full-load copper loss of 90 W. What is the full-load efficiency of the transformer if the power factor of the load is unity?

12.18 A 350 kVA single-phase transformer has a full-load copper loss of 3.5 kW. Calculate the copper loss in the transformer when it delivers (a) 300 kVA, (b) 250 kW at 0.75 power factor.

12.19 The number of turns on the primary and secondary winding, respectively, of a single-phase transformer are 605 and 55. If the primary winding is supplied at 3.3 kV, determine (a) the no-load secondary voltage, (b) the primary winding current and its power factor when the secondary current is 300 A at 0.8 power factor lagging, the no-load primary current being 7.5 A at 0.2 power factor lagging.

Draw the phasor diagram of the transformer to scale.

12.20 Modify the program in Listing 12.1 so that, at the completion of each option, you can either return to the main menu, or terminate the program.

Summary of important facts

The e.m.f. equation for a transformer is

$$\mathbf{E = 4.44\, N f \Phi_m = 4.44\, N f B_m a}$$

where

E is the e.m.f. induced in the winding
N is the number of turns on the winding
f is the supply frequency in Hz
Φ_m is the maximum flux in the core
B_m is the maximum flux density in the core
a is the cross-sectional area of the core

In an *ideal transformer* in which all the magnetic flux leaving the primary winding links with the secondary winding, and the windings have no resistance, the following holds good:

$$\frac{V_2}{V_1} = \frac{I_1}{I_2} = \frac{N_2}{N_1}$$

where

V_1 is the voltage applied to the primary winding
V_2 is the voltage induced in the secondary winding
I_1 is the primary winding current
I_2 is the secondary winding current
N_1 is the number of turns on the primary winding
N_2 is the number of turns on the secondary winding.

If $V_2 < V_1$, the transformer is said to have a **step-down voltage ratio**, and if $V_2 > V_1$, the transformer is said to have a **step-up voltage ratio**. A transformer having a step-down voltage ratio has a **step-up current ratio** (i.e. $I_2 > I_1$), and vice versa.

In an ideal transformer, the *number of turns per volt on each winding is the same*, that is

$$\frac{V_1}{N_1} = \frac{V_2}{N_2}$$

and there is *ampere-turn balance between the windings*, or

$$V_1 I_1 = V_2 I_2$$

The two equations above *also apply to auto-transformers.*

Since the heat generated by a transformer is largely dependent on the load current, **the rating of a transformer is given in VA** or an SI multiple of VA, i.e. kVA, MVA, etc.

The efficiency of a transformer is given by

$$\textbf{efficiency} = \frac{\textbf{output power}}{\textbf{input power}}$$

$$= \frac{\textbf{output power}}{\textbf{output power} + \textbf{total loss}}$$

$$= \frac{\textbf{input power} - \textbf{total loss}}{\textbf{input power}}$$

and the *total loss* in the transformer is

total loss = copper loss + iron loss

The above equations for efficiency give a **per-unit** (p.u.) result; if this value is multiplied by 100, the result is the **per-cent** (%) efficiency.

Maximum efficiency occurs in a transformer when

iron loss = copper loss

The copper loss at load current I_2 is given by

$$\textbf{copper loss} = \textbf{full-load copper loss} \times \left[\frac{I_2}{I_{FL}}\right]^2$$

$$= \textbf{full-load copper loss} \times \left[\frac{kVA_2}{kVA_{FL}}\right]^2$$

where I_{FL} is the full-load current, kVA_2 is the load kVA at current I_2, and kVA_{FL} is the full-load kVA.

Maximum power is transferred from a source to a resistive load when the internal resistance of the source is equal to the 'reflected' value of the load resistance. This is known as 'impedance matching', and occurs when

$$R_1 = \left[\frac{N_1}{N_2}\right]^2 \times R_L$$

where

R_1 is the internal resistance or output resistance of the source

N_1 is the number of turns on the primary winding

N_2 is the number of turns on the secondary winding

R_L is the resistance of the load.

13 Three-phase systems

13.1 Introduction

A **three-phase supply system** is produced by three sets of windings on an alternator, the windings being displaced from one another around the alternator rotor by 120°.

Three-phase systems have a number of advantages over single-phase systems including the fact that, for a given amount of power transmitted, the total power loss is less than in three single-phase systems. Moreover, the total volume of conductor required for a given power transmitted is less than in the equivalent single-phase system. For these and other reasons, power is transmitted by three-phase systems.

Also, the torque produced by a single-phase motor is pulsating rather than rotating; that is, an ideal single-phase motor has no starting torque! In order to cause the rotor of a single-phase motor to begin to rotate, it is necessary (at starting) to convert it into a two-phase motor to produce a starting torque. On the other hand, a three-phase motor produces a starting torque, whose speed can be controlled by relatively straightforward methods.

In the United Kingdom, the three phases are known as the *R* (Red), *Y* (Yellow) and *B* (Blue) phases or as phases 1, 2 and 3; in the United States (and some other countries) the notation is *A*, *B* and *C*.

By the end of this chapter, the reader will be able to

- Draw waveform and phasor diagrams for three-phase systems.
- Understand the meaning of balanced and unbalanced supplies and loads.
- Solve problems involving balanced and unbalanced star-connected systems.
- Solve problems involving balanced delta-connected systems.
- Carry out power, volt-ampere and reactive volt-ampere calculations on balanced loads.
- Use software to solve three-phase problems.

13.2 Three-phase waveforms

A basic three-phase alternator is shown in Figure 13.1. The three windings on the rotor of the alternator are known as the **red** (*R*), **yellow** (*Y*) and **blue** (*B*) phases, the windings being physically displaced from one another by 120°.

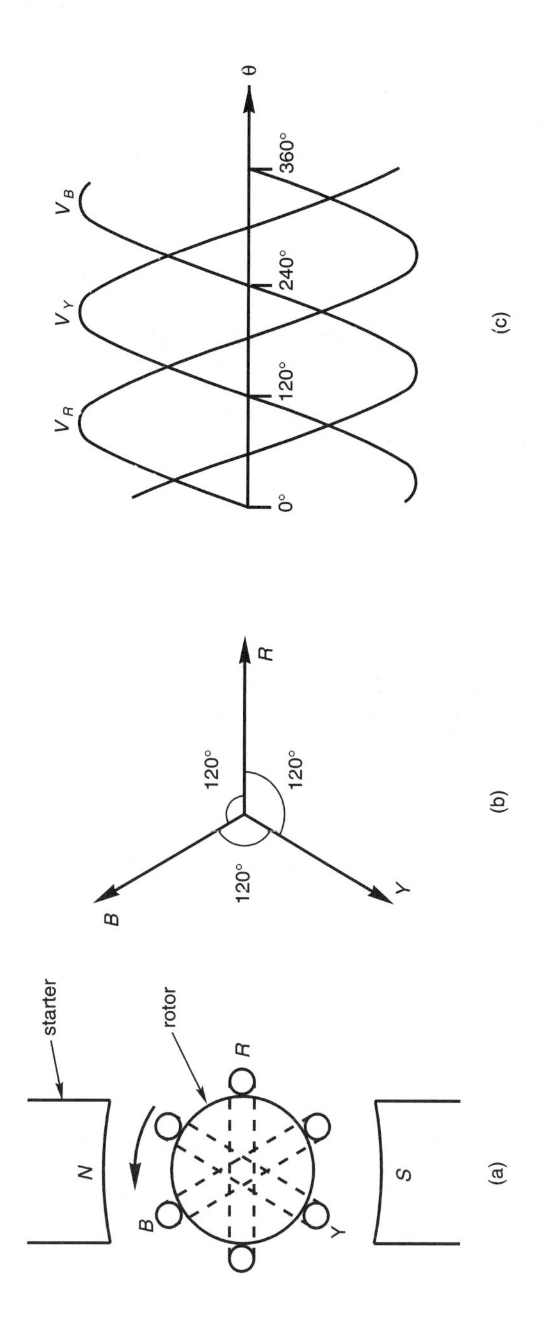

Figure 13.1 (a) A simplified three-phase alternator; (b) the voltage phasors; (c) the waveform diagram

It is generally the case that we draw the R-phase in the reference (horizontal) direction, as in Figure 13.1(b). However, the reader is warned that it is sometimes convenient to use another phasor as the reference phasor.

Positive Phase Sequence (P.P.S.)
The phasor diagram for the normal voltage sequence or **positive phase sequence** is as shown in Figure 13.1(b), in which the Y-phase lags behind the R-phase by 120°, and the B-phase leads the R-phase by 120°. The waveform diagrams are shown in Figure 13.1(c).

13.3 Phase and line voltage

The **phase voltage**, V_P, is the voltage induced in a *phase winding* of an alternator. The **line voltage** or **line-to-line voltage**, V_L, is the voltage between a pair of lines in the system.

Depending on the method of connecting the windings, there is a definite relationship between the phase and line voltages. These are described later in the chapter.

13.4 Phase and line voltage of a balanced star-connected system

In the 'star' connection, the 'start' of each end of the three phase windings are connected together (see Figure 13.2(a)) at the **neutral point** (N), which is usually connected to earth (hence the name 'neutral' point).

In a **balanced supply system** the magnitude of each of the phase voltages is the same. That is

$$V_{RN} = V_{YN} = V_{BN} = V_P$$

where V_P is the *phase voltage* of the system. The phase angle between each of the phase voltages is 120°.

The voltage between any pair of lines is the **line voltage** (or *line-to-line voltage*), V_L. The voltage between lines R and Y is given by

$$\begin{aligned} V_{RY} &= \text{voltage of line } R \text{ with respect to line } Y \\ &= V_{RN} - V_{YN} \end{aligned}$$

and

$$\begin{aligned} V_{BR} &= \text{voltage of line } B \text{ with respect to line } R \\ &= V_{BN} - V_{RN} \\ V_{YB} &= \text{voltage of line } Y \text{ with respect to line } B \\ &= V_{YN} - V_{BN} \end{aligned}$$

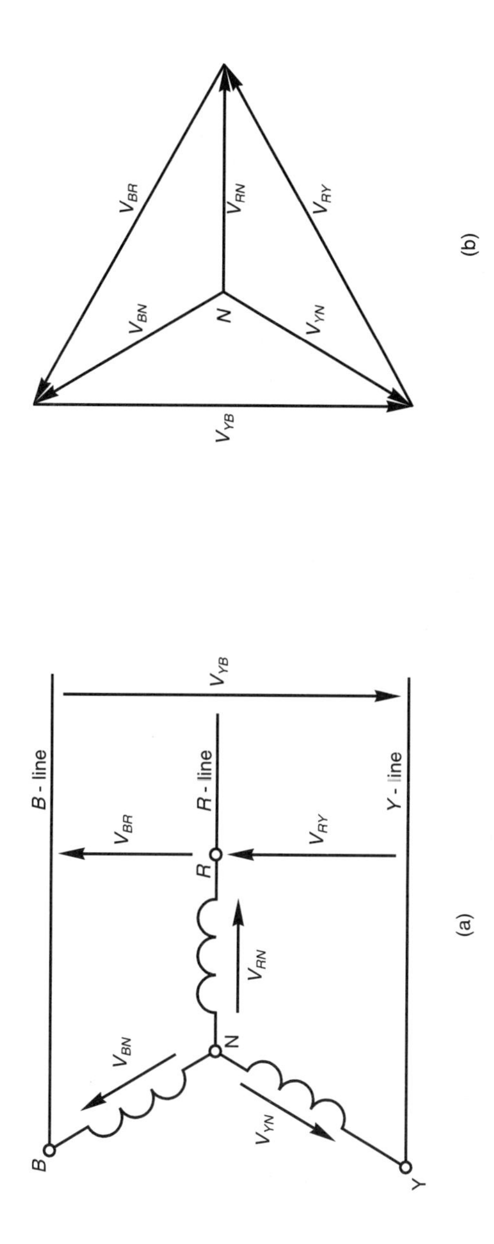

Figure 13.2 Rebalanced star-connection system: (a) star-connection; (b) its phasor diagram

It can be shown that, in a balanced system, the magnitude of the line voltage is

$$V_L = V_{RY} = V_{BR} = V_{YB} = \sqrt{3} \times V_P$$

Whenever engineers refer to, say, a 440 V three-phase system, or to a 3.3 kV three-phase system, they always mean the line voltage of a balanced system.

Worked Example 13.1

Determine the phase voltage of a 440 V, three-phase star-connected system.

Solution

Since $V_L = 440$ V, then

$$V_P = V_L/\sqrt{3} = 440/\sqrt{3} = 254 \text{ V}.$$

13.5 Phase and line current in a balanced star-connected system

Since the current in one phase of a star-connected alternator must flow into the line connected to it, then

line current = phase current

If both the supply and load are balanced (see also section 13.6 for a definition of a balanced load), then the magnitude of the phase current, I_P, is equal to the magnitude of the line current, I_L. That is

$$I_L = I_P.$$

13.6 Three-phase, four-wire, star-connected system

A three-phase, four-wire, star-connected system is shown in Figure 13.3. The wire connecting the neutral point, N, of the supply is connected to the **star point**, S, of the load. If the resistance of the neutral wire is zero, then the potential of the neutral point and the star point are equal to one another.

If we apply Kirchhoff's current law to the star point of the load (or to the neutral point of the supply), we have

$$I_N = I_R + I_Y + I_B$$

A three-phase load is said to be balanced if

1. the magnitude of the impedance of each phase of the load is the same, and
2. the phase angle of each phase of the load is the same.

If this is not the case, the load is **unbalanced**.

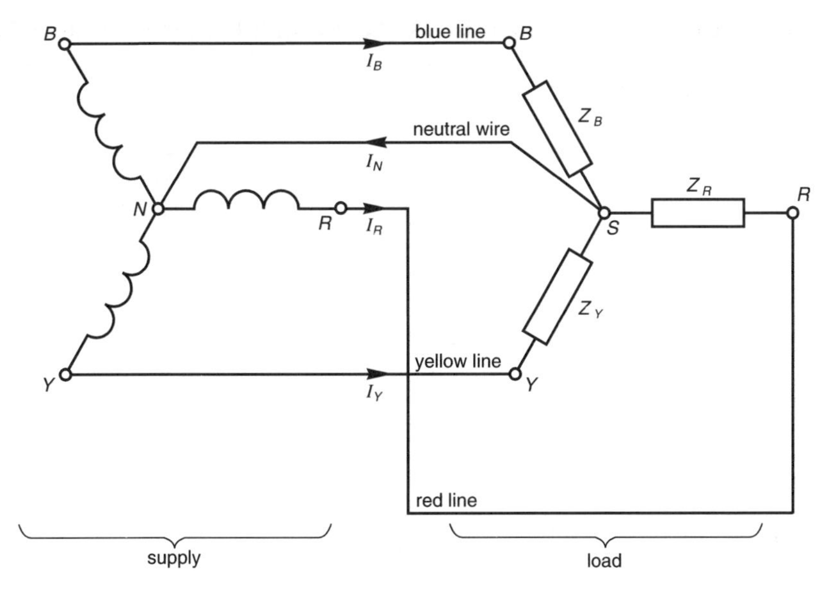

Figure 13.3 Three-phase, four-wire, star-connected system

From a point of view of calculations, *we can regard each phase of a three-phase, four-wire, star-connected load as being directly connected to one phase of the supply*. That is Z_R is energised by V_{RN}, Z_Y is energised by V_{YN}, and Z_B is energised by V_{BN} (see also section 13.8).

13.7 Balanced supply and balanced star-connected load

When both the supply and the load are balanced, **the current in the neutral wire is zero**. In this event, *the neutral wire can be removed*, thereby saving 25 per cent of the cost of the conductor material. This is illustrated in Worked Example 13.2. In the majority of supply systems, it is assumed that the load is balanced, so that a **three-phase, three-wire supply** can be used.

Worked Example 13.2

A 173.2 V balanced three-phase, four-wire supply is connected to a balanced star-connected load having a resistance of 100 Ω per phase. Calculate the current in each line and in the neutral wire.

Solution

The phase voltage is

$$V_P = V_L/\sqrt{3} = 173.2/\sqrt{3} = 100 \text{ V}$$

The current in each phase is

$$I_P = 100 \text{ V}/100\,\Omega = 1 \text{ A}$$

That is, the current in the R-line is 1 A in phase with V_RN, the current in the Y-line is 1 A in phase with V_YN, and the current in the B-line is 1 A in phase with V_BN. The resolved components of these currents are

	horizontal component	*vertical component*
I_R	1 A	0 A
I_Y	−0.5 A	−0.866 A
I_B	−0.5 A	0.866 A
Sum	0 A	0 A

That is, the current in the neutral wire is

$$I_N = \sqrt{\left(\begin{bmatrix}\text{horizontal}\\\text{component}\end{bmatrix}^2 + \begin{bmatrix}\text{vertical}\\\text{component}\end{bmatrix}^2\right)}$$

$$= \sqrt{(0^2 + 0^2)} = 0$$

As mentioned above, the neutral wire in a system with a balanced supply and load is redundant, because it carries no current.

13.8 Summary of line and phase values for a balanced star-connected system

	Line values	*Phase values*
Voltages:	$V_L = \sqrt{3}V_P$	$V_P = V_L/\sqrt{3}$
Currents:	$I_L = I_P$	$I_P = I_L$

13.9 Balanced supply and unbalanced load

When the load is unbalanced, in general, a current will flow in the neutral wire, and a three-phase, four-wire supply is necessary if the phase voltage at the load is to be maintained constant.

Worked Example 13.3

The following resistors are connected in star to a balanced 762 V, three-phase, four-wire supply:

Red phase: 20 Ω
Blue phase: 10 Ω
Yellow phase: 15 Ω

Determine the current in each line and in the neutral wire. Draw the phasor diagram showing the line and neutral currents.

Solution

The phase voltage of the system is

$$V_P = 762/\sqrt{3} = 440 \text{ V}$$

Using Ohm's law, the magnitude of the current in each line (and phase) is

$$\begin{aligned} I_R &= V_{RN}/Z_{RN} = V_{RN}/R_{RN} = 440/20 = 22 \text{ A} \\ I_B &= V_{BN}/Z_{BN} = V_{BN}/R_{BN} = 440/10 = 44 \text{ A} \\ I_Y &= V_{YN}/Z_{YN} = V_{YN}/R_{YN} = 440/15 = 29.33 \text{ A} \end{aligned}$$

Since the loads are purely resistive, the current in each line will be in phase with the associated voltage, that is I_R will be in phase with V_{RN}, I_B will be in phase with V_{BN}, etc.

One method of determining the value of the neutral current is to *draw the phasor diagram to scale*, and measure the magnitude and phase angle of I_N after adding the three line currents together, as shown in Figure 13.4. To do this we add one pair of current phasors together, say $(I_R + I_B)$, then add the result to I_Y. The result is

$$I_N = I_R + I_Y + I_B$$

As mentioned earlier, any graphical solution is approximate, and does not give a precise result.

Alternatively, we can resolve the phasors into their horizontal and vertical components (as described in Chapter 8), and add the results as follows:

	horizontal component		*vertical component*	
I_R	$22 \cos 0^\circ$	$= 22$ A	$22 \sin 0^\circ$	$= 0$ A
I_B	$44 \cos 120^\circ$	$= -22$ A	$44 \sin 120^\circ$	$= 38.1$ A
I_Y	$29.33 \cos(-120^\circ)$	$= -14.66$ A	$29.33 \sin(-120^\circ)$	$= -25.4$ A
Sum		-14.66 A		12.7 A

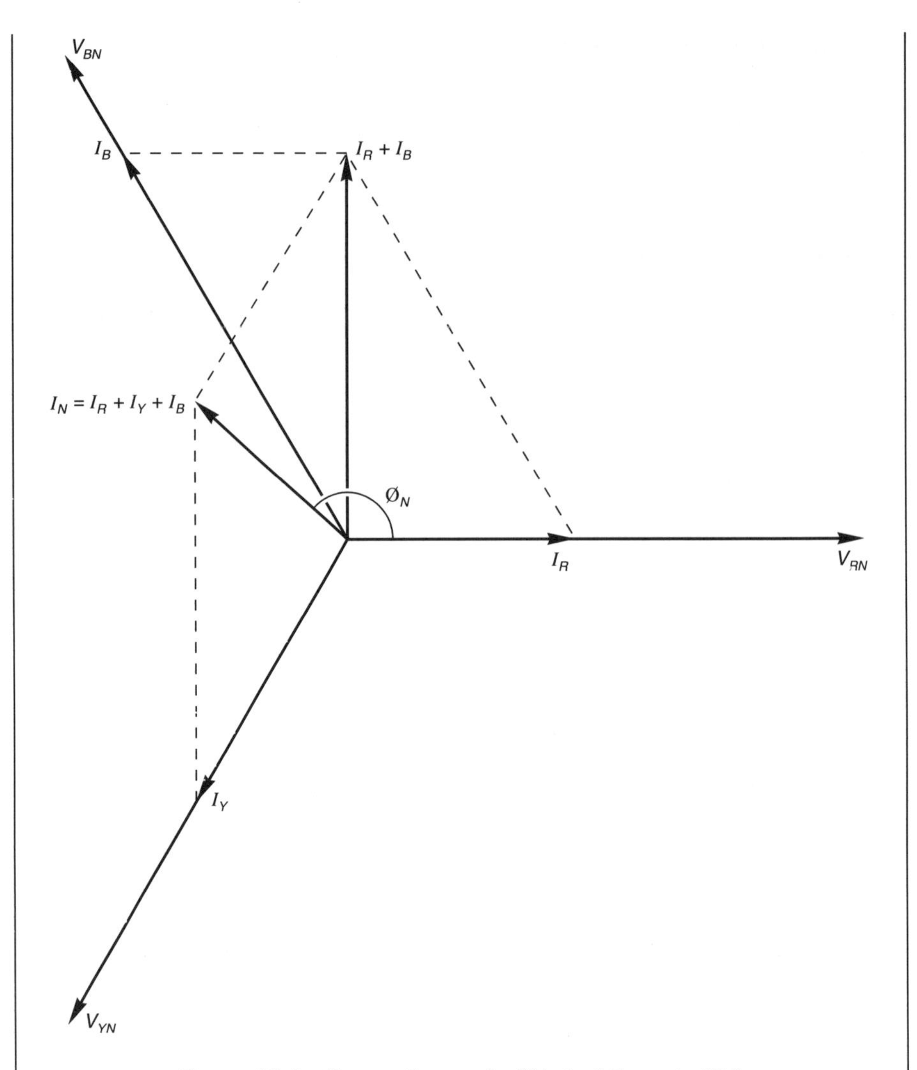

Figure 13.4 Phasor diagram for Worked Example 13.3

The magnitude of the current in the neutral wire is therefore

$$I_N = \surd((-14.66)^2 + 12.7^2) = 19.4 \text{ A}$$

Since the horizontal component of I_N is negative, and the vertical component is positive, I_N lies in the second quadrant. Assuming that the reader is using a calculator to determine the phase angle, the actual value is given by

$$\begin{aligned}\phi_N &= \text{phase angle shown by calculator} + 180^\circ \\ &= \arctan(12.7/(-14.66)) + 180^\circ \\ &= -40.9^\circ + 180^\circ = 139.1^\circ.\end{aligned}$$

Worked Example 13.4

The following impedances are connected in star to a 550 V balanced supply. Determine the current in each line and in the neutral wire.

Red phase: a resistance of 10 Ω in series with an inductive reactance of 15 Ω

Yellow phase: a resistance of 25 Ω

Blue phase: A resistance of 30 Ω in series with a capacitive reactance of 50 Ω

Solution

The solution follows a similar pattern to that of Worked Example 13.2, with the exception that the current in each phase will be at some angle with respect to the phase voltage. The magnitude of the phase voltage is

$$V_P = 550/\sqrt{3} = 317.5\text{ V}$$

The impedance and phase angle for each phase of the load is calculated as follows, the circuit diagram being shown in Figure 13.5(a).

R-phase

$$\begin{aligned} Z_R &= \sqrt{(R_R{}^2 + X_L{}^2)} = \sqrt{(10^2 + 15^2)} = 18.03\ \Omega \\ \phi_R &= \arctan(X_L/R_R) = \arctan(15/10) \\ &= 56.3^\circ (I_R \text{ lagging } V_{RN}) \end{aligned}$$

Y-phase

$$\begin{aligned} Z_Y &= R_Y = 25\ \Omega \\ \phi_Y &= 0^\circ (I_Y \text{ in phase with } V_{YN}) \end{aligned}$$

B-phase

$$\begin{aligned} Z_B &= \sqrt{(R_B{}^2 + X_C{}^2)} = \sqrt{(30^2 + 50^2)} = 58.31\ \Omega \\ \phi_B &= \arctan(X_C/R_B) = \arctan(50/30) \\ &= 59^\circ (I_B \text{ leading } V_B N) \end{aligned}$$

The line currents are calculated as follows:

$$\begin{aligned} I_R &= V_P/Z_R = 317.5/18.03 \\ &= 17.61\text{ A (lagging behind } V_{RN} \text{ by } 56.3^\circ) \\ I_Y &= V_P/Z_Y = 317.5/25 \\ &= 12.7\text{ A (in phase with } V_{YN}) \\ I_B &= V_P/Z_B = 317.5/58.31 \\ &= 5.45\text{ A (leading} V_{BN} \text{ by } 59^\circ) \end{aligned}$$

The corresponding phasor diagram is shown in Figure 13.5(b).

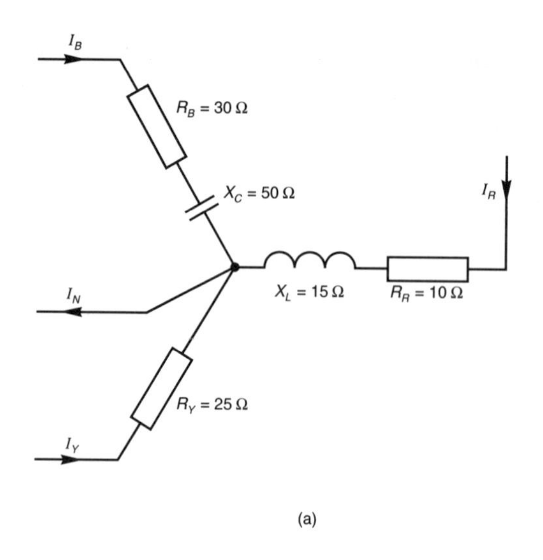

(a)

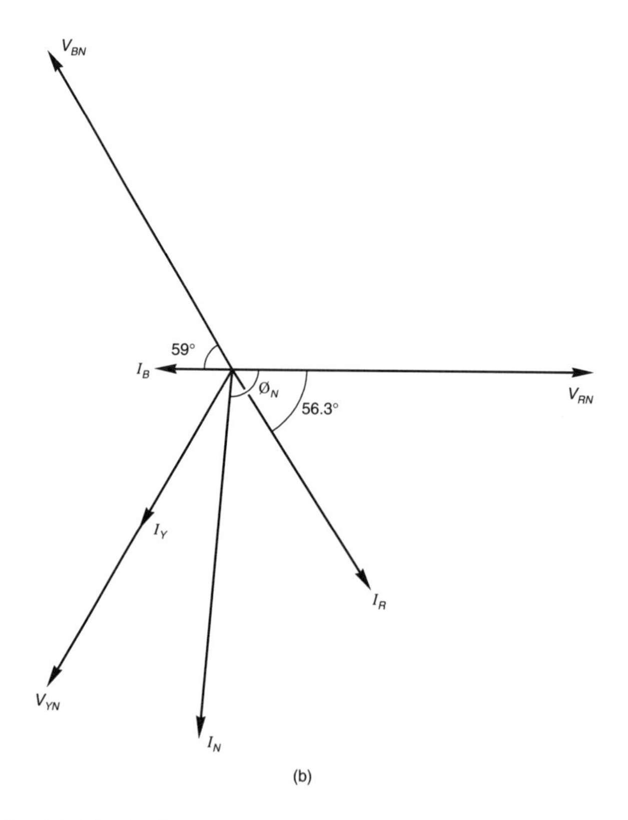

(b)

Figure 13.5 (a) Circuit diagram for Worked Example 13.4; (b) phasor diagram

The neutral current, I_N, can either be determined by graphical methods, or by calculation as shown below. The phase angle of each current with respect to the reference direction is as follows:

I_R : phase angle $= -56.3°$
I_Y : phase angle $= -120°$
I_B : phase angle $= 120° + 59° = 179°$

The horizontal and vertical components of I_R are calculated below

$$\text{horizontal component} = I_R \cos \phi_R$$
$$= 17.61 \cos(-56.3°) = 9.77 \text{ A}$$
$$\text{vertical component} = I_R \sin \phi_R$$
$$= 17.61 \sin(-56.3°) = -14.65 \text{ A}$$

and it can be shown that (the reader should verify these values)

$$\text{horizontal component of } I_Y = I_R \cos \phi_Y = -6.35 \text{ A}$$
$$\text{vertical component of } I_Y = I_R \sin \phi_Y = -11 \text{ A}$$
$$\text{horizontal component of } I_B = I_B \cos \phi_B = -5.45 \text{ A}$$
$$\text{vertical component of } I_B = I_B \sin \phi_B = 0.1 \text{ A}$$

hence

$$\text{horizontal component of } I_N$$
$$= \text{horizontal component of } (I_R + I_Y + I_B)$$
$$= 9.77 - 6.35 - 5.45 = -2.03 \text{ A}$$

therefore

$$\text{magnitude of } I_N = \sqrt{((-2.03)^2 + (-25.55)^2)}$$
$$= 25.63 \text{ A}$$

Since both the vertical and horizontal components of I_N are negative, I_N lies in the third quadrant. Assuming that the reader is using a calculator to calculate the phase angle, then

$$\phi_N = \text{angle given for arctan } \frac{-25.55}{-2.03} - 180°$$
$$= 85.46° - 180° = -94.54°.$$

13.10 Delta connection or mesh connection

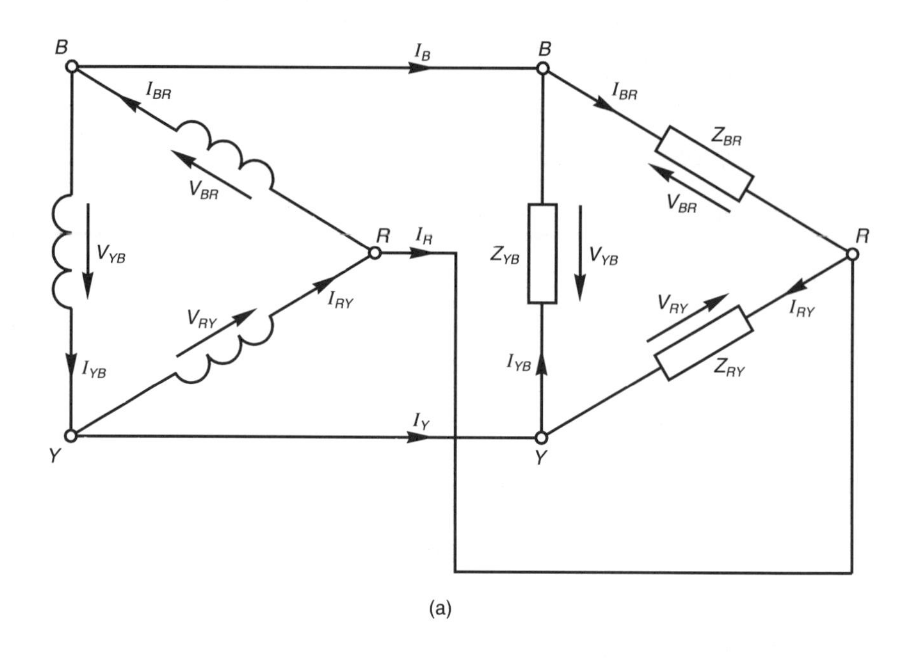

(a)

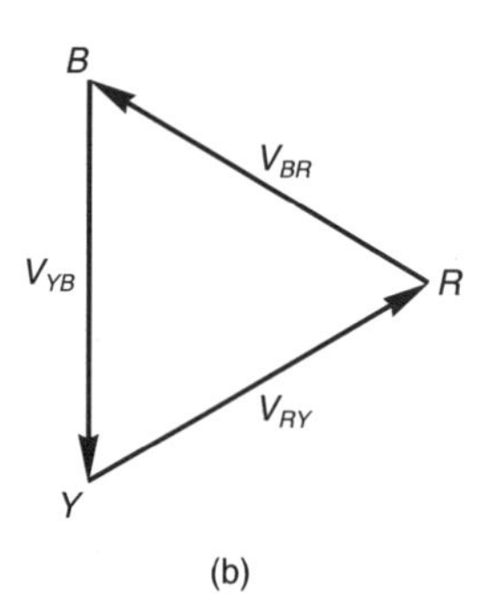

(b)

Figure 13.6 (a) A delta- or mesh-connected system; (b) phasor diagram of its voltages

When an alternator is **delta connected** or **mesh connected**, the 'end' of one winding is connected to the 'start' the next winding, as shown in Figure 13.6(a). In this case, each 'phase' winding of the alternator is connected between a pair of lines so that, in a *balanced system*

line voltage = phase voltage

or

$$V_L = V_P$$

Once again, we can treat each phase of the system as a single-phase problem, since each phase of the load is directly connected to one phase of the supply. It is then a matter of applying Kirchhoff's current law to the nodes of the system to determine the line current. (**Note**: A delta-connected system is a three-phase, three-wire system.

In the case of a **balanced delta-connected system**

Line current $= \sqrt{3} \times$ **phase current**

or

$$I_L = \sqrt{3} \times I_P$$

13.11 Summary of line and phase values for a balanced delta-connected system

	Line values	*Phase values*
Voltage:	$V_L = V_P$	$V_P = V_L$
Current:	$I_L = \sqrt{3} \times I_P$	$I_P = I_L/\sqrt{3}$

Worked Example 13.5

A delta-connected a.c. motor (which is a balanced load) draws a line current of 40 A. What is the phase current?

Solution

From the above relationships the phase current is

$$I_P = I_L/\sqrt{3} = 40/\sqrt{3} = 23.09 \text{ A}.$$

Worked Example 13.6

Three 20 Ω resistors are connected in delta to a balanced 240 V three-phase supply. What is the phase current and the line current?

Solution

Since 240 V is applied to each resistor, the phase current is

$$I_P = 240/20 = 12 \text{ A}$$

and the corresponding line current is

$$I_L = \sqrt{3}I_P = \sqrt{3} \times 12 = 20.8 \text{ A}.$$

13.12 Power, VA and VAr consumed by a balanced three-phase load

Power consumption

Total power, $P = 3 \times$ power consumed by one phase

In a star-connected system, $V_P = V_L/\sqrt{3}$ and $I_P = I_L$, hence

$$P = 3V_P I_P \cos\phi = 3\frac{V_L}{\sqrt{3}} I_L \cos\phi = \sqrt{3} V_L I_L \cos\phi$$

In a delta-connected system, $V_P = V_L$ and $I_P = I_L/\sqrt{3}$, hence

$$P = 3V_P I_P \cos\phi = 3V_L \frac{I_L}{\sqrt{3}} \cos\phi = \sqrt{3} V_L I_L \cos\phi$$

That is, for either star or delta connection

$$\boldsymbol{P = \sqrt{3} V_L I_L \cos\phi}$$

VAr consumption

Total VAr, $Q = 3 \times$ *VAr* consumed by one phase

Now, for both balanced star and delta connections

$$V_P I_P = \sqrt{3} V_L I_L$$

hence the *total VAr consumed* is

$$\boldsymbol{Q = \sqrt{3} V_L I_L \sin\phi}$$

VA consumption

$$\textbf{Total VA}, \boldsymbol{S} = 3 \times VA \text{ consumed by one phase}$$
$$= \boldsymbol{3V_P I_P = \sqrt{3} V_L I_L}$$

It also follows that

$$\boldsymbol{S = \sqrt{(P^2 + S^2)}}.$$

Worked Example 13.7

A 15 kW, 440 V, three-phase a.c. motor has an efficiency of 80 per cent and a power factor of 0.6 lagging when delivering one-half of its rated output. If the windings are connected (a) in star, (b) in delta, determine the phase current and the phase voltage in each case.

Solution

For the stated condition

$$\text{mechanical output power} = 15/2 = 7.5 \text{ kW}$$

$$\text{electrical input power} = \frac{\text{output power}}{\text{efficiency}}$$
$$= 7.5 \text{ kW}/0.8 = 9.375 \text{ kW}$$

Since the motor is a balanced load, then

$$P = \sqrt{3}V_L I_L \cos\phi$$

hence

$$I_L = \frac{P}{\sqrt{3}V_L \cos\phi} = \frac{9.375 \times 10^3}{\sqrt{3} \times 440 \times 0.6} = 20.5 \text{ A}$$

(a) *Star connection*

$$I_P = I_L = 20.5 \text{ A}$$

(b) *Delta connection*

$$I_P = I_L/\sqrt{3} = 11.84 \text{ A}.$$

13.13 Computer listing

Listing 13.1 Three-phase calculations for balanced systems

```
10 CLS '**** CH13-1 ****
20 PRINT "THREE-PHASE CALCULATIONS FOR BALANCED SYSTEMS": PRINT
30 PRINT
40 PRINT "Select ONE of the following, and determine:": PRINT
50 PRINT "1. the line voltage of a star-connected system."
60 PRINT "2. the phase voltage of a star-connected system."
70 PRINT "3. the line current of a delta-connected system."
80 PRINT "4. the phase current of a delta-connected system."
90 PRINT "5. Three-phase VA, power and VAr.": PRINT
100 INPUT "Enter you selection here (1 - 5) :", S
110 IF S < 1 OR S > 5 THEN GOTO 10
120 PRINT
130 IF S = 2 THEN GOTO 200
140 IF S = 3 THEN GOTO 230
150 IF S = 4 THEN GOTO 260
160 IF S = 5 THEN GOTO 290
170 CLS : PRINT "LINE VOLTAGE CALCULATION (STAR CONNECTION)": PRINT
180 INPUT "Phase voltage (V) = ", VP: PRINT
190 PRINT "Line voltage = "; SQR(3) * VP; "V": GOTO 390
200 CLS : PRINT "PHASE VOLTAGE CALCULATION (STAR CONNECTION)": PRINT
210 INPUT "Line voltage (V) = ", VL: PRINT
220 PRINT "Phase voltage = "; VL / SQR(3); "V": GOTO 390
230 CLS : PRINT "LINE CURRENT CALCULATION (DELTA CONNECTION)": PRINT
240 INPUT "Phase current (A) = ", IP: PRINT
250 PRINT "Line current = "; SQR(3) * IP; "A": GOTO 390
```

```
260 CLS : PRINT "PHASE CURRENT CALCULATION (DELTA CONNECTION)": PRINT
270 INPUT "Line current (A) = ", IL: PRINT
280 PRINT "Phase current = "; IL / SQR(3); "A": GOTO 390
290 CLS : PRINT "THREE-PHASE VA, POWER AND VAr CALCULATION.": PRINT
300 INPUT "Line voltage (V) = ", VL
310 INPUT "Line current (A) = ", IL
320 INPUT "Phase angle (degrees) = ", PHI: PRINT
330 PHI = PHI * 3.14159 / 180
340 VA = SQR(3) * VL * IL
350 PRINT "Volt-amperes = "; VA; "VA"
360 PRINT "Power = "; VA * COS(PHI); "W"
370 PRINT "Volt-amperes reactive = "; VA * SIN(PHI); "VAr"
380 PRINT "Power factor = "; COS(PHI)
390 PRINT
400 PRINT "Do you wish to:": PRINT
410 PRINT "1. return to the menu, or"
420 PRINT "2. leave the program?": PRINT
430 INPUT "Enter your selection here (1 or 2): ", S
440 IF S < 1 OR S > 2 THEN GOTO 390
450 IF S = 1 THEN GOTO 10
460 END
```

Computer listing 13.1 offers five options, namely

1. given the phase voltage of a balanced star-connected system, determine the line voltage
2. given the line voltage of a balanced star-connected system, determine the phase voltage
3. given the phase current of a balanced delta-connected system, determine the line current
4. given the line current of a balanced delta-connected system, determine the phase current
5. for a balanced system, and given the line voltage, line current and phase angle, calculate the three-phase power, VA and VAr.

The above options provide sufficient power to solve many problems. The reader should note that in item 5 (power, VA and VAr), no precaution is taken to restrict its use to a system which has a phase angle which is greater than $+90°$ or less than $-90°$.

Exercises

Where possible, the reader should draw the phasor diagram for each exercise to scale.

13.1 A three-phase 440 V system supplies a balanced load of 15 kVA. Determine the line current.

13.2 A three-phase alternator supplies a balanced three-phase load of 50 kVA. If the line current is 64.15 A, calculate the line voltage.

13.3 A balanced three-phase load dissipates 25 kW at a power factor of 0.8. If the line current is 41 A, what is the line voltage?

13.4 A three-phase, star-connected alternator supplies a line current of 150 A to a balanced load at a line voltage of 400 V. Determine (a) the phase voltage of the system, (b) the apparent power supplied to the load in kVA.

13.5 A balanced delta-connected load draws a phase current of 30 A when the phase voltage is 440 V. Determine the value of the line current.

If the alternator which supplies the current is star-connected, what is the phase voltage of the alternator?

13.6 A three-phase, star-connected alternator delivers 5.0 MW at a power factor of 0.9 lagging, the line voltage being 11 kV. Calculate (a) the alternator output in MVA, (b) the reactive power consumed by the load in MVAr, (c) the line current.

13.7 A balanced delta-connected load is supplied at 450 V, and consumes a line current of 20 A at a power factor of 0.9 lagging. Determine (a) the apparent power and the active power consumed by the load, (b) the phase current.

13.8 A balanced three-phase load takes a line current of 40 A at a line voltage of 1100 V. If the power consumed is 58 kW, calculate the kVA and kVAr supplied to the load. What is the power factor of the load?

13.9 A three-phase star-connected alternator delivers a line current of 173.2 A to a balanced delta-connected load at a line voltage of 346.4 V. Calculate (a) the phase voltage of the alternator, (b) the phase current supplied by the alternator, (c) the phase current in the load. (d) What kVA is consumed by the load?

13.10 A 12 kW, 500 V, three-phase, 50 Hz delta-connected a.c. motor has a full-load power factor of 0.85 and an efficiency of 87 per cent. Determine (a) the line current taken by the motor, (b) the phase current of the motor.

13.11 The ends of the three windings on a small alternator are brought out to six slip-rings, and connected to a balanced load. On full-load the phase current is 30 A and the phase voltage is 300 V. If the windings are connected (i) in star, (ii) in delta, determine (a) the line voltage, (b) the line current, (c) the kVA output.

13.12 A pump is driven by a 12 kW, 440 V, three-phase, 50 Hz a.c. motor which operates on full-load. The efficiency and power factor of the motor are 85 per cent and 0.9, respectively.

Calculate the phase current and the phase voltage when the stator windings are connected (a) in star, (b) in delta.

13.13 A d.c. motor drives a 500 V, three-phase, star-connected alternator, which supplies a 15 kW, three-phase, delta-connected induction motor which operates at full-load.

If the d.c. motor-alternator combination has an overall efficiency of 82 per cent, the induction motor has an efficiency of 90 per cent and a power factor of 0.85, determine (a) the phase voltage of the alternator, (b) the line and phase current of the induction motor, (c) the input power to the d.c. motor. Ignore losses in the system.

13.14 A 500 V balanced three-phase supply is connected to three inductors connected in delta. Each inductor has a resistance of 20 Ω and an inductive reactance of 15 Ω. Determine (a) the current flowing in each inductor, (b) the line current, (c) the power dissipated by the load, (d) the power factor of the load.

13.15 Three 30μF capacitors are connected in delta to a 440 V, 50 Hz, balanced three-phase supply. Calculate (a) the current in each capacitor and (b) the line current.

13.16 Three similar coils are connected in star to a 400 V balanced three-phase system. If the line current is 15 A and the overall power factor is 0.8 lagging, calculate (a) the resistance, inductive reactance and impedance of each coil, and (b) the total power consumed.

If the coils are re-connected in delta to the same supply, determine the line current and total power consumed.

13.17 A load consists of three star-connected impedances, each impedance has a value of 10 Ω and phase angle 60° lagging. If the phase voltage is 250 V, determine the line voltage, the line current and the total power consumed. Draw the phasor diagram to scale.

13.18 A three-phase, four-wire, 440 V system supplies a star-connected load comprising the following impedances.

R-phase: a resistance of 10 Ω
Y-phase: a coil of impedance 15 Ω and phase angle of 30° lagging
B-phase: an impedance of 10 Ω and phase angle 30° lead

Determine the magnitude and phase angle of each line current, and of the neutral wire current. Calculate also the total power consumed by the load. Draw the phasor diagram to scale.

13.19 A symmetrical star-connected, three-phase alternator supplies an unbalanced, four-wire, star-connected load as follows:

R-phase: a resistance of 100 Ω
Y-phase: a resistance of 100 Ω in series with an inductive reactance of 30 Ω
B-phase: a resistance of 75 Ω in series with a capacitive reactance of 130 Ω

If the phase voltage of the alternator is 100 V, determine the current in each line and in the neutral wire, and calculate the total power consumed. Draw the phasor diagram to scale.

Summary of important facts

A **balanced** or **symmetrical** three-phase supply comprises three voltages *which have the same magnitude and are phase displaced from one another by 120°*. The normal phase sequence or **positive phase sequence** in the United Kingdom is R (red phase), Y (yellow phase), B (blue phase) (or phases 1, 2 and 3); the United States these are called the A, B and C phases.

The **phase voltage**, V_P, of a system is the voltage induced in one phase of the alternator, and the **line voltage**, V_L, is the voltage between a pair of lines. *In a symmetrical star-connected system*

$$\boldsymbol{V_L = \sqrt{3}V_P}$$

and if a *balanced star-connected load* is connected to the symmetrical supply, then the **line current**, I_L, is related to the **phase current**, I_P, as follows:

$$\boldsymbol{I_L = I_P}$$

In a symmetrical **delta-connected** or **mesh-connected** system

$$\boldsymbol{V_L = V_P}$$

and

$$\boldsymbol{I_L = \sqrt{3}I_P}$$

When a star-connected load is connected to a three-phase, four-wire supply, the current in the neutral wire is

$$\boldsymbol{I_N = \textbf{phasor sum of } (I_R + I_Y + I_B)}$$

When both the supply and the load are balanced then $I_N = 0$. This is usually the case in many industrial loads, *making it possible to remove the neutral wire from the system*; that is we can use a **three-phase, three-wire supply**. When the load is connected in delta, it is only necessary to use a three-phase, three-wire supply.

For a balanced supply and a balanced load (either star or delta)

$$\textbf{total power consumed, } \boldsymbol{P = \sqrt{3}V_L I_L \cos\phi}$$

$$\textbf{total VA consumed, } \boldsymbol{S = \sqrt{3}V_L I_L}$$

$$\textbf{total VAr consumed, } \boldsymbol{Q = \sqrt{3}V_L I_L \sin\phi}$$

where ϕ is the phase angle of the load, and

$$\boldsymbol{S = \sqrt{(P^2 + Q^2)}}$$

14 Transistor amplifiers

14.1 Introduction

In this chapter we look at the basic design and operation of transistor amplifiers. Whilst it may seem that this is an exact scientific process, there remains an element of experience or 'art' in the selection of component values. Nothing, in fact, is precise or exact, and there is always an element of uncertainty about the operating values associated with a transistor. For example, one of the objects of transistor circuit design is to remove the effect of, for instance, the variation in current gain between transistors in a production batch.

By the end of this chapter, the reader will be able to

- Understand the difference between bipolar junction transistors and field-effect transistors.
- Understand the meaning and operation of common-emitter, common-base and common-collector bipolar transistor amplifiers.
- Appreciate the elements of a field-effect transistor.
- Make calculations on amplifier circuits.
- Draw the load line of a transistor amplifier, and carry out calculations using the load line.
- Use computer software for transistor amplifier bias circuit design.

14.2 Types of transistor

There are two broad classifications of transistor, namely the **bipolar junction transistor** (BJT) and the **field-effect transistor** (FET).

A BJT has three regions in its structure, namely the **emitter region** (which emits charge carriers), the **base region** (which controls the flow of charge carriers), and the **collector region** (which collects charge carriers).

In a *p-n-p* BJT, one *p*-region is the emitter and the other is the collector; the *n*-region is the base region. In an *n-p-n* BJT, one *n*-region is the emitter and the other the collector; the *p*-region is the base region. In this chapter we will concentrate on the use of *p-n-p* transistors; in the case of *n-p-n* transistors, the supply voltage polarities and the direction of the current flow is reversed when compared with *p-n-p* types.

FETs are divided into two groups, namely **junction-gate FETs** (JUGFETS) and **metal-oxide-semiconductor FETs** (MOSFETS). Further subdivisions of FET types are possible, but we ignore them here.

An essential difference between BJTs and FETs is that the input resistance of a FET is very much higher than that of a BJT. This has a number of advantages, including the fact that the power consumption of FETs is very low.

14.3 p-n-p transistor configurations

The three basic configurations in which *p-n-p* BJTs are used are shown in Figure 14.1, and are

(a) **the common-emitter configuration**
(b) **the common-base configuration**
(c) **the common-collector configuration** or **emitter follower**.

All three are used in electronic circuits, the common-emitter being the most popular.

In addition to the elements shown in Figure 14.1, a **bias circuit** must also be provided, and we look at some of these later. It may also be necessary to include components which improve the **thermal stability** of the circuit.

14.4 Common-emitter bias circuits

We take a look at three types of bias circuit, namely

1. **fixed bias**
2. **collector bias**
3. **potential divider bias.**

We will look in detail at the d.c. bias conditions here.

The principal object of a bias circuit is to bias the operating point of the transistor so that it operates in a precise area of the transistor characteristic. The four main classifications of bias are

Class A: current flows in the load during the whole period of the input signal cycle.
Class AB: current flows in the load for more than one-half of the input signal cycle, but less than the complete cycle.
Class B: current flows in the load for one-half of the input signal cycle
Class C: current flows in the load for less than one-half of the input signal cycle.

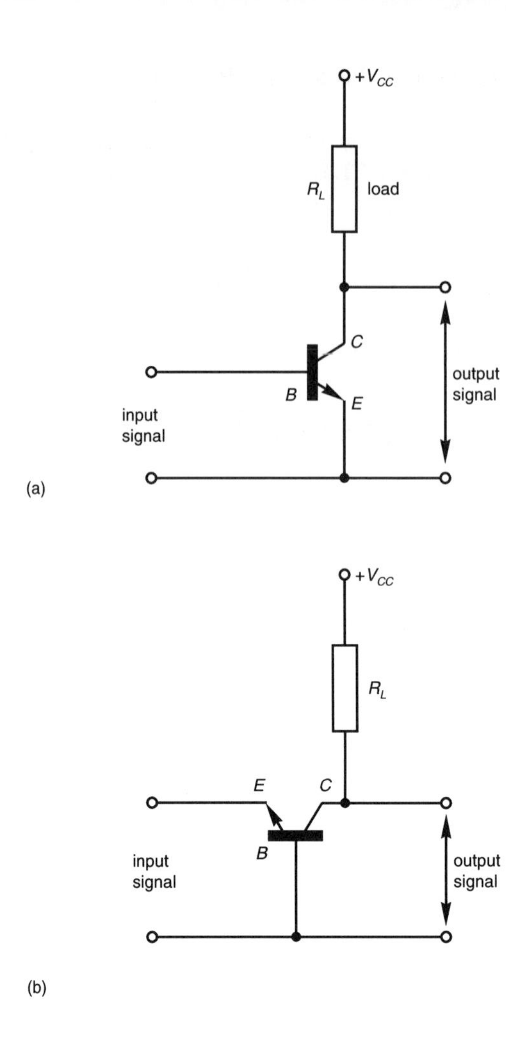

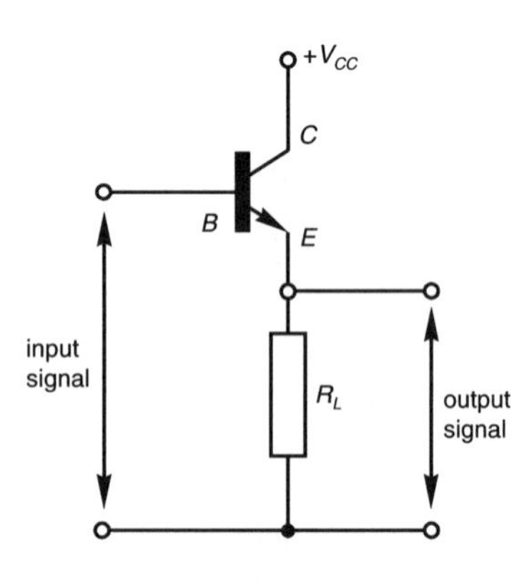

Figure 14.1 Basic circuits for (a) p-n-p common-emitter; (b) p-n-p common-base; (c) p-n-p common-collector (emitter-follower) amplifiers

(a) Common-emitter: fixed bias circuit

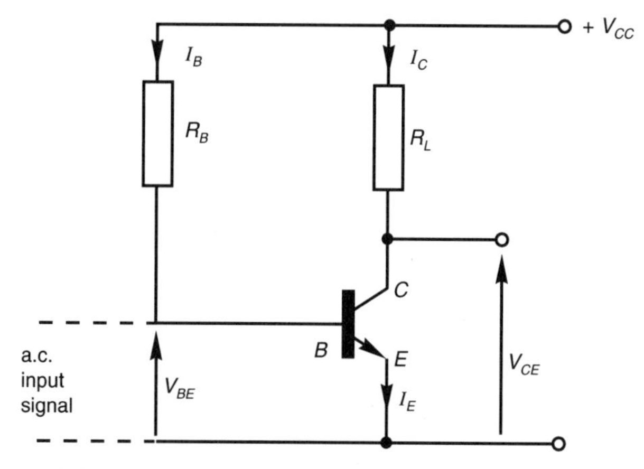

Figure 14.2 Fixed bias circuit for a common-emmitter p-n-p amplifier

The basic circuit is shown in Figure 14.2, in which

$R_B =$ bias resistor
$R_L =$ collector load resistor
$V_{CC} =$ collector supply voltage
$V_{CE} =$ d.c. (quiescent) collector-emitter voltage
$V_{BE} =$ d.c. (quiescent) base-emitter voltage
$I_B =$ quiescent base current
$I_C =$ quiescent collector current
$I_E =$ quiescent emitter current $= I_B + I_C$.

In a practical silicon transistor, V_{BE} has a value of about 0.6–0.7 V, and V_{CE} should have a value of about $V_{CC}/2$. The reader will note the use of the expression 'a value of about . . .'; this implies that, in practice, there is a small element of uncertainty relating to these values.

We shall also refer to certain *parameters* of the transistor itself, namely the **input resistance**, the **output resistance** and the **current gain**. These are defined for the *transistor only* (not for the complete circuit) as follows:

input resistance, $\boldsymbol{R_{\text{in}}} = \dfrac{V_{BE}}{I_B}\ \Omega$

output resistance, $\boldsymbol{R_{\text{out}}} = \dfrac{V_{CE}}{I_C}\ \Omega$

current gain $= \dfrac{I_C}{I_B}$ (dimensionless).

Worked Example 14.1

For a circuit of the type in Figure 14.2, determine the value of (a) the base bias resistor R_B, (b) V_{CE}, (c) I_E, (d) R_{in}, (e) R_{out} and (f) the current gain of the transistor if $R_L = 4.7\ \text{k}\Omega$, $I_C = 1$ mA, $I_B = 10\mu$A, $V_{BE} = 0.6$ V and $V_{CC} = 10$ V.

Solution

(a) Since $V_{BE} = 0.6$ V, the voltage across R_B is

$$V_{CC} - V_{BE} = 10 - 0.6 = 9.4\ \text{V}$$

Also, since $I_B = 10\mu$A, then

$$\begin{aligned} R_B &= 9.4\text{V}/I_B = 9.4\ \text{V}/10\mu\text{A} \\ &= 940 \times 10^3\ \Omega \text{ or } 940\ \text{k}\Omega \end{aligned}$$

(b) With a quiescent collector current of 1 mA, the p.d. across R_L is

$$I_C R_L = (1 \times 10^{-3}) \times 4700 = 4.7\ \text{V}$$

hence

$$V_{CE} = 10 - 4.7 = 5.3\ \text{V}$$

(c) The quiescent emitter current is

$$I_E = I_B + I_C = 10\ \mu\text{A} + 1\ \text{mA} = 1.01\ \text{mA}$$

(d) The input resistance of the transistor is

$$R_{in} = V_{BE}/I_B = 0.6\ \text{V}/10\ \mu\text{A} = 60\,000\ \Omega \text{ or } 60\ \text{k}\Omega$$

(e) The output resistance of the transistor is

$$R_{out} = V_{CE}/I_C = 5.3\ \text{V}/1\ \text{mA} = 5300\ \Omega \text{ or } 5.3\ \text{k}\Omega$$

(f) The current gain of the transistor is

$$I_C/I_B = 1\ \text{mA}/10\ \mu\text{A} = 100$$

(b) Common-emitter: collector bias circuit

The bias circuit of the amplifier is shown in Figure 14.3, in which the terms and definitions used are identical to those for Figure 14.2. However, this circuit derives its base bias via R_B from its collector. This provides *improved thermal stability* when compared with the circuit for Figure 14.2; that is, it has a better performance when the ambient temperature changes.

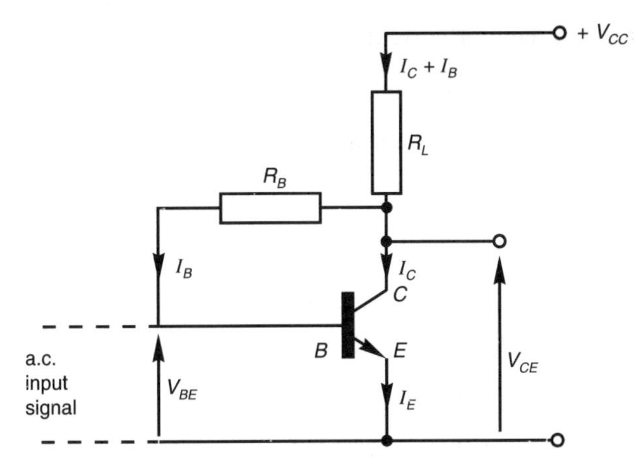

Figure 14.3 Collector bias circuit for a p-n-p common-emmitter amplifier

Worked Example 14.2

Calculate the values specified in Worked Example 14.1, but for the case where collector bias is used (see Figure 14.3). Use the values specified in Worked Example 14.1, that is $R_L = 4.7k\Omega$, $I_C = 1$ mA, $I_B = 10$ μA, $V_{BE} = 0.6$ V and $V_{CC} = 10$ V.

Solution

(a) In this case, the voltage across R_B is the difference between V_{CE} and V_{BE}. However, the current flowing through R_L is

$$I_C + I_B = 1 \text{ mA} + 10 \text{ μA} = 1.01 \text{ mA}$$

and the voltage drop across R_L is

$$(I_C + I_B)R_L = (1.01 \times 10^{-3}) \times 4.7 \times 10^3 = 4.747 \text{ V}$$

hence

$$V_{CE} = V_{CC} - 4.747 = 10 - 4.747 = 5.253 \text{ V}$$

The voltage across R_B is

$$5.253 - 0.6 = 4.653 \text{ V}$$

hence

$$R_B = 4.653/I_B = 4.653/10 \text{ μA}$$
$$= 465300 \text{ Ω or } 465.3 \text{ kΩ}$$

(b) The quiescent collector voltage is (see above)

$$V_{CE} = 5.253 \text{ V}$$

(c) The quiescent emitter current is

$$I_E = I_C + I_B = 1.01 \text{ mA}$$

(d) The input resistance of the transistor is

$$R_{\text{in}} = V_{BE}/I_B = 0.6/10 \times 10^{-6} = 60 \text{ k}\Omega$$

(e) The output resistance of the transistor is

$$R_{\text{out}} = V_{CE}/I_C = 5.253/1 \times 10^{-3}$$
$$= 5253 \ \Omega \text{ or } 5.253 \text{ k}\Omega$$

(f) The current gain of the transistor is

$$I_C/I_B = 1 \text{ mA}/10 \ \mu\text{A} = 100.$$

(c) Common-emitter: potential divider bias

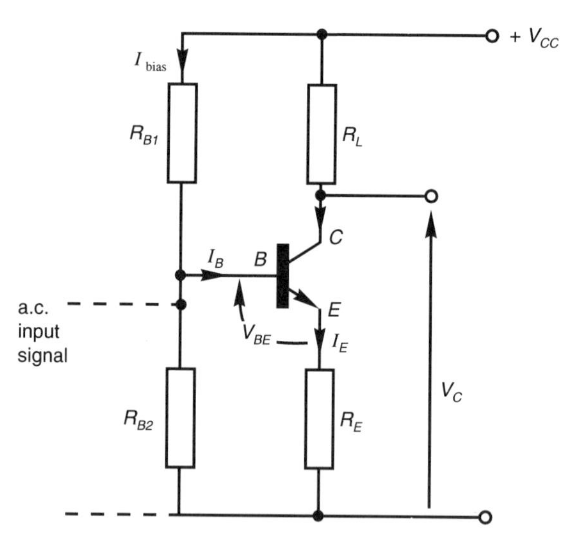

Figure 14.4 Potential divider bias circuit for a common-emitter p-n-p amplifier

This bias circuit is shown in Figure 14.4, and gives improved thermal stability over circuits described hitherto, and is a very popular circuit. For the circuit to give good thermal stability, we need to ensure that

1. the quiescent voltage across R_E is about 10 per cent of V_{CC}, and
2. the quiescent current entering the bias chain, I_{bias}, is about five times greater than the base current, I_B.

Once again, the reader is asked to note the use of the expression 'is about. . .'; this implies that the value is subject to variation.

Worked Example 14.3

In a circuit employing a base bias circuit of the type in Figure 14.4, the collector supply voltage is 10 V, the quiescent base-emitter voltage of the transistor is 0.6 V, the quiescent collector and base currents are 1 mA and 10μA, respectively, the emitter resistor (R_E) is 1 kΩ, and the collector load resistor is 4.7 kΩ. If the d.c. current entering the bias chain is five times the base current, determine (a) the value of R_{B1} and R_{B2} and (b) the quiescent collector voltage.

Solution

(a) The emitter current is

$$I_E = I_C + I_B = 1\text{ mA} + 10\text{ μA} = 1.01\text{ mA}$$

and the voltage across R_E is

$$R_E I_E = (1 \times 10^3) \times (1.01 \times 10^{-3}) = 1.01\text{ V}$$

The quiescent voltage of the base region is

$$V_B = R_E I_E + 0.6 = 1.61\text{ V}$$

which is also the voltage across R_{B2}. The current flowing in R_{B1} is $5I_B$, of which I_B flows into the base. It follows that $4I_B$ flows in R_{B2}, hence

$$\begin{aligned} R_{B2} &= V_B/4I_B = 1.61/(4 \times 10 \times 10^{-6}) \\ &= 40\,250\ \Omega \text{ or } 40.25\text{ k}\Omega \end{aligned}$$

The voltage across R_{B1} is $(10 - 1.61) = 8.39$ V, and the current flowing through it is $5I_B = 50$ μA, therefore

$$R_{B1} = 8.39\text{ V}/50\text{ μA} = 167\,800\ \Omega \text{ or } 167.8\text{ k}\Omega$$

(b) The quiescent collector voltage is

$$V_{CC} - I_C R_L = 10 - (1 \times 10^{-3} \times 4.7 \times 10^3) = 5.3\text{ V}.$$

14.5 Common-emitter class A amplifier

A simple class A transistor amplifier using fixed bias is shown in Figure 14.5. The transistor is biased so that the maximum negative input signal voltage does not cause the transistor to fail to carry current, and the maximum positive signal excursion does not cause the transistor to saturate. As a general rule of thumb, the quiescent collector voltage should be in the region of $V_{CC}/2$. The following are shown in Figure 14.5.

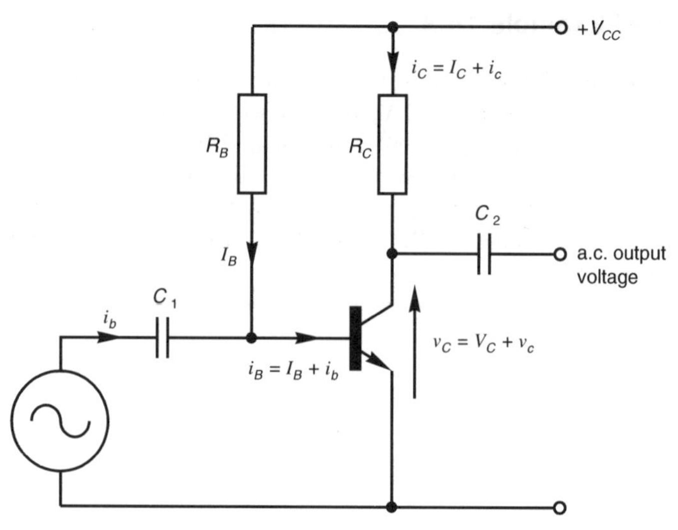

Figure 14.5 A common-emitter class A amplifier

R_C = collector load resistor
v_1 = a.c. signal voltage (the input voltage)
i_b = a.c. input current
I_B = base bias (quiescent) current
$i_B = I_B + i_b$ = total current flowing into the base
i_c = a.c. component of the collector current
I_C = quiescent collector current (due to I_B)
$i_C = I_C + i_c$ = total collector current (due to i_B)
v_c = a.c. component of collector voltage
V_C = quiescent collector voltage
$v_C = V_C + v_c$ = total collector voltage.

Capacitor C_1 is a **blocking capacitor**, whose function it is to prevent current from V_{CC} from flowing into the a.c. signal source. Capacitor C_2 is another blocking capacitor, whose function it is to prevent current from V_{CC} from flowing into the output circuit.

Worked Example 14.4

The collector characteristic of an *n-p-n* transistor is linear between the points given in Table 14.1.

The transistor is used in a circuit of the type in Figure 14.5, with a collector load of 1.5 kΩ and a supply voltage of 8 V. Using a load line, determine

(a) the total power dissipated by the circuit
(b) the quiescent collector power dissipation if the quiescent base current is 25 μA

(c) the current gain, h_{fe}, of the transistor at the quiescent point.
(d) If the peak a.c. input current is 5 μA estimate the peak-to-peak change in the collector voltage.

Table 14.1 Table for Worked Example 14.4

Base current (μA)	*Collector current* (mA) *for a collector voltage of*	
	2V	10V
10	0.9	1.7
20	1.8	2.8
30	2.8	4.2
40	3.9	5.5

Solution

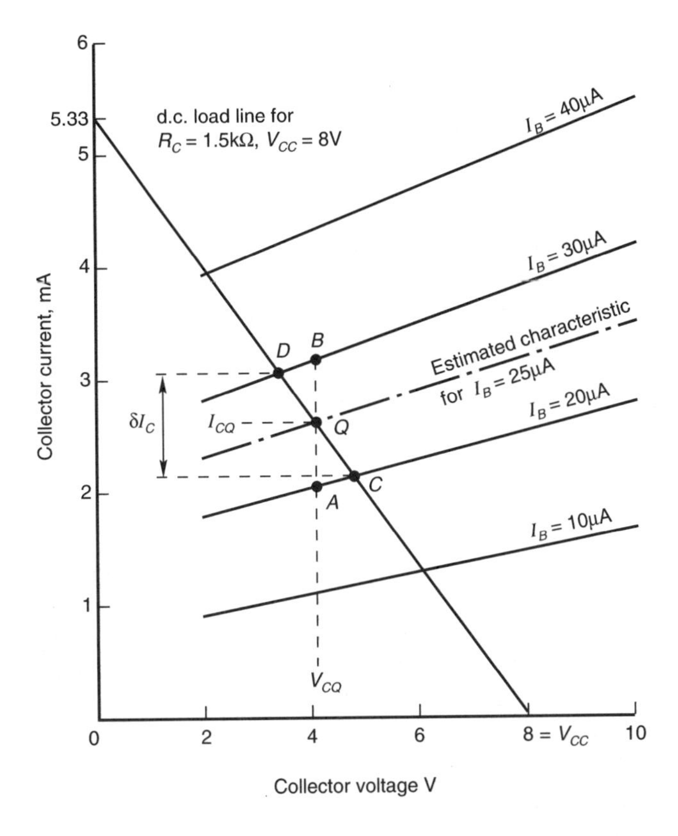

Figure 14.6 Graph for Worked Example 14.4

(a) The point at which the transistor operates on the output characteristics (the **operating point**), is given by the intersection of the output characteristic for the quiescent base current with what is known as the **d.c. load line** for the collector load resistance (also see below). This is illustrated in Figure 14.6.

The collector characteristics for the four values of base current in Table 14.1 are shown in full line in Figure 14.6. Since the specified quiescent base current is 25μA, we can estimate the position of the load line as being mid-way between those for 20μA and 30μA, and is shown in broken line in the figure.

The *d.c. load line* for $R_L = 1.5\ \text{k}\Omega$ and $V_{CC} = 8$ V is drawn on the characteristics, and commences at $V_{CC} = 8$ V when $i_C = 0$, and terminates when $i_C = V_{CC}/R_C = 8\ \text{V}/1.5\ \text{k}\Omega = 5.33$ mA when $v_C = 0$. The gradient of the d.c. load line is: $-1/R_C = -1/1.5\ \text{k}\Omega = -0.667\ \text{mA}/V$.

The quiescent (no-signal) operating point of the transistor lies at point Q on the characteristic, where the load line intersects with the $i_B = 25\ \mu\text{A}$ collector characteristic. This occurs at a collector voltage of $V_{CQ} = 4.1$ V, and a quiescent collector current of $i_{CQ} = 2.6$ mA.

The quiescent (no-signal) collector power dissipation *of the circuit* is

$$\begin{aligned} V_{CC}(i_{CQ} + i_{BQ}) &= 8\ \text{V} \times (2.6 + 0.25)\ \text{mA} \\ &= 21\ \text{mW} \end{aligned}$$

(b) The quiescent (no-signal) collector power dissipation *of the transistor* is

$$V_{CQ} i_{CQ} = 4.1\ \text{V} \times 2.6\ \text{mA} = 10.66\ \text{mW}$$

(c) The *current gain*, h_{fe}, of the transistor at the operating point is the ratio of the change in collector current to the change in base current. This is given by the change in collector current between A and B on the characteristics in Figure 14.6. The collector current at A is 2.06 mA (corresponding to a base current of 20 μA), and the collector current at point B is 3.18 mA (corresponding to a base current of 30 μA). Hence

$$h_{fe} = \frac{(3.18 - 2.06)\ \text{mA}}{(30 - 20)\ \mu\text{A}} = 112$$

(d) When the transistor is used in an amplifier with resistor R_C in circuit, the operating point is constrained to move along the load line as the base current changes. That is when the base current is 20μA, the transistor operates at point C on the characteristics, and when the base current is 30 μA the operating point is at point D. The corresponding values of collector current are 2.15 mA and 3.05 mA (corresponding to points C and D, respectively). That is *the effective current gain of the amplifier* is

$$\frac{(3.05 - 2.15)\ \text{mA}}{(30 - 20)\ \mu\text{A}} = 90$$

From the results, we see that *the current gain of the amplifier is less than that of the transistor itself* (the reader will find it an interesting exercise to argue why this is the case).
We can see that the peak-to-peak change in collector current is

$$3.05 - 2.15 = 0.9 \text{ mA}$$

and the peak-to-peak change in collector voltage is

$$0.9 \text{ mA} \times 1.5k\Omega = 1.35 \text{ V}.$$

which corresponds to an r.m.s. output voltage of

$$1.35/(2\sqrt{2}) = 0.48 \text{ V}.$$

14.6 The a.c. load line or dynamic load line

The circuit for a fixed-bias, common-emitter, class A amplifier is shown in figure 14.7(a), with the d.c. load line and a.c. load line (see below) being superimposed on the output characteristics of the transistor in Figure 14.7(b). The d.c. load line has been described in Worked Example 14.4, and the position of the quiescent point, Q, obtained in the manner described.

The reactance of the blocking capacitors C_1 and C_2 at the lowest operating frequency have a very low value and, so far as the operation of the amplifier is concerned, they can be regarded as being a.c. 'short-circuits'.

When the external load, R_L, is connected to the output terminals of the amplifier, we can predict the operation of the amplifier from the **a.c. load line** or **dynamic load line**. This is a straight line passing through point Q (see Figure 14.7(b)), and has a slope of $1/R_L{}'$, where

$$\begin{aligned} R_L{}' &= R_C \text{ in parallel with } R_L \\ &= R_C R_L/(R_C + R_L) \end{aligned}$$

The use of the a.c. load line is illustrated in Worked Example 14.5.

Worked Example 14.5

The transistor in Figure 14.7(a) has the characteristics in Table 14.2, which are linear between the points given.

The collector load resistance is 1.0 kΩ, that of R_L is 4 kΩ, and the collector supply voltage is 9 V. If the quiescent collector voltage is 4.5 V, determine the quiescent value of the collector and base current of the transistor. Estimate the value of base bias resistor, R_B, given that the quiescent base-emitter voltage is 0.6 V.

Plot the d.c. and a.c. load lines on the output characteristics of the transistor, and determine the r.m.s. value of the output voltage if the r.m.s. value of the a.c. component of the base current is 14.14μA. Calculate also the value of h_{fe} of the transistor at the operating point, together with the current

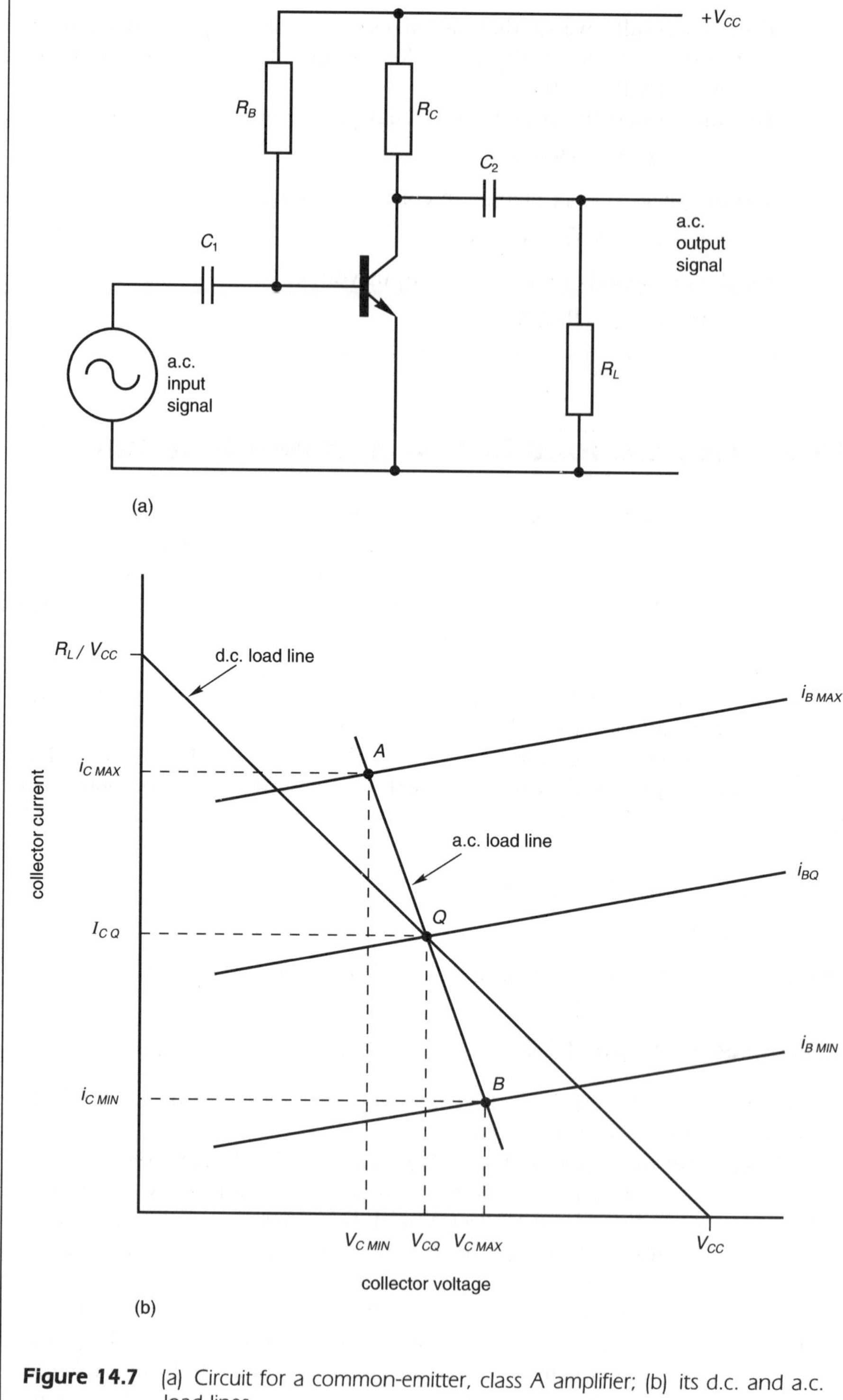

Figure 14.7 (a) Circuit for a common-emitter, class A amplifier; (b) its d.c. and a.c. load lines

gain of the amplifier. The reactance of the blocking capacitors is very low at the operating frequency.

Table 14.2 Table for Worked Example 14.5

Base current (μA)	*Collector current* (mA) *for a collector voltage of*	
	1.0V	8.5V
40	1.9	2.65
60	2.9	3.95
80	3.95	5.2
100	4.95	6.45
120	5.9	7.7

Solution

It is left as an exercise for the reader to plot the output characteristics and the d.c. load line of the amplifier, and it will be seen that the load line cuts the output characteristic when $V_{CQ} = 4.5$ V and $i_{BQ} = 80$ μA. It will also be found that the quiescent collector current, I_{CQ}, is 4.5 mA.

The value of the base bias resistor is determined as follows:

$$R_B = (V_{CC} - V_{CQ})/i_{BQ} = (9 - 0.6)/80 \times 10^{-6}$$
$$= 105\,000\ \Omega \text{ or } 105\ \text{k}\Omega$$

The a.c. load, $R_L{}'$, has a value of $R_C(1\ \text{k}\Omega)$ in parallel with $R_L(4\ \text{k}\Omega)$, or

$$R_L{}' = 1 \times 4/(1 + 4) = 0.8\ \text{k}\Omega$$

The a.c. load line therefore has a slope of $-1/0.8 = -1.25$ mA/V, and passes through the quiescent point Q.

This line passes through $(i_{CQ} - 1.25) = 3.25$ mA at a voltage of $(V_{CQ} + 1) = 5.5$ V, and through $(i_{CQ} - 2.5) = 2$ mA at a voltage of $(V_{CQ} + 2) = 6.5$ V, etc. Using this information, it is possible to plot the a.c. load line without too much difficulty, as shown in Figure 14.8.

A sinusoidal base current of r.m.s. value 14.14 μA corresponds to a peak-to-peak base current of $2\sqrt{2} \times 14.14 = 40$ μA. That is, the minimum base current is $(80 - 40/2) = 60$ μA, and the maximum base current is $(80 + 40/2) = 100$ μA. This causes the transistor operating point to swing between points A and B on the a.c. load line. The corresponding collector voltage at point A is 3.7 V , and 5.3 V at point B, giving a peak-to-peak collector voltage swing of $(5.3 - 3.7) = 1.6$ V, or a corresponding r.m.s. output voltage of

$$1.6/(2\sqrt{2}) = 0.57\ \text{V}$$

The value of h_{fe} for the transistor is calculated from

$$h_{fe} = i_{CQ}/i_{BQ} = 4.5\ \text{mA}/80\ \mu\text{A} = 56.25.$$

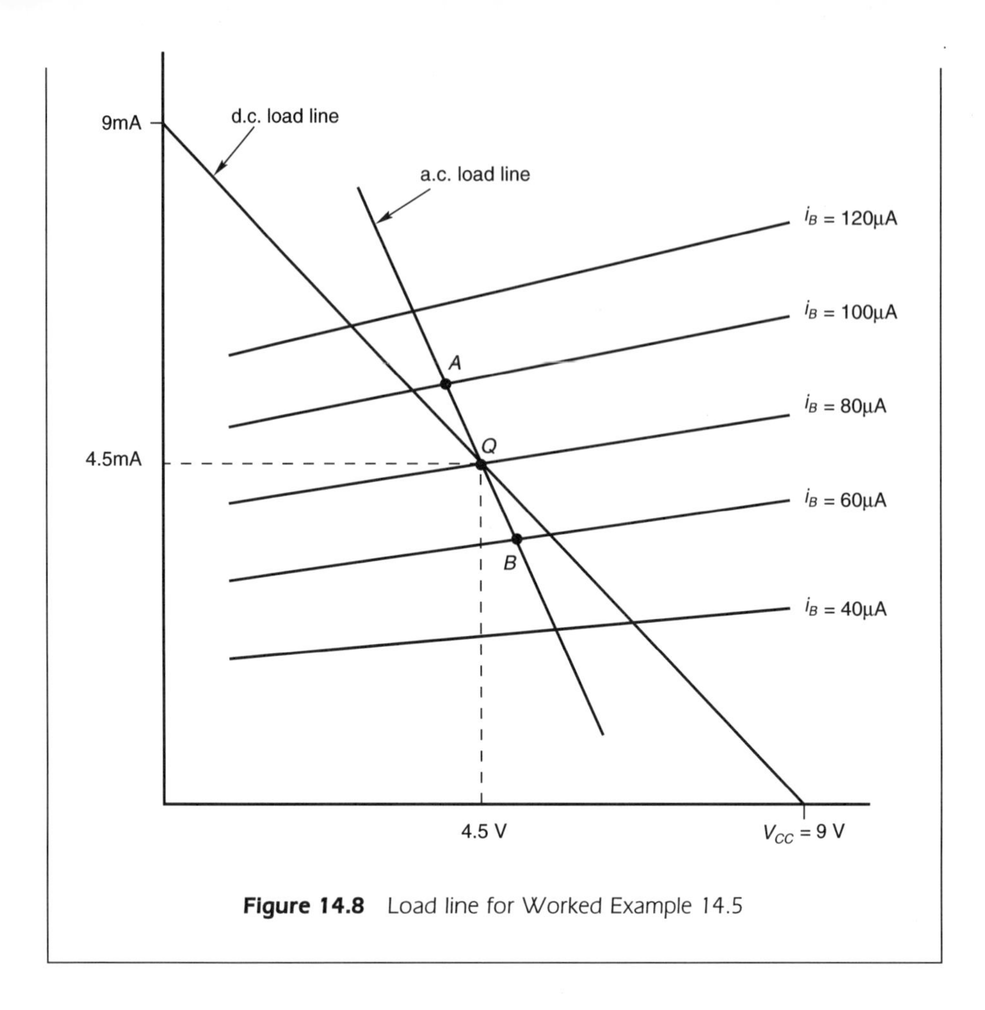

Figure 14.8 Load line for Worked Example 14.5

14.7 The emitter follower circuit

The basic emitter follower circuit is shown in Figure 14.9, where

C_1 and C_2 are blocking capacitors
R_B is the base bias resistor
R_E is the emitter resistor.

The value of the base bias resistor can be calculated by applying Kirchhoff's voltage law to the input circuit as follows (see also Worked Example 14.6):

$$V_E + V_{BE} + I_B R_B = V_{CC}$$

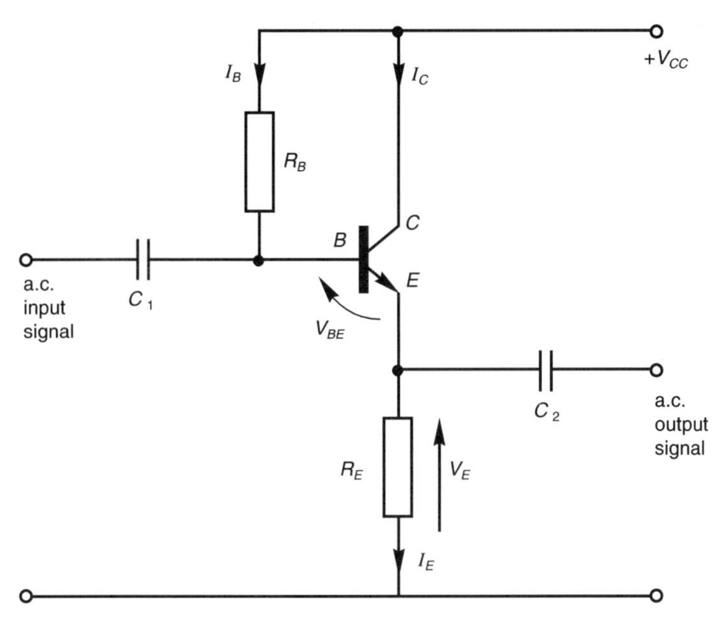

Figure 14.9 p-n-p emitter follower circuit

Also, since $V_E = R_E(I_B + I_C)$, then

$$R_E(I_B + I_C) + V_{BE} + I_B R_B = V_{CC}$$

or

$$R_E I_E + V_{BE} + I_B R_B = V_{CC}$$

Depending on the required circuit conditions, it is often desirable to have a quiescent voltage of about $V_{CC}/2$ across R_E, that is $V_E \approx V_{CC}/2$.

Worked Example 14.6

Estimate suitable values for R_B and R_E in the emitter follower circuit in Figure 14.9, given that $V_{CC} = 10V$, $I_{CQ} = 10$ mA, $I_{BQ} = 100$ µA and $V_{BE} = 0.6$ V.

Solution

Now, $I_E = I_C + I_B = 10$ mA + 100 µA = 1.1 mA, hence

$$V_E \approx V_{CC}/2 = 5 \text{ V} = I_E R_E$$

or

$$R_E = 5/(10.1 \times 10^{-3}) = 495\ \Omega$$

Also

$$R_E I_E + V_{BE} + I_B R_B = V_{CC}$$

or

$$(495 \times 10.1 \times 10^{-3}) + 0.6 + (100 \times 10^{-6} \times R_B) = 10$$

then

$$R_B = (10 - 5 - 0.6)/100 \times 10^{-6} = 44000\ \Omega \text{ or } 44\ \text{k}\Omega$$

14.8 Field-effect transistors

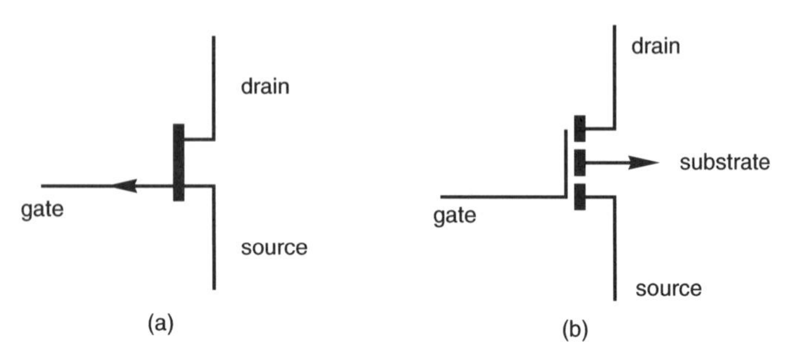

Figure 14.10 Field-effect transistor: (a) p-channel JUGFET; (b) p-channel MOSFET

There are two basic types of **field-effect transistor** (FET), namely the **junction-gate FET** (JUGFET) and the **metal-oxide-semiconductor FET** (MOSFET), symbols for *p*-channel devices being shown in Figure 14.10. MOSFETs are widely used in digital circuits, and have a very low power consumption and their gate current is very low indeed.

14.9 Computer listing

Listing 14.1 Common-emitter bias calculations

```
10 CLS '*** CH14-1 ***
20 PRINT "COMMON EMITTER BIAS CALCULATIONS": PRINT
30 PRINT "Select ONE of the following:": PRINT
40 PRINT "1. FIXED BIAS and COLLECTOR BIAS design.": PRINT
50 PRINT "2. POTENTIAL DIVIDER BIAS design:": PRINT
60 INPUT "Enter your selection here (1 or 2): ", Sel
70 IF Sel < 1 OR Sel > 2 THEN GOTO 10
80 IF Sel = 2 THEN GOTO 400
90 CLS : PRINT "FIXED BIAS or COLLECTOR BIAS design": PRINT
100 INPUT "Collector supply voltage (V) = ", Vcc
110 INPUT "Load resistance (k ohm) = ", RL
```

```
120 INPUT "Quiescent base-emitter voltage (V) = ", Vbe
130 INPUT "Quiescent base current (microamperes) = ", Ib
140 INPUT "Quiescent collector current (mA) = ", Ic: PRINT
150 Vce1 = Vcc - RL * Ic
160 Vce2 = Vcc - RL * (Ic + (Ib / 10 ^ 3))
170 Rbias1 = (Vcc - Vbe) / (Ib / 10 ^ 3)
180 Rbias2 = (Vce2 - Vbe) / (Ib / 10 ^ 3)
190 Rin = Vbe / (Ib / 10 ^ 3)
200 Rout1 = Vce1 / Ic
210 Rout2 = Vce2 / Ic
220 Ie = Ic + (Ib / 10 ^ 3)
230 PRINT "Input resistance = "; Rin; "k ohms"
240 PRINT "Quiescent emitter current = "; Ie; "mA"
250 PRINT "Current gain = "; Ic / (Ib / 10 ^ 3): PRINT
260 PRINT "For FIXED BIAS design"
270 PRINT TAB(4); "Output resistance = "; Rout1; "k ohms"
280 PRINT TAB(4); "Quiescent collector-emitter voltage = "; Vce1
290 PRINT TAB(4); "Bias resistor = "; Rbias1; "k ohms": PRINT
300 PRINT "For COLLECTOR BIAS design"
310 PRINT TAB(4); "Output resistance = "; Rout2; " k ohms"
320 PRINT TAB(4); "Quiescent collector-emitter voltage = "; Vce2
330 PRINT TAB(4); "Bias resistor = "; Rbias2; " k ohms"
340 END
400 CLS : PRINT "POTENTIAL DIVIDER BIAS design.": PRINT
410 INPUT "Collector supply voltage (V) = ", Vcc
420 INPUT "Quiescent base-emitter voltage (V) = ", Vbe
430 INPUT "Quiescent base current (microamperes) = ", Ib
440 INPUT "Quiescent collector current (mA) = ", Ic
450 INPUT "Emitter resistor (ohms) = ", Re
460 INPUT "Load resistance (k ohm) = ", RL: PRINT
470 Ie = Ic + (Ib / 10 ^ 3)
480 Ve = (Ie / 10 ^ 3) * Re
490 VR2 = Ve + Vbe
500 PRINT "Please enter the ratio of the direct current entering"
510 PRINT "the bias chain (i.e., RB1 and RB2) to the quiescent"
520 PRINT "base current. A nominal value of 5 is suitable, but"
530 PRINT "you MUST ENTER the ratio here.": PRINT
540 INPUT "                Ratio = ", N
550 RB2 = VR2 / ((N - 1) * Ib / 10 ^ 6)
560 VR1 = Vcc - VR2
570 RB1 = VR1 / (N * Ib / 10 ^ 6): PRINT
580 PRINT "RB1 = "; RB1 / 1000; "k ohms"
590 PRINT "RB2 = "; RB2 / 1000; "k ohms"
600 PRINT "Collector voltage = "; Vcc - Ic * RL
610 PRINT "Emitter voltage = "; (Ie / 1000) * Re
620 END
```

The program in Computer listing 14.1 offers two options, namely the design of

1. fixed bias and collector bias circuits, and
2. potential divider bias circuits.

Option 1 asks you to supply the value of V_{CC}, together with the load resistance in the collector circuit, the quiescent base-emitter voltage, the quiescent collector current, and the quiescent base current. Having been supplied with this data, it calculates the input resistance of the transistor, the quiescent base current, the current gain of the transistor, the output resistance of the transistor, the quiescent collector voltage, and the value of the bias resistor.

Option 2 asks for the value of V_{CC}, the quiescent base-emitter voltage of the transistor, together with the collector and base current. You also need to supply the value of the resistance connected to the collector and the emitter, and the ratio of the current entering the potential divider chain to the base current. Having been supplied with this data, the computer calculates the value of each of the resistors in the bias chain, together with the quiescent collector and emitter voltage.

Exercises

14.1 Describe the essential difference between bipolar junction transistors and field-effect transistors.

14.2 A common-emitter amplifier uses the fixed-bias circuit in Figure 14.2. If the quiescent collector current is 10 mA, the base current is 110 μA, $V_{CC} = 12$ V, $R_L = 600\ \Omega$ and $V_{BE} = 0.65$ V, determine (a) the value of the base bias resistor, (b) the value of V_{CE}, (c) the input resistance of the transistor, (d) the emitter current, (e) the output resistance of the transistor, (f) the current gain.

14.3 Re-design the bias circuit for the amplifier in Problem 14.2 for collector bias (see Figure 14.3) given the same operating conditions.

14.4 A common-emitter amplifier with fixed bias has the following values: $V_{CC} = 10$ V, $V_{BE} = 0.55$ V, $I_C = 5$ mA, current gain = 83.33, and $V_{CE} = 5$ V. Determine (a) the quiescent value of base current, (b) the collector load resistance, (c) the input resistance of the transistor, (d) the output resistance of the transistor, (e) the value of the bias resistor.

14.5 A common-emitter amplifier with collector bias has the following values: $V_{CC} = 8$ V, $R_L = 800\ \Omega$, $V_{BE} = 0.6$ V, quiescent base current = 68.8 μA, current gain = 80. Determine (a) the quiescent value of the collector current, (b) the quiescent collector voltage, (c) the input resistance of the transistor, (d) the output resistance of the transistor, (e) the value of the bias resistor.

14.6 A common-emitter amplifier with a potential-divider bias circuit of the type in Figure 14.4 has a collector supply voltage of 15 V, its quiescent base-emitter voltage being 0.55 V. If the quiescent collector current is 5 mA, the emitter resistor is 200 Ω, the collector load resistance is 1.4 kΩ, and the ratio of the current entering the bias circuit to the bias current is 5. Determine (a) the value of the resistors in the bias chain, (b) the quiescent collector voltage and (c) the quiescent emitter voltage.

14.7 A common-emitter amplifier with a potential divider bias circuit has the following values: $V_{CC} = 10$ V, $V_{BE} = 0.6$ V, $I_B = 40$ μA, $R_E = 300$ Ω, $R_C = 1$ kΩ, and $V_C = 6$ V. If the ratio of the current entering the bias chain to the base bias current is 4, determine (a) the quiescent collector current, (b) the value of the resistors in the bias chain, (c) the quiescent emitter voltage.

14.8 An emitter follower circuit of the type in Figure 14.7 has an emitter resistor of 100 Ω, and a collector supply voltage of 10 V. If the quiescent collector current is 50 mA, the current gain of the transistor is 80, and the quiescent base-emitter voltage is 0.55 V, determine the value of the base bias resistor.

14.9 The collector characteristic of a transistor connected in the common-emitter mode are linear between the following points:

v_{CE} (V)	2	25
i_C (mA)	4.5	5.5 when $i_B = 50$ μA
i_C (mA)	14.0	16.0 when $i_B = 150$ μA
i_C (mA)	23.5	26.5 when $i_B = 250$ μA

Draw the collector characteristic, and the construct the load line for a collector load of 1 kΩ and a collector supply voltage of 30 V.

If the peak-to-peak current swing in the base circuit is 200μA, about a quiescent base current of 150μA, determine

(a) the r.m.s. value of the a.c. component of collector current
(b) the r.m.s. value of the voltage developed across the load
(c) the r.m.s. value of the a.c. power developed in the load
(d) the value of h_{fe} at the quiescent point.

Summary of important facts

The regions of a **bipolar junction transistor** (BJT) are the **emitter** (where charge carriers are emitted), the **base** (where the flow of charge carriers is controlled), and the **collector** (where charge carriers are collected).

The principal types of bias circuit used in BJT circuits are

1. fixed bias (see Figure 14.2)
2. collector bias (see Figure 14.3)
3. potential divider bias (see Figure 14.4).

Fixed bias is the simplest form of circuit, and collector bias provides an element of **thermal stability**, that is it gives improved performance against ambient temperature variation. Potential divider bias can provide very good protection against ambient temperature change.

The **common-emitter amplifier circuit** (see Figure 14.5) has a blocking capacitor at its input, and one at its output. The purpose of the blocking capacitors is to allow the passage of a.c. signals respectively in to and out of, the circuit, whilst blocking the flow of d.c. from the supply. The capacitors have a low reactance at the lowest frequency of operation, so that there is very little voltage drop in them. The operation of a common-emitter amplifier can be predicted using the **load-line method** (see Figure 14.6).

The **emitter-follower circuit** (see Figure 14.7) has a voltage gain of practically unity, but has a very high input impedance and a low output impedance. For this reason, it is used as a unity-gain buffer amplifier.

The regions of a **field effect transistor** (FET) are the **source** (the source of charge carriers), the **gate** (which is used to control the flow of charge carriers), and the **drain** (where charge carriers are drained from the transistor).

15 Operational amplifier circuits

15.1 Introduction

Operational amplifiers (abbreviated to *op-amps*) form a very important part of the subject matter of electronics, not only because they are used in all types of electronic systems, but also because they are vital to power engineers in the field of the control of power systems. By the end of this chapter, the reader will be able to understand and appreciate

- Phase inverting amplifiers.
- Summing amplifiers.
- Non-inverting amplifiers.
- Differential input (instrumentation) amplifiers.
- Integrators.
- Computer software for operational amplifier circuits.

15.2 Features of an ideal operational amplifier

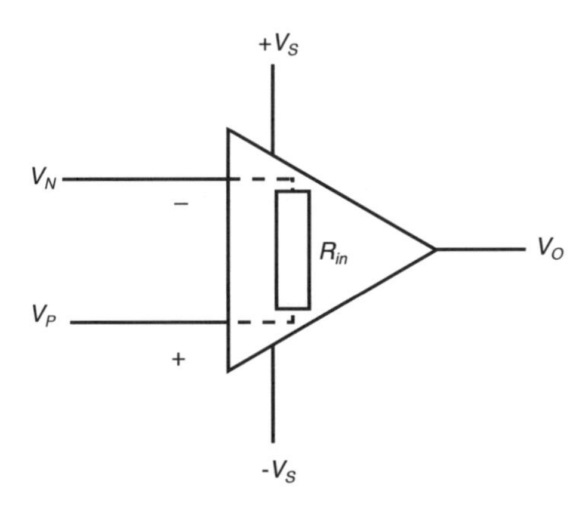

Figure 15.1 Circuit representation of an op-amp

An ideal operational amplifier (see Figure 15.1) is a high-gain, direct coupled voltage amplifier. The op-amp has two supply voltages, one being positive $(+V_S)$ with respect to ground, the other being negative $(-V_S)$. The output signal, V_O, from the amplifier is

$$V_O = A(V_P - V_N)$$

where V_P is known as the **non-inverting input signal** (or the '+' input), V_N is the **inverting input signal** (or the '−' input), and A is the **forward gain** or **d.c. gain** of the amplifier.

The forward gain of the amplifier is very high (typically 10^5 or 10^6), so that the **differential input signal** $(V_P - V_N)$ is very small. Since the differential input signal is so small it can, for most purposes, be thought of as being *virtually zero* (do not get confused, its value is *not zero*, otherwise the amplifier output would also be zero!).

The output voltage from the amplifier is in-phase with any signal applied to the non-inverting input (the '+' input), and is anti-phase to any signal applied to the inverting input (the '−' input). Ideally, if the same signal is simultaneously applied to both inputs, the output voltage is zero.

Another parameter which we sometimes have to account for is the **differential input resistance**, R_{in}, of the amplifier. This is the resistance between terminals V_P and V_N; its value can be very high (particularly in FET op-amps), so that the input current drawn by the op-amp may be very small.

In an **ideal op-amp**, we can assume the following values:

forward gain = infinity
differential input voltage = zero
differential input resistance = infinity
input current = zero.

15.3 An inverting amplifier

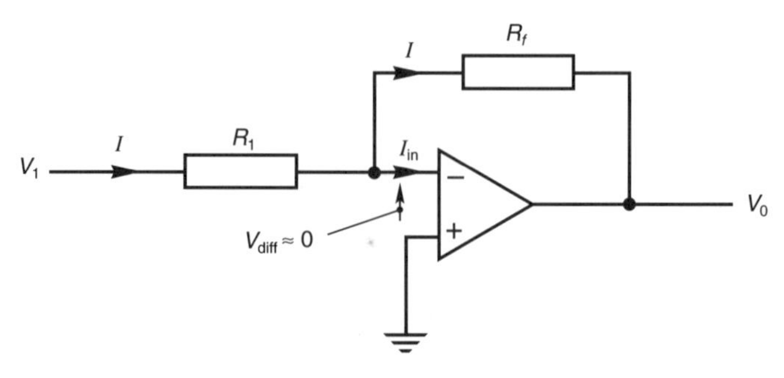

Figure 15.2 The inverting amplifier circuit

The basic configuration of an inverting amplifier using an op-amp is shown in Figure 15.2. If we use an ideal op-amp, it is safe to assume that the differential input voltage, V_{diff}, is zero. That is, I_{in} is also zero, so that the current, I, flowing in R_1 must flow through the feedback resistor R_f. The input current is

$$I = \frac{V_1 - V_{diff}}{R_1} \approx \frac{V_1}{R_1}$$

and the current flowing through the feedback resistor is

$$I = \frac{V_{diff} - V_O}{R_f} \approx -\frac{V_O}{R_f}$$

hence

$$\frac{V_1}{R_1} = -\frac{V_O}{R_f}$$

so that the overall voltage gain of the amplifier is

$$\frac{V_O}{V_1} = -\frac{R_f}{R_1}$$

The negative sign associated with the voltage gain tells us that the circuit acts as an inverting amplifier.

Worked Example 15.1

Determine the voltage gain of an inverting amplifier of the type in Figure 15.2 if $R_1 = 100\ \text{k}\Omega$ and $R_f = 500\ \text{k}\Omega$. What is the output voltage if $V_1 = 1.5$ V?

Solution

The voltage gain is

$$\frac{V_O}{V_1} = -\frac{R_f}{R_1} = -\frac{500 \times 10^3}{100 \times 10^3} = -5$$

If $V_1 = 1.5$ V, then

$$V_O = -5 \times 1.5 = -7.5\ \text{V}$$

(**Note**: The amplifier must not be operated so that we try to obtain a higher output voltage than the supply voltage (either $+V_S$ or $-V_S$), otherwise the amplifier *saturates*.)

15.4 A phase inverting summing amplifier

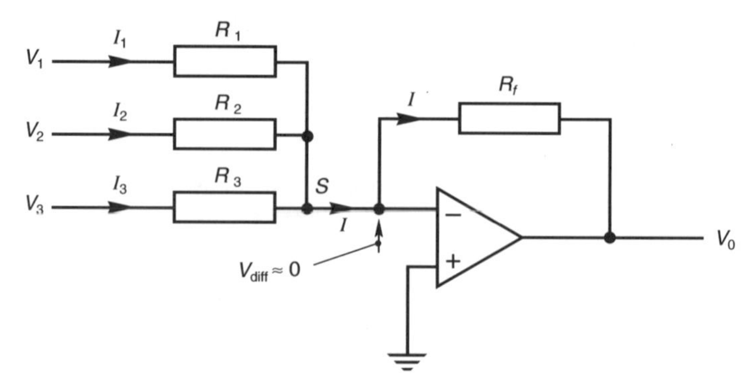

Figure 15.3 The inverting summing amplifier

The basic op-amp configuration for adding several signals together is shown in Figure 15.3. Since the op-amp is considered to be ideal, $V_{\text{diff}} \approx 0$, that is the **summing junction**, S, is assumed to be a **virtual earth point**. The total current flowing from the three input sources is

$$I_1 + I_2 + I_3 = \frac{V_1}{R_1} + \frac{V_2}{R_2} + \frac{V_3}{R_3} = I$$

Also, since V_{diff} is assumed to be zero, it follows that I flows through R_f, that is

$$I = \frac{V_{\text{diff}} - V_O}{R_f} \approx -\frac{V_O}{R_f}$$

therefore

$$-\frac{V_O}{R_f} = \frac{V_1}{R_1} + \frac{V_2}{R_2} + \frac{V_3}{R_3}$$

The output voltage from the summing amplifier is

$$V_O = -\left[V_1 \frac{R_f}{R_1} + V_2 \frac{R_f}{R_2} + V_3 \frac{R_f}{R_3}\right].$$

Worked Example 15.2

A voltage of 1 V and another of 0.5 V are added together in a phase inverting summing amplifier. Determine the output voltage if an ideal op-amp is used (a) with input resistors of 1 MΩ and a feedback resistor of 1 MΩ, (b) if the 1 V signal has an input resistor of 0.5 MΩ, the 0.5 V signal has an input resistor of 0.8 MΩ, and the feedback resistor is 1.5 MΩ.

Solution

(a) The output voltage is

$$V_O = -\left[V_1 \frac{R_f}{R_1} + V_2 \frac{R_f}{R_2}\right]$$
$$= -\left[1 \frac{10^6}{10^6} + 0.5 \frac{10^6}{10^6}\right] = -1.5 \text{ V}$$

(see also the solution of Worked Example 20.7 in Chapter 20)

(b) In this case, the output voltage is

$$V_O = -\left[1 \times \frac{1.5 \times 10^6}{0.5 \times 10^6} + 0.5 \times \frac{1.5 \times 10^6}{0.8 \times 10^6}\right]$$
$$= -3.9375 \text{ V}.$$

15.5 A non-inverting amplifier

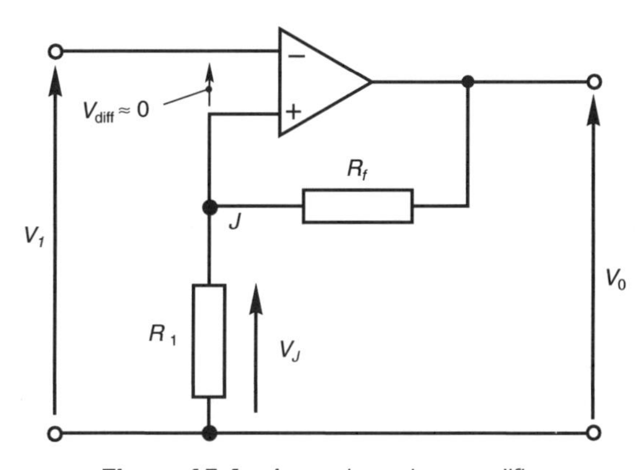

Figure 15.4 A non-inverting amplifier

A basic non-inverting amplifier using an op-amp is shown in Figure 15.4. Once again, the op-amp is ideal so that $V_{\text{diff}} \approx 0$, hence the summing junction J is at the same potential as that of the non-inverting input, that is $V_J \approx V_1$. Using the equation for the potential division between two resistors, the potential at junction J is

$$V_J = \frac{R_1}{R_1 + R_f} V_O$$

or

$$V_O = \frac{R_1 + R_f}{R_1} V_J \approx \frac{R_1 + R_f}{R_1} V_1$$

Since the polarity of V_O is the same as that of V_1, the amplifier is non-inverting. The voltage gain of the amplifier in Figure 15.4 is therefore

$$\frac{V_O}{V_1} = \frac{R_1 + R_f}{R_1}.$$

Worked Example 15.3

If, in a circuit of the type in Figure 15.4, $V_1 = 0.5\text{ V}$, $R_1 = 5\text{ k}\Omega$ and $R_f = 100\text{ k}\Omega$, determine the value of the output voltage if an ideal op-amp is used.

Solution

The output voltage is given by

$$V_O = \frac{R_1 + R_f}{R_1} \times V_1 = \frac{5000 \times 100\,000}{5000} \times 0.5$$
$$= 10.5\text{ V}$$

(see also Worked Example 20.8 in Chapter 20).

15.6 A differential input (instrumentation) amplifier

In many cases we need to measure the potential difference between two points in a circuit, both of which are 'live' with respect to ground (zero voltage). There are many examples of this in practice, including instrumentation amplifiers, and oscilloscopes with differential (or difference) inputs. The basis of a differential input amplifier or instrumentation amplifier is shown in Figure 15.5.

If, in the circuit in Figure 15.5

$$\frac{R_{2A}}{R_{1A}} = \frac{R_{2B}}{R_{1B}}$$

then the **differential voltage gain** of the amplifier between the differential input voltage $(V_2 - V_1)$ and the output is

$$\frac{V_O}{V_2 - V_1} = \frac{R_2}{R_1}$$

By altering the ratio R_2/R_1, we can alter the forward gain of the amplifier (see also Worked Example 15.4).

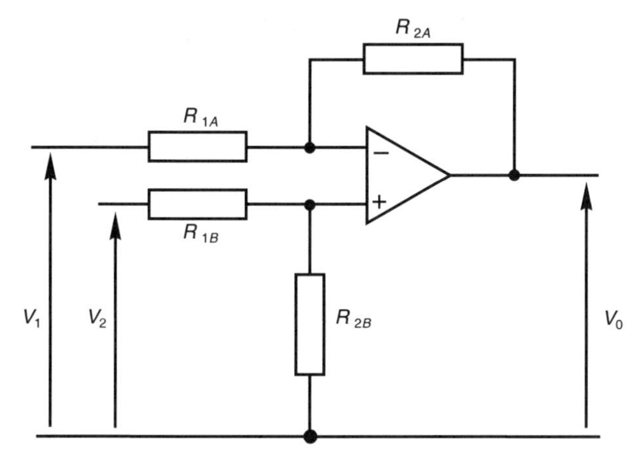

Figure 15.5 Differential amplifier or instrumentation amplifier

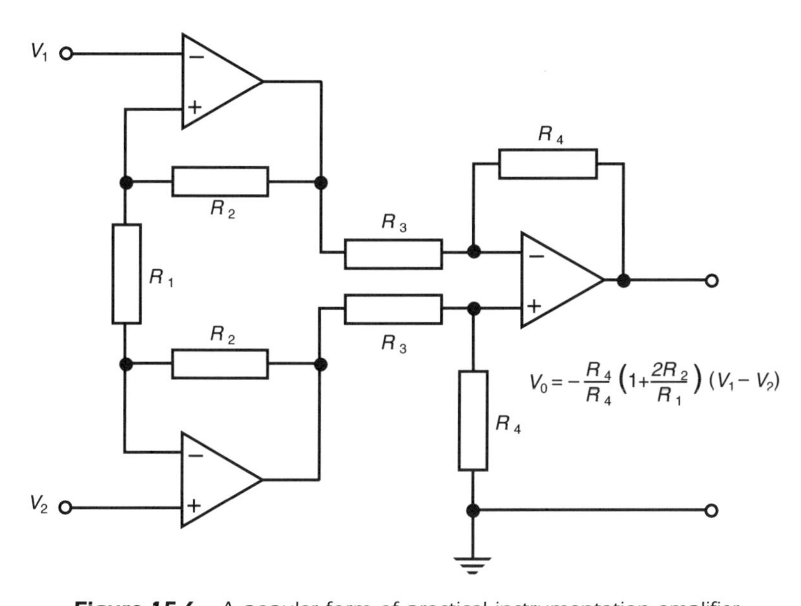

Figure 15.6 A popular form of practical instrumentation amplifier

The circuit in Figure 15.6 is a popular form of instrumentation amplifier, whose gain can be adjusted simply by altering the value of R_1 alone (the other values either being fixed, or used as range-changing resistors). The output voltage from the amplifier is

$$V_O = -\frac{R_4}{R_3}\left[1+\frac{2R_2}{R_1}\right](V_1 - V_2)$$

$$= \frac{R_4}{R_3}\left[1+\frac{2R_2}{R_1}\right](V_2 - V_1)$$

Worked Example 15.4

A simple difference amplifier of the type in Figure 15.5 has inputs of $V_1 = 3$ V and $V_2 = 1.5$ V. If (a) $R_{1A} = R_{1B} = 10$ kΩ, $R_{2A} = R_{2B} = 100$ kΩ, (b) $R_{1A} = R_{1B} = 20$ kΩ, $R_{2A} = R_{2B} = 50$ kΩ, determine the output voltage.

Solution

(a) Since, in Figure 15.5, $R_{2A}/R_{1A} = 10$, and $R_{2B}/R_{1B} = 10$, the overall gain of the amplifier to the differential input signal is 10, thc output voltage is

$$V_O = 10(V_2 - V_1) = 10(1.5 - 3) = -15 \text{ V}$$

(b) In this case $R_{2A}/R_{1A} = 2.5$, and $R_{2B}/R_{1B} = 2.5$, hence

$$V_O = 2.5(V_2 - V_1) = 2.5(1.5 - 3) = -3.75 \text{ V}.$$

Worked Example 15.5

A differential amplifier of the type in Figure 15.6 is required to have a differential gain ranging from 10 to 100 simply by adjusting the value of R_1. The differential gain of the final stage is to be unity by making $R_3 = R_4$. If $R_2 = 50$ kΩ, determine the minimum and maximum value of R_1 to give the required overall gain.

Solution

The overall differential voltage gain of the circuit in Figure 15.6 is

$$\frac{V_0}{V_2 - V_1} = \frac{R_4}{R_3}\left[1 + \frac{2R_2}{R_1}\right] = 1 + \frac{2R_2}{R_1}$$

$$= 1 + 2\frac{50\,000}{R_1} = 1 + \frac{100\,000}{R_1}$$

For a differential voltage gain of 10, then

$$10 = 1 + 100\,000/R_{1(\text{min gain})}$$

or

$$R_{1(\text{min gain})} = 100\,000/9 = 11\,111 \ \Omega \text{ or } 11.111 \text{ k}\Omega$$

and for a differential voltage gain of 100 then

$$100 = 1 + 100\,000/R_{1(\text{max gain})}$$

or

$$R_{1(\text{max gain})} = 100\,000/99 = 1010 \ \Omega \text{ or } 1.01 \text{ k}\Omega.$$

15.7 An electronic integrator circuit

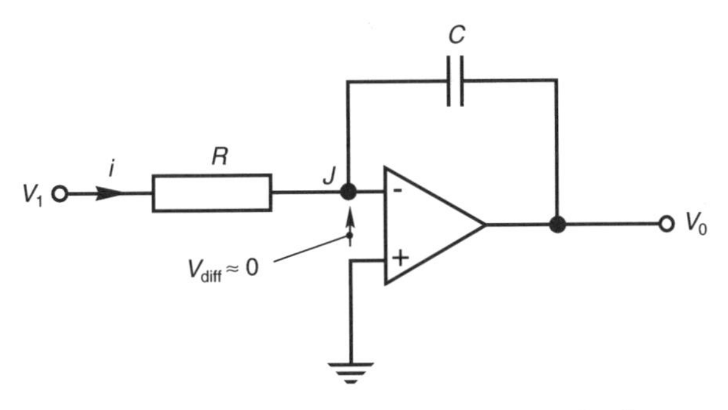

Figure 15.7 An electronic integrator circuit

If we increase the voltage across a capacitor by dv volts, the current flowing through the capacitor is

$$i = Cdv/dt$$

If the capacitor is connected across an ideal operational amplifier, as shown in Figure 15.7(b), the current flowing in R must also flow in C (remember, the input current to the amplifier itself is practically zero). It therefore follows that, since the voltage at junction J is zero (or nearly so), then

$$i = CdV_0/dt$$

or

$$dV_0 = -\frac{1}{C} i \, dt$$

Integrating both sides of the equation gives

$$V_0 = \frac{1}{C}\int i \, dt$$

In the case of Figure 15.7

$$i = \frac{V_1 - V_{\text{diff}}}{R} \approx \frac{V_1}{R}$$

Since current i flows into the phase inverting terminal of the op-amp, then output V_0 must be negative, hence

$$V_0 = -\frac{1}{C}\int \frac{V_1}{R} \, dt = -\frac{1}{CR}\int V_1 \, dt$$

That is, the output voltage is the time integral of the input voltage.

This suggests that if V_1 is a constant voltage, the output voltage will fall at a constant rate of $1/CR$ V/s. That is, if $C = 1$ μF and $R = 1$ MΩ, then

$$1/CR = 1/(10^{-6} \times 10^{6}) = 1 \text{ V/s}$$

Unfortunately there is a limit to this since, if V_1 is constant, the amplifier output will very quickly reach saturation voltage!

Worked Example 15.6

An op-amp used in an integrator circuit has supply voltages of +10 V and −10 V. If $R = 1\ \text{M}\Omega, C = 0.5\ \mu\text{F}$ and the input d.c. voltage is +0.4 V, determine (a) the output voltage 3 s after the input signal is applied (assuming that the capacitor is initially discharged), (b) the time taken for the amplifier to saturate.

Solution

(a) Since the input voltage is constant, the output voltage changes at a constant rate of

$$-\frac{\text{input signal voltage}}{R_C} = -\frac{0.4}{1 \times 10^6 \times 0.5 \times 10^{-6}}$$
$$= -0.8\ \text{V/s}$$

After 3 s the output voltage is

$$\text{time} \times \text{rate of change of voltage} = 3 \times (-0.8)$$
$$= -2.4\ \text{V}$$

(b) The time taken for the output voltage to reach −10 V is

$$\frac{-10\ \text{V}}{\text{rate of change of output voltage}} = \frac{-10}{0.8} = 12.5\ \text{s.}$$

15.8 Computer listings

Two computer listings are included here. Listing 15.1 is for a phase inverting summing amplifier of the type in Figure 15.3, having up to four input resistors. Listing 15.2 deals with an integrator circuit which has a constant input voltage.

In the program in Listing 15.1, you are asked to supply the number of input resistors used, the input voltage applied to each resistor, the value of each input resistor and of the feedback resistor. The program evaluates the current in each resistor and, in line 170, it totals the current in the resistors. Finally, the program gives all the information originally supplied, together with the output voltage.

Listing 15.2 deals with a basic integrator circuit with a constant voltage applied to the input. You are asked to supply the value of R and C in the integrator circuit, together with the input signal voltage and the supply voltage to the op-amp. You must also give the period over which the signal voltage is to be integrated. The integration period is divided into ten equal intervals, and the program gives the output voltage over the integration period. If the amplifier output voltage exceeds the op-amp supply voltage, you are given a warning that the amplifier has saturated.

Listing 15.1 Phase inverting summing amplifier

```
10 CLS '**** CH 15-1 ****
20 DIM D(12): FOR Y = 1 TO 12: D(Y) = 0: NEXT Y: I = 0
30 PRINT "PHASE INVERTING SUMMING AMPLIFIER": PRINT
40 PRINT "The summing amplifier has a feedback resistor, Rf,"
50 PRINT "and up to 4 input resistors. The open-loop gain of the"
60 PRINT "amplifier is infinity, and the input resistance is"
70 PRINT "infinity.": PRINT
80 INPUT "Value of Rf (Megohms) = ", Rf: PRINT
90 INPUT "Number of input resistors (max. 4) = ", N
100 IF N < 1 OR N > 4 THEN CLS: GOTO 90
110 PRINT
120 FOR Y = 1 TO N
130     PRINT TAB(5); "For resistor "; Y
140     INPUT "value (Megohms) = ", D(Y)
150     INPUT "input voltage (V) = ", D(Y + 4)
160     D(Y + 8) = D(Y + 4) / D(Y) '** current in resistor **
170     I = I + D(Y + 8) '** sum of currents **
180 NEXT Y
190 CLS
200 PRINT "Rf = "; Rf; "Megohms": PRINT
210 PRINT "Resistor No."; TAB(15); "Value (Megohms)"; TAB(40); "Voltage"
220 FOR Y = 1 TO N
230     PRINT Y; TAB(15); D(Y); TAB(40); D(Y + 4)
240 NEXT Y
250 PRINT : PRINT : Vout = -Rf * I
260 PRINT "Output voltage = "; Vout; "V": PRINT
270 PRINT "NOTE: If the calculated value of the ouput voltage exceeds"
280 PRINT "      the supply voltage to the amplifier, the amplifier"
290 PRINT "      output will saturate."
300 END
```

Listing 15.2 Operational amplifier integrator

```
10 CLS '**** CH 15-2 ****
20 PRINT "OPERATIONAL AMPLIFIER INTEGRATOR": PRINT
30 INPUT "Input resistor R (Megohms) = ", R
40 INPUT "Feedback capacitor C (microfarads) = ", C
50 INPUT "Input voltage Vi (V) = ", Vi: PRINT
60 INPUT "Amplifier D.C. supply voltage = ", Vs: PRINT
70 slope = -Vi / (R * C)
80 PRINT "NOTE: The amplifier will saturate after a time of"
90 PRINT TAB(5); ABS(Vs / slope); "seconds.": PRINT
100 INPUT "Length of integration period (sec) = ", T
110 CLS
```

```
120 PRINT "R (Megohm) = "; R; ", C (microfarad) = "; C
130 PRINT "Input voltage (V) ="; Vi; ", supply voltage (V) = "; Vs
140 PRINT "Integration period (sec) = "; T: dt = T / 10: PRINT
150 PRINT "Time (s)"; TAB(15); "Output voltage (V)"
160 FOR Y = 0 TO 10
170     PRINT Y * dt; TAB(15); slope * (Y * dt)
180 NEXT Y
190 IF ABS(slope * (Y * dt)) > Vs THEN PRINT : PRINT "Amplifier has saturated"
200 END
```

Exercises

15.1 An inverting amplifier of the type in Figure 15.2 has $R_1 = 27\ \text{k}\Omega$ and $R_f = 560\ \text{k}\Omega$. Calculate the overall voltage gain.

15.2 If the resistor in Exercise 15.1 have a 5 per cent tolerance, what is (a) the maximum gain, (b) the minimum gain?

15.3 Determine the output voltage from an inverting summing amplifier of the type in Figure 15.3 if $V_1 = 1.2$ V, $V_2 = -0.8$ V, $V_3 = 0.5$ V, $R_1 = 82\ \text{k}\Omega$, $R_2 = 15\ \text{k}\Omega$, $R_3 = 27\ \text{k}\Omega$ and $R_f = 47\ \text{k}\Omega$.

15.4 A 3-input inverting summing amplifier of the type in Figure 15.3 has an output voltage of 0.667 V. If $V_1 = 0.2$ V, $V_2 = 0.4$ V, $R_1 = 10\ \text{k}\Omega$, $R_2 = 20\ \text{k}\Omega$, $R_3 = 15\ \text{k}\Omega$ and $R_f = 100\ \text{k}\Omega$, determine the value of V_3.

15.5 A non-inverting amplifier of the type in Figure 15.4 has the following values: $R_1 = 10\ \text{k}\Omega$, $R_f = 100\ \text{k}\Omega$. What is the overall voltage gain of the amplifier?

15.6 Given that the tolerance of the resistors in Example 15.5 is 5 per cent, determine (a) the maximum, (b) the minimum gain of the amplifier.

15.7 If a non-inverting amplifier of the type in Figure 15.4 has an input voltage of 1 V and an output voltage of 8 V, and $R_f = 80\ \text{k}\Omega$, what is the value of R_1?

15.8 A differential amplifier of the type in Figure 15.5 has an output voltage of 8 V. If R_2 is 100 kΩ and the magnitude of the differential input voltage is 2 V, calculate the value of R_1.

15.9 An ideal integrator of the type in Figure 15.7 has an input resistor of 100 kΩ, and an integrating capacitor of 0.1 μF. If the input voltage is 15 mV, determine the output voltage after a time of (a) 3 s, (b) 5 s.

If the amplifier is supplied at ±10 V, how long will it take for the amplifier to saturate?

15.10 An ideal integrator has an output voltage of −2.553 V after a period of 10 s (the capacitor being initially uncharged). If the input voltage is 12 mV, and the integrating capacitor is 0.1 μF, what is the value of the resistor in the integrator circuit?

15.11 Prove that the magnitude of the voltage gain of the amplifier in Figure 15.6 is

$$\frac{R_4}{R_3}\left[1+\frac{2R_2}{R_1}\right].$$

Summary of important facts

Operational amplifiers (*op-amps*) are key components in the design of electronic circuits. An op-amp amplifies the signal between its two input terminals, namely the **inverting input** and the **non-inverting input** terminal. The inverting input is so named because the output signal is of opposite polarity or opposite phase angle to the voltage applied to that input terminal.

An op-amp has several important features, namely it has a very high input resistance between its input terminals (known as the *differential input resistance*), so that it draws very little input current, and its forward gain is very high. An *ideal op-amp* has infinite input resistance, and infinite forward gain.

Typical circuits using the op-amp include

1. an inverting amplifier (section 15.3)
2. an inverting summing amplifier (section 15.4)
3. a non-inverting amplifier (section 15.5)
4. a differential input (instrumentation) amplifier (section 15.6)
5. an integrator (section 15.7).

Using the above circuits, many configurations are possible.

16 Oscillators

16.1 Introduction

Electronic oscillators are a group of *positive feedback amplifier circuits*, in which a proportion of the output signal is fed back and added to the input signal. In fact, the signal which is fed back forms the only input to the oscillator! In this way, the output oscillation is self-sustaining.

There is a group of oscillators which produce a *sinusoidal oscillation*, others producing a *non-sinusoidal output*. Included in the latter are square (rectangular), pulse and triangular (ramp) wave generators. By the end of this chapter, the reader will be able to

- Calculate the frequency of a wide range of sinusoidal oscillator.
- Calculate the frequency of an astable multivibrator.
- Solve problems involving op-amp oscillators.
- Use software to calculate the frequency of oscillators.

16.2 General principles of oscillators

Sinusoidal oscillators usually (but not always) contain inductors and capacitors connected so that the circuit resonates at a particular frequency. Some sinusoidal oscillators contain resistors and capacitors connected in a way to control the frequency of oscillation.

Oscillators are positive feedback circuits, which are arranged so that the *overall loop gain of the circuit is unity under oscillatory conditions.*

A **relaxation oscillator** is a circuit which generates **non-sinusoidal waveforms**. The timing elements are resistors and capacitors, and the periodic time of the wave is controlled by R and C. The circuit gradually charges the capacitors through the resistors, and then rapidly discharges them.

16.3 R-C phase-shift oscillators

A basic form of *discrete component* ***R-C*** **phase-shift oscillator** is shown in Figure 16.1. Components R_{B1}, R_{B2}, R_E and C_E are part of the bias circuit, which ensure that the transistor is biased to the correct operating point.

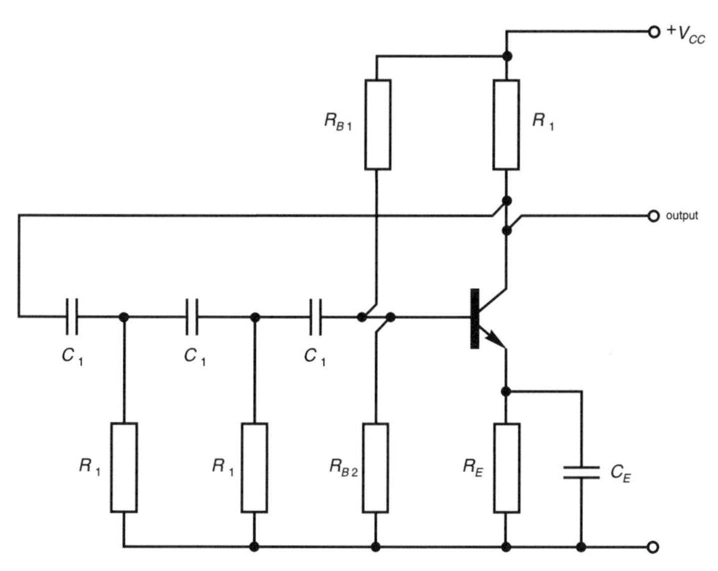

Figure 16.1 Discrete component R-C phase-shift oscillator

The components which control the frequency of oscillation are the three resistors shown as R_1, and the three capacitors shown as C_1. The theoretical frequency of oscillation is

$$f_O = \frac{1}{2\pi R_1 C_1 \sqrt{6}} \text{ Hz}$$

To provide the correct gain for sustained oscillations, the *current gain of the transistor must be at least 29*, but to ensure that oscillation will take place under all operating conditions, the current gain should be at least 60.

The way in which this circuit provides positive feedback is as follows: Each of the R_1C_1 circuits provides 60° phase shift at the oscillatory frequency to the signal fed back to the input, so that the overall phase shift is 180°. This, in addition to the 180° phase shift introduced by the transistor, ensures that the sinusoidal base current is in phase with the sinusoidal collector current.

Worked Example 16.1

An *R-C* phase shift oscillator of the type in Figure 16.1 operates at a frequency of 3 kHz. If $C_1 = 0.01\ \mu\text{F}$, what is the value of R_1 used in the circuit?

Solution

From the equation $f_O = 1/(2\pi R_1 C_1 \sqrt{6})$, it follows that

$$\begin{aligned} R_1 &= 1/(2\pi f_O C_1 \sqrt{6}) \\ &= 1/(2\pi \times 3000 \times 0.01 \times 10^{-6} \times \sqrt{6}) \\ &= 2166\ \Omega \text{ or } 2.166\ \text{k}\Omega. \end{aligned}$$

16.4 The Wein-bridge oscillator

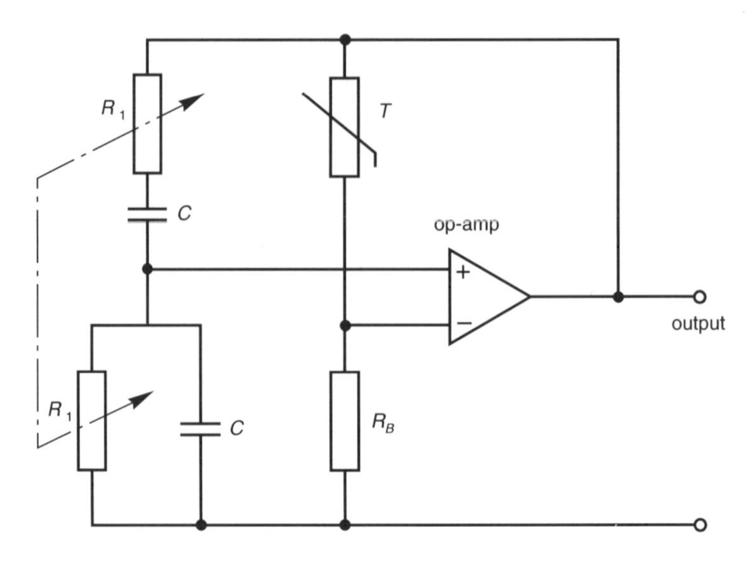

Figure 16.2 An op-amp Wien-bridge oscillator

The resistance-capacitance circuit used in this oscillator (the series R_1C, and the parallel R_1C circuits) is very similar to the electrical bridge circuit known as the *Wein bridge* and, for this reason, it is given the name the **Wein-bridge oscillator**.

Under oscillatory conditions, the voltage at the non-inverting input of the op-amp is in phase with the output voltage, giving rise to sustained oscillations.

The function of *thermistor* T, and resistor R_B, is to provide oscillations with a very stable amplitude. If the amplitude of the output voltage varies, the resistance of T changes, thereby altering the proportion of the output signal applied to the phase inverting input. This has the effect of controlling the amplitude of the output signal.

By using a pair of *ganged resistors* R_1, the frequency can be altered smoothly over a wide range. Alternatively, the frequency range can be altered in steps by simultaneously changing the capacitance of the capacitors C. The frequency of oscillation is

$$fo = \frac{1}{2\pi R_1 C}\ \text{Hz}$$

Worked Example 16.2

What is the oscillatory frequency of a Wein-bridge oscillator of the type in Figure 16.2 if $R_1 = 20\ \text{k}\Omega$ and $C = 0.05\ \mu\text{F}$?

Solution

The oscillatory frequency is

$$f_O = 1/(2\pi R_1 C) = 1/(2\pi \times 20 \times 10^3 \times 0.05 \times 10^{-6})$$
$$= 159.1 \text{ Hz}$$

16.5 The twin-T oscillator

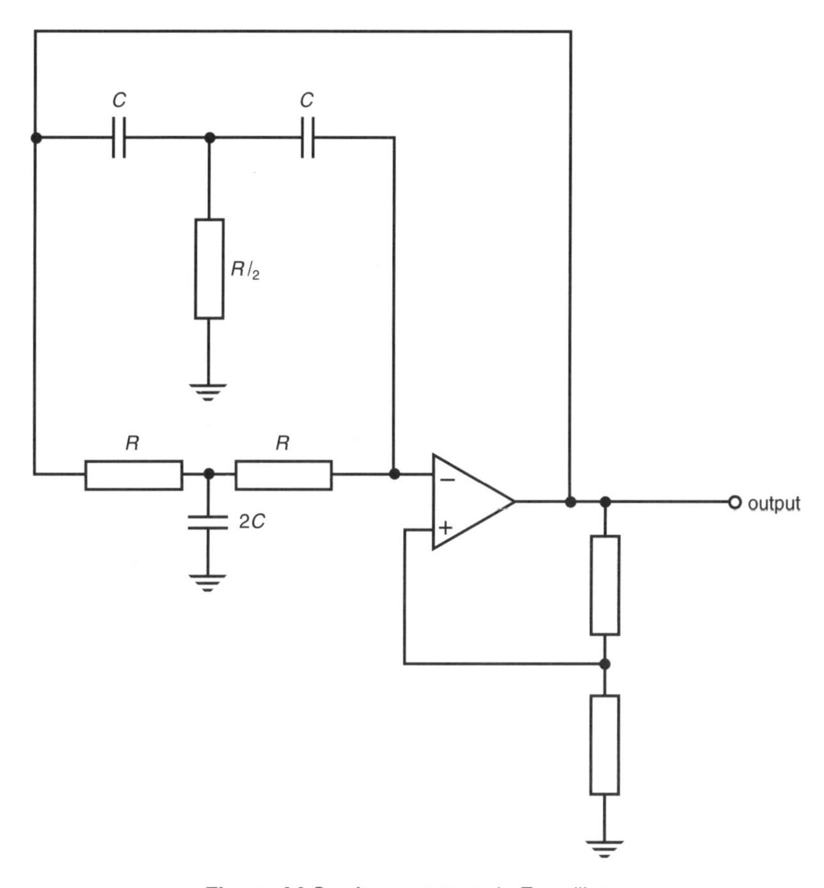

Figure 16.3 An op-amp twin-T oscillator

The basis of an **op-amp twin-T oscillator** is shown in Figure 16.3, and the principle is generally similar to that of the Wein-bridge oscillator. The frequency of operation is

$$f_0 = \frac{1}{2\pi RC} \text{ Hz}$$

16.6 The Colpitts oscillator

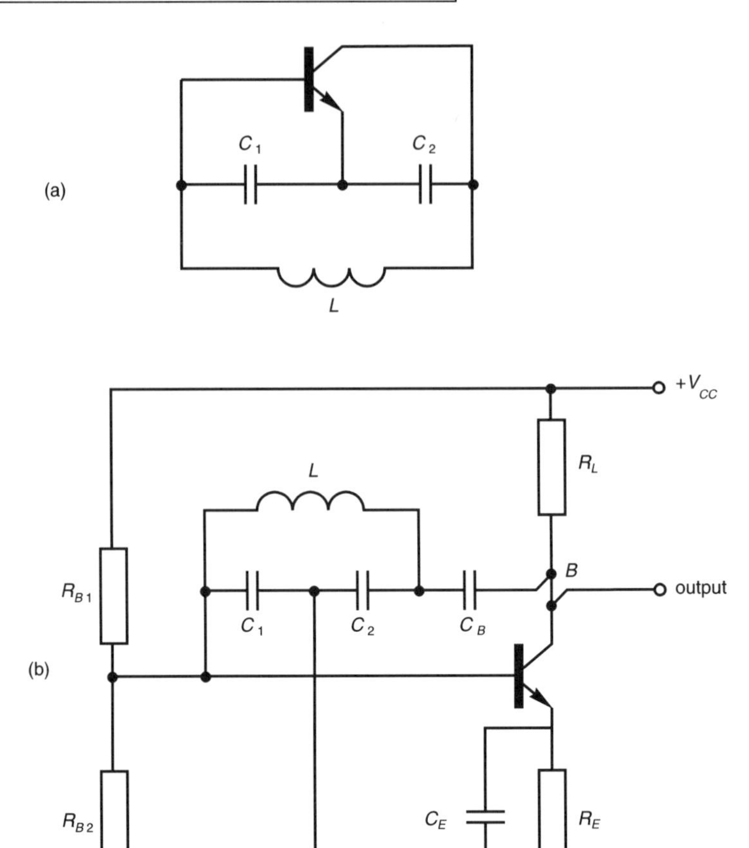

Figure 16.4 (a) Basis of the Colpitts oscillator (b) a practical form of discrete component circuit

The **Colpitts oscillator**, shown in Figure 16.4, is an *L-C* oscillator capable of producing audio- and radio-frequency oscillations. The basis of the circuit is shown in Figure 16.4(a), in which the resonant circuit comprises L in parallel with the series combination of C_1 and C_2. The resonant frequency of the circuit is

$$\omega_0 = \sqrt{\left[\frac{1}{L}\left[\frac{1}{C_1} + \frac{1}{C_2}\right]\right]} \text{ rad/s}$$

and

$$f_0 = \omega_0/2\pi \text{ Hz}$$

As a general rule, the reactance of C_1 at the resonant frequency is much less than that of C_2, so that

$$\omega_0 \approx 1/\surd(LC_2)$$

Control of the oscillatory frequency is primarily provided by C_2 and, for sustained oscillations to occur, the ratio C_2/C_1 must be greater than the current gain of the transistor. Altering the value of C_1 effectively controls the magnitude of the output voltage.

A practical discrete component circuit is shown in Figure 16.4(b). In addition to the capacitors C_1 and C_2, and inductor L, the following components are required:

R_L – needed to develop the output voltage
R_{B1}, R_{B2}, C_E and R_E – bias and thermal stability components
C_B – blocking capacitor which has a low reactance at the oscillatory frequency.

Worked Example 16.3

A Colpitts oscillator of the type in Figure 16.4(b) has in its resonant circuit an inductor of 20 μH, $C_1 = 10$ nF and $C_2 = 250$ pF. Determine the frequency of oscillation.

Solution

The oscillatory frequency is

$$\omega_O = \sqrt{\left[\frac{1}{L}\left[\frac{1}{C_1} + \frac{1}{C_2}\right]\right]}$$

$$= \sqrt{\left[\frac{1}{2 \times 10^{-6}}\left[\frac{1}{10 \times 10^{-9}} + \frac{1}{250 \times 10^{-12}}\right]\right]}$$

$$= \sqrt{(2.05 \times 10^{15})}$$

$$= 45.28 \times 10^6 \text{ rad/s} \equiv 7.206 \text{ MHz}$$

(**Note**: The simplified solution is

$$\omega_0 = 1/\sqrt{(LC_2)} = 44.72 \times 10^6 \text{ rad/s} \equiv 7.12 \text{ MHz.})$$

16.7 The Hartley oscillator

The basis of the **Hartley oscillator** is shown in Figure 16.5(a), in which the resonant circuit is the pair of series-connected inductors L_1 and L_2, which is in parallel with C. The two inductors may, or may not, be mutually coupled by mutual inductance M. If no mutual coupling exists, the resonant frequency of the circuit is

$$\omega_0 = 1/\sqrt{((L_1 + L_2)C)} \text{ rad/s}$$

and $f_0 = \omega_0/2\pi$ Hz.

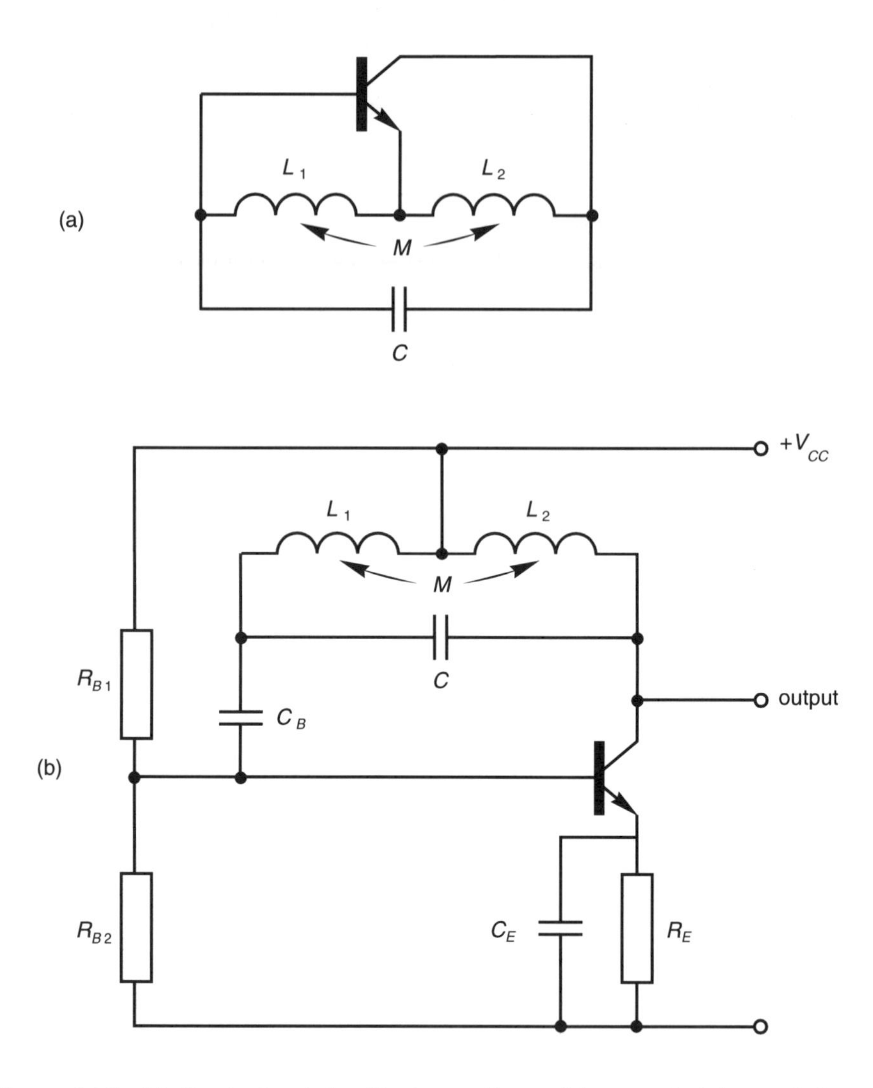

Figure 16.5 (a) The basis of the Hartley oscillator; (b) a practical form of discrete component circuit

The condition for oscillations to be sustained in the transistor circuit is $h_{fe} > L_1/L_2$. In the event that mutual coupling, M, exists between the coils, the resonant frequency is

$$\omega_0 = 1/\sqrt{((L_1 + L_2 + 2M)C)} \text{ rad/s}$$

A practical discrete component version of the circuit is shown in Figure 16.5(b). In addition to the elements in the resonant circuit, the following are required:

R_{B1}, R_{B2}, R_E, C_E – bias and thermal stability components,
C_B – blocking capacitor of low reactance at the oscillatory frequency.

The reader should note that L_1 is connected between V_{CC} and C_B. For sustained oscillations to occur, the current gain of the transistor should be greater than $(L_1 + M)/(L_2 + M)$.

Worked Example 16.4

A Hartley oscillator of the type in Figure 16.5(b) has an oscillatory frequency of 190.2 kHz. If $L_1 = 400$ μH and $L_2 = 1000$ μH, and no mutual inductance exists between them, determine the value of C.

Solution

The equation for the resonant frequency is

$$\omega_0 = 1/\sqrt{((L_1 + L_2)C)}$$

Transposing for C gives

$$\begin{aligned} C &= 1/((L_1 + L_2)\omega_0^2) \\ &= 1/\left((400 \times 10^{-6} + 1000 \times 10^{-6})(2\pi \times 190.2 \times 10^{-3})^2\right). \\ &= 500 \times 10^{-12} F \text{ or } 500 \text{ pF.} \end{aligned}$$

16.8 The tuned collector oscillator

A practical form of discrete component **tuned collector oscillator** is shown in Figure 16.6, in which L and C form the oscillatory elements, and the frequency of oscillation is

$$\omega_0 = 1/\sqrt{(LC)} \text{ rad/s}$$

The output from the circuit is taken from a winding which is magnetically coupled to L, as is the feedback signal to the base circuit. The other elements in the circuit are as follows:

$R_{B1}, R_{B2}, R_{E2}, C_E$ – bias and thermal stability components,
R_{E1} – a resistor which controls the initial gain.

The initial loop gain of the circuit should lie between about 2 and 5, and R_{E1} is included in the emitter circuit to give some negative feedback to reduce the gain to a reasonable value.

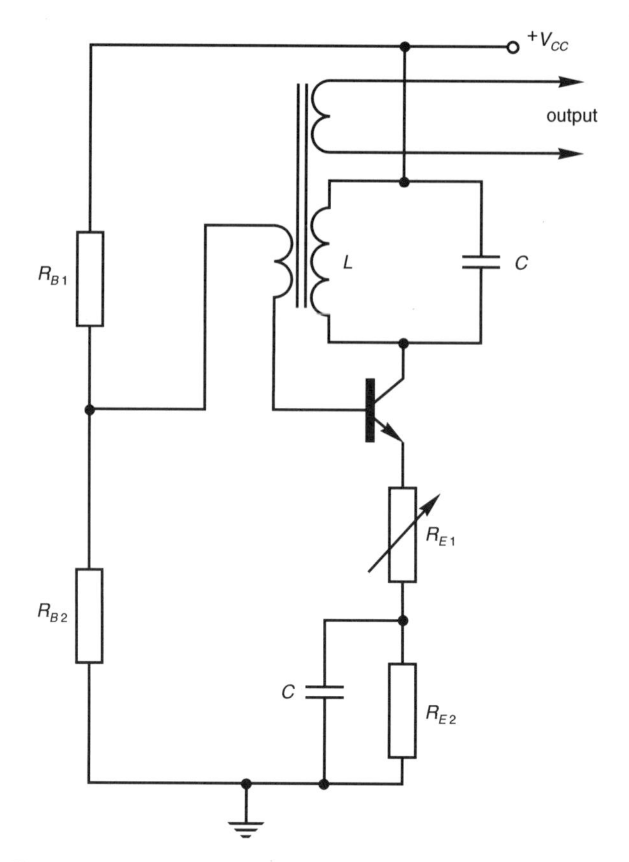

Figure 16.6 A discrete component tuned collector oscillator

16.9 The astable multivibrator

There are many forms of astable multivibrator circuit, and we look in detail at one of the most popular, which is shown in Figure 16.7. This circuit generates 'square waves' (or nearly so) at outputs V_{O1} and V_{O2}. The pulse period (in s) of the square wave V_{O1} is

$$T_1 \approx 0.7 R_4 C_2$$

and the pulse period of V_{O2} is

$$T_2 \approx 0.7 R_3 C_1$$

The ratio T_1/T_2 is known as the **mark-to-space ratio** of the waveform. The periodic time of the output wave is

$$T = T_1 + T_2$$

and its frequency is

$$f = 1/T \text{ Hz}$$

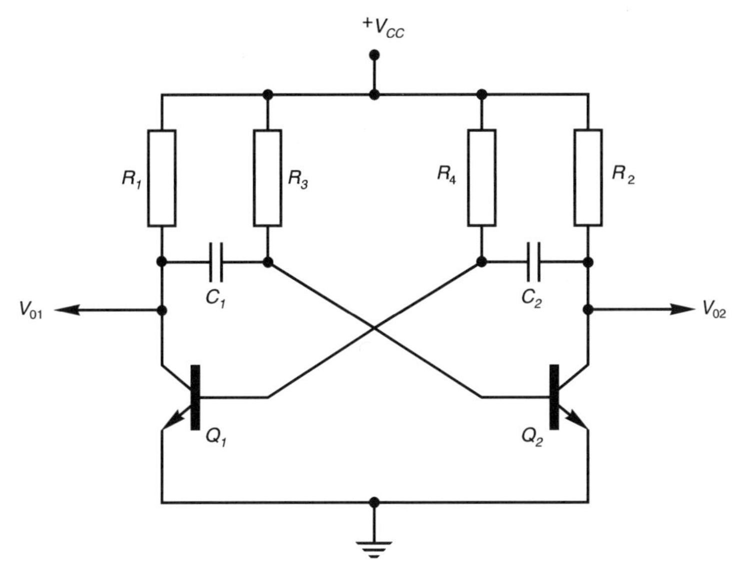

Figure 16.7 A popular astable multivibrator circuit

If $R_3 = R_4$ and $C_1 = C_2$, the mark-to-space ratio is unity, and the output signal at V_{O1} and V_{O2} is a square wave. In this case the periodic time of the wave is

$$T = 0.7R_4C_2 + 0.7R_3C_1 = 1.4R_4C_2$$

and

$$f = 1/T = 1/(1.4R_4C_2)\ \text{Hz}$$

In this case $T_1 = T_2$, and the oscillator is known as a **symmetrical astable multivibrator**.

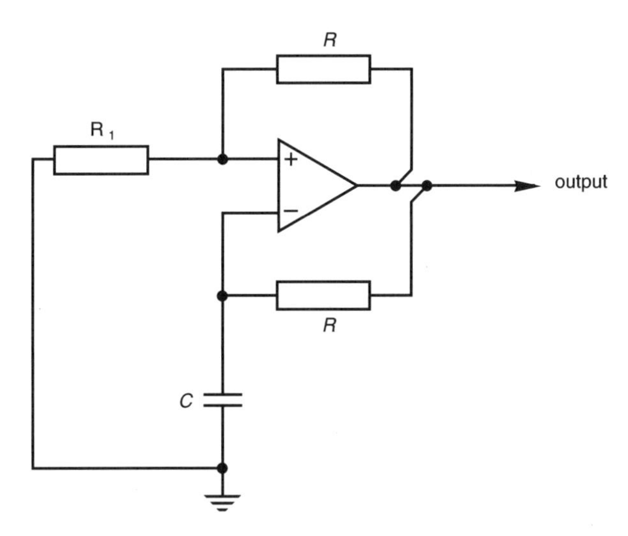

Figure 16.8 An op-amp symmetrical astable multivibrator

The circuit of an op-amp symmetrical multivibrator is shown in Figure 16.8, in which the lower *R-C* circuit is the effective timing circuit of the oscillator.

Worked Example 16.5

A symmetrical astable multivibrator in Figure 16.7 uses $R_3 = R_4 = 18\ \text{k}\Omega$ and $C_1 = C_2 = 0.01\ \mu\text{F}$. What is the frequency of oscillation, and the periodic time of the output wave.

Solution

Since the output will be a true square wave, then

$$f = 1/(1.4R_C)$$
$$= 1/\left(1.4 \times 18 \times 10^3 \times 0.01 \times 10^{-6}\right) = 3968\ \text{Hz}$$

and

$$T = 1/f = 1/3968 = 0.00025\ \text{s or } 0.25\ \text{ms}.$$

16.10 Computer listing

Listing 16.1 Oscillation frequency of popular oscillator circuits

```
10 CLS '**** CH16-1 ****
20 PI = 3.14159: R = 0: C = 0: L = 0: M = 0
30 PRINT "OSCILLATION FREQUENCY OF POPULAR OSCILLATOR CIRCUITS"
40 PRINT : PRINT "Select ONE of the following:": PRINT
50 PRINT " 1. R-C phase-shift oscillator.": PRINT
60 PRINT " 2. Wein Bridge and Twin-T oscillators.": PRINT
70 PRINT " 3. Colpitts oscillator.": PRINT
80 PRINT " 4. Hartley oscillator.": PRINT
90 PRINT " 5. Tuned Collector oscillator.": PRINT
100 PRINT " 6. Symmetrical Astable Multivibrator.": PRINT
110 INPUT "Enter your selection here (1 - 6): ", S
120 IF S < 1 OR S > 6 THEN GOTO 10
130 CLS
140 IF S = 2 THEN GOTO 230
150 IF S = 3 THEN GOTO 270
160 IF S = 4 THEN GOTO 390
170 IF S = 5 THEN GOTO 510
180 IF S = 6 THEN GOTO 550
```

```
190 PRINT "*** R-C phase-shift oscillator ***": PRINT
200 INPUT "R (ohms) = ", R
210 INPUT "C (F) = ", C
220 w = 1 / (C * R * SQR(6)): GOTO 590
230 PRINT "*** Wein Bridge and Twin-T oscillators ***": PRINT
240 INPUT "R (ohms) = ", R
250 INPUT "C (F) = ", C
260 w = 1 / (R * C): GOTO 590
270 PRINT "*** Colpitts oscillator ***"
280 PRINT "In the Colpitts oscillator, the frequency is primarily"
290 PRINT "controlled by capacitor C2, which is connected between"
300 PRINT "the collector and the emitter of the transistor."
310 PRINT : INPUT "Capacitance C2 (F) = ", C2
320 INPUT "Other capacitance C1 (F) = ", C1
330 INPUT "Inductance L (H) = ", L
340 IF C1 = 0 THEN w = SQR(1/(C2 * L)): GOTO 380
350 w = SQR(((1 / C1) + (1 / C2)) / L)
360 PRINT : PRINT "For sustained oscillations to occur"
370 PRINT "then hfe > "; C2 / C1
380 GOTO 590
390 PRINT "*** Hartley oscillator ***": PRINT
400 PRINT "The Hartley oscillator has two inductors L1 and"
410 PRINT "L2, where L2 is in the collector circuit, which may"
420 PRINT "or may not be mutually coupled by"
430 PRINT "mutual inductance M.": PRINT
440 INPUT "L1 (H) = ", L1
450 INPUT "L2 (H) = ", L2
460 INPUT "Mutual inductance M (H) = ", M
470 INPUT "Capacitance C (F) = ", C
480 w = 1 / SQR((L1 + L2 + (2 * M)) * C): PRINT
490 PRINT "NOTE: For sustained oscillations to occur"
500 PRINT "then hfe > "; (L1 + M) / (L2 + M): GOTO 590
510 PRINT "*** Tuned Collector oscillator ***": PRINT
520 INPUT "Inductance L (H) = ", L
530 INPUT "Capacitance C (F) = ", C
540 w = 1 / SQR(L * C): GOTO 590
550 PRINT "*** SYMMETRICAL ASTABLE MULTIVIBRATOR ***": PRINT
560 INPUT "Resistor R (ohms) = ", R
570 INPUT "Capacitor C (F) = ", C
580 w = 2 * PI / (1.4 * R * C)
590 PRINT
600 PRINT "Frequency = "; w / (2 * PI); "Hz"
610 PRINT "          = "; w; "rad/s"
620 PRINT "Periodic time = "; 2 * PI / w; "seconds"
630 PRINT "              = "; 2000 * PI / w; "milliseconds"
650 END
```

Computer listing 16.1 is a menu-driven package which determines the frequency of

1. an R-C phase-shift oscillator (lines 190 to 220)
2. Wein-bridge and Twin-T oscillators (lines 230 to 260)
3. a Colpitts oscillator (lines 270 to 380)
4. a Hartley oscillator (lines 390 to 500)
5. a Tuned Collector oscillator (lines 510 to 540)
6. a Symmetrical Astable Multivibrator (lines 550 to 580).

The program uses the equations given in this chapter, and lines 590 to 630 print the resulting frequency and periodic time of oscillation. As an aside, the program also determines the minimum value of current gain of the transistor for both the Colpitts and Hartley oscillators.

Exercises

16.1 Determine the oscillatory frequency of an *R-C* phase-shift oscillator of the type in Figure 16.1 if $R_1 = 4.7$ kΩ and $C_1 = 0.001$ μF.

16.2 If the frequency of oscillation of an *R-C* phase-shift oscillator is 196.9 Hz, and $R_1 = 3.3$ kΩ, calculate the value of C_1.

16.3 What is the oscillatory frequency of a Wein-bridge oscillator of the type in Figure 16.2 if $R_1 = 10$ kΩ and $C = 0.01$ μF?

16.4 A twin-T oscillator using $R = 15$ kΩ produces a frequency of 10.6 kHz. What is the value of C?

16.5 In a Colpitts oscillator, $C_1 = C_2 = 500$ pF and $L = 15$ μH. Determine the frequency of oscillation and the minimum value of h_{fe} required for the transistor.

16.6 If the frequency of a Colpitts oscillator is 5.68 MHz, and $L = 4$ μH, and $C_2 = 200$ pF, determine the value of C_1 (see Figure 16.4).

16.7 A Hartley oscillator produces a frequency of 225.1 kHz. If $L_1 = L_2 = 500$ μH (see Figure 16.5), and there is no mutual coupling between the inductors, determine the value of C.

16.8 A Hartley oscillator uses the following components: $L_1 = 250$ μH, $L_2 = 1200$ μH (there being no mutual coupling), and $C = 1000$ pF, calculate the oscillatory frequency and the minimum value of h_{fe} of the transistor in the circuit.

16.9 A tuned collector oscillator uses $L = 1000$ μH and $= 10$ nF. What is the frequency of oscillation?

16.10 A symmetrical astable multivibrator of the type in Figure 16.7 uses the following components: $R_3 = R_4 = 10$ kΩ, and $C_1 = C_2 = 0.01$ μF. Determine the frequency of oscillation.

16.11 An astable multivibrator contains the following components: $R_3 = 100$ kΩ, $R_4 = 47$ kΩ, $C_1 = 0.01$ μF and $C_2 = 0.05$ μF. Calculate the periodic time of the output wave, and the mark-to-space ratio of the wave.

Summary of important facts

The output signal from an oscillator may either be **sinusoidal** or **non-sinusoidal**. The latter are generally required for special purposes such as for timing and pulse generators, or for oscilloscope time bases, etc.

Oscillators are **positive feedback amplifiers**, in which either a part or all of the output signal is fed back to the input. The signal fed back is sufficient to maintain the output in an oscillatory state. That is, the **loop gain of the amplifier is unity**.

Many oscillators contain resistors and capacitors (***R-C*** **oscillators**) in the feedback circuits, and others contain inductors and capacitors (***L-C*** **oscillators**).

The ***R-C*** **phase-shift oscillator** (Figure 16.1) has an oscillatory frequency of

$$f_0 = 1/(2\pi R_1 C_1 \sqrt{6})\ \text{Hz}$$

and the current gain of the transistor used in the circuit must be at least 29.

The **Wein-bridge oscillator** (Figure 16.2) is an *R-C* oscillator, and has an oscillatory frequency of

$$f_0 = 1/(2\pi R_1 C)\ \text{Hz}$$

The **Twin-T oscillator** (Figure 16.3) is yet another *R-C* oscillator, and is generally similar to the Wein-bridge oscillator, which oscillates at

$$f_0 = 1/(2\pi RC)\ \text{Hz}$$

The **Colpitts oscillator** (Figure 16.4) is an *L-C* oscillator, and its angular oscillatory frequency is

$$\omega_0 = \sqrt{\left[\frac{1}{L}\left[\frac{1}{C1} + \frac{1}{C_2}\right]\right]}\ \text{rad/s}$$

This oscillator (and the Hartley oscillator) is capable of producing very high frequencies.

The **Hartley oscillator** (Figure 16.5) is an *L-C* oscillator whose oscillatory frequency is

$$\omega_0 = 1/\sqrt{((L_1 + L_2)C)}\ \text{rad/s}$$

If the inductors are mutually coupled, the oscillatory frequency is

$$\omega_0 = 1/\sqrt{((L_1 + L_2 + 2M)C)}\ \text{rad/s}$$

Yet another *L-C* oscillator is the **tuned collector oscillator** (Figure 16.6), in which the feedback signal is obtained by mutual coupling between the collector circuit and the base circuit. Its frequency of oscillation is

$$\omega_0 = 1/\sqrt{(LC)}\ \text{rad/s}$$

The **astable multivibrator** (Figure 16.7) is an *R-C* **relaxation oscillator**, which generates square waves (or nearly so) by charging a capacitor through a resistor, and then quickly discharging it. For the circuit in Figure 16.7, the periodic time of each pulse controlled by each transistor is

$$T_1 = 0.7R_4C_4 \qquad T_2 = 0.7R_3C_1$$

and the periodic time of the output square wave is

$$T = T_1 + T_2 = 0.7(R_4C_4 + R_3C_1)$$

The **mark-to-space ratio** (or on-to-off ratio) of the square wave is given by T_1/T_2.

If the mark-to-space ratio is unity, then the output is a **symmetrical square wave** (and the circuit is described as a *symmetrical multivibrator*). If the mark-to-space ratio is not unity, then the circuit is described as an *unsymmetrical multivibrator*.

17 Attenuators and filters

17.1 Introduction

An **attenuator** reduces the magnitude of the voltage or current entering its terminals by the a constant amount *at all frequencies*, and is usually constructed using pure resistors. Attenuators (or attenuator pads) are used, for example, at the input of an electronic instrument in order to reduce a voltage or current to a value which can be handled by the instrument.

A **filter** is a circuit which attenuates certain frequencies, whilst allowing others to pass with very little attenuation.

Whilst we can specify the numerical amount of attenuation introduced by an attenuator or filter, the attenuation is usually expressed in **decibels** (dB), which is a logarithmic ratio. By the end of this chapter, the reader will be able to

- Convert a numerical 'gain' or 'attenuation' into its decibel value, and vice versa.
- Understand the terminology associated with attenuators and filters.
- Design passive attenuator circuits.
- Design passive and active *R-C* filter networks.
- Use software to deal with calculations relating to conversion between. numerical and decibel values, attenuator calculations, together with T-to-π and π-to-T resistive attenuators.

17.2 The decibel

The **Bel** (unit symbol B) is the ratio of two **power values**, P_1 and P_2 as follows:

$$\text{Power ratio in bels} = \log_{10}\frac{P_2}{P_1}$$

The bel is an inconveniently large unit and, in practice, we use the **decibel** (dB), where

$$1\text{ B} = 10\text{ dB or }1\text{ dB} = 0.1\text{ B}$$

If two power values, P_1 and P_2, differ by n dB, then

$$n = 10 \log_{10} \frac{P_2}{P_1} \text{ dB}$$

or the power ratio P_2/P_1 is

$$\frac{P_2}{P_1} = 10^{n/10}$$

Since we are dealing with a logarithmic ratio, then

if $P_2 > P_1$, the dB ratio is positive
if $P_2 = P_1$, the dB ratio is zero
if $P_2 < P_1$, the dB ratio is negative

That is, a reduction in the power value , i.e. an *attenuation*, results in a *negative dB ratio*.

Worked Example 17.1

If the output power from, and input power to, an electronic system are respectively (a) 2 W and 0.1 W, (b) 10 mW and 200 mW, (c) 100 W and 0.1 kW, calculate the power ratio in decibels.

Solution

(a) In this case $P_2 = 2$ W and $P_1 = 0.1$ W, then

$$n = 10 \log_{10} \frac{P_2}{P_1} = 10 \log_{10} = \frac{2}{0.1} 10 \log_{10} 20$$
$$= 13.01 \text{ dB}$$

That is, there is a 'gain' of 13.01 dB.

(b) Here $P_2 = 10 \text{ mW} = 10 \times 10^{-3}$ W and $P_1 = 200 \text{ mW} = 200 \times 10^{-3}$ W, hence

$$n = 10 \log_{10} \frac{P_2}{P_1} = 10 \log_{10} \frac{10 \times 10^{-3}}{200 \times 10^{-3}}$$
$$= 10 \log_{10} 0.05 = 10 \times (-1.301) = -13.01 \text{ dB}$$

That is, we have an *attenuation* of 13.01 dB.

(c) Since $P_1 = 0.1 \text{ kW} = 100$ W, the dB ratio is

$$n = 10 \log_{10} \frac{P_2}{P_1} = 10 \log_{10} \frac{100}{100} = 10 \log_{10} 1$$
$$= 0 \text{ dB}$$

That is the power 'gain' is zero.

Worked Example 17.2

If the dB ratio produced by a system is (a) 20 dB, (b) 0 dB, (c) −30 dB, determine the overall numerical power ratio.

Solution

(a) The overall numerical power ratio is

$$\frac{P_2}{P_1} = 10^{n/10} = 10^{20/10} = 10^2 = 100$$

That is the output power is 100 times greater than the input power

(b) The numerical ratio is

$$10^{n/10} = 10^{0/10} = 10^0 = 1$$

In this case the output power is equal to the input power.

(c) The numerical ratio in this case is

$$10^{n/10} = 10^{-30/10} = 10^{-3} = 0.001$$

or the output power is 0.1 per cent of the input power.

(a) Voltage and current decibel ratios

The power consumed in a resistor is

$$I^2R = \left[\frac{V}{R}\right]^2 \times R = V^2/R$$

If the voltages which produce power P_1 and P_2 in a resistor are V_1 and V_2, respectively, and if we can assume that P_1 and P_2 are consumed in an equal value of resistance, R, then the decibel ratio is

$$n = 10\log_{10}\frac{P_2}{P_1} = 10\log_{10}\frac{V_2{}^2/R}{V_1{}^2/R}$$

$$= 10\log_{10}\left[\frac{V_2}{V_1}\right]^2 = 20\log_{10}\frac{V_2}{V_1}\ \text{dB}$$

It may similarly be shown that the dB ratio of two currents is

$$n = 20\log_{10}\frac{I_2}{I_1}\ \text{dB}$$

It follows that

the numerical voltage ratio $= 10^{n/20}$

and

the numerical current ratio $= 10^{n/20}$

Table 17.1 Table of decibel values

Numerical ratio	*dB power ratio*	*dB voltage and current ratio*
20:1	+13	+26
16:1	+12	+24
10:1	+10	+20
8:1	+9	+18
4:1	+6	+12
2:1	+3	+6
1:1	0	0
0.5:1	−3	−6
0.25:1	−6	−12
0.125:1	−9	−18
0.1:1	−10	−20
0.0625:1	−12	−24
0.05:1	−13	−26

A table of dB power together with voltage and current ratios for selected numerical ratios are given in Table 17.1.

Worked Example 17.3

(a) If a circuit provides an attenuation of 10 dB, and the input voltage is 8 V, what is the output voltage?
(b) The output current from a circuit is 10 mA, and it has a gain of 5 dB, what is the input current?
(c) What is the numerical gain of the circuit in part (b)?

Solution

(a) The dB ratio of the attenuator is

$$n = 20\log_{10}(V_2/V_1)$$

and, since the 'gain' is −10 dB, then

$$-10 = 20\log_{10}(V_2/8)$$

or

$$-0.5 = \log_{10}(V_2/8)$$

hence

$$\frac{V_2}{8} = 10^{-0.5} = 0.316$$

therefore

$$V_2 = 8 \times 0.316 = 2.528 \text{ V}$$

(b) the current ratio of the circuit is

$$n = 20\log_{10}(I_2/I_1)$$

or

$$5 = 20\log_{10}(10 \times 10^{-3}/I_1)$$

That is

$$\frac{10 \times 10^{-3}}{I_1} = 10^{0.25} = 1.78$$

therefore

$$I_1 = 10 \times 10^{-3}/1.78 = 5.62 \times 10^{-3}\ \text{A or } 5.62\ \text{mA}.$$

(c) The numerical current gain is calculated from

$$\text{current gain} = I_2/I_1 = 10 \times 10^{-3}/5.62 \times 10^{-3} = 1.78$$

or it can be calculated as follows:

$$\text{current ratio} = 10^{n/10} = 10^{5/20} = 10^{0.25} = 1.78.$$

17.3 Using decibel ratios

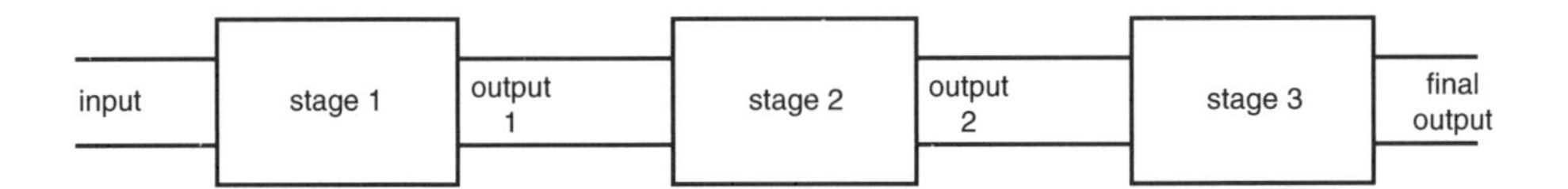

Figure 17.1 Cascaded stages

When several circuit stages are cascaded as shown in Figure 17.1, the output of stage 1 is 'multiplied' by the 'gain' of stage 2 before it is passed on, and so on. The net gain of the circuit is given by the product of the 'gain' of each of the stages.

Calculations are simplified if each stage gain is expressed in decibels, when the overall gain is the *sum of the individual gains*.

Worked Example 17.4

(a) A circuit has three stages having a decibel 'gain', respectively, of 10 dB, −6 dB and 4 dB. What is the overall dB voltage gain?
(b) What is the overall numerical gain of the circuit in part (a)?
(c) If the output voltage from the circuit in section (a) is 10 mV, what is the input voltage?

Solution

(a) The overall dB voltage gain is

$$n = 10 - 6 + 4 = 8 \text{ dB}$$

(b) The numerical gain of the circuits is

$$10^{n/20} = 10^{8/20} = 10^{0.4} = 2.51$$

(c) We can calculate the input voltage by either of the following methods:

(i) $$n = 20 \log_{10}(V_2/V_1)$$

or

$$8 = 20 \log_{10}(10 \times 10^{-3}/V_1)$$

therefore

$$\frac{10 \times 10^{-3}}{V_1} = 10^{8/20} = 10^{0.4} = 2.51$$

or $V_1 = 10 \times 10^{-3}/2.51 = 3.98 \times 10^{-3} V$ or 3.98 mV

(ii) Since the numerical gain is 2.51 (see part (b)), then

$$\frac{10 \text{ mV}}{V_1} = 2.51$$

or $V_1 = 10 \text{ mV}/2.51 = 3.98 \text{ mV}$.

17.4 T- and π-attenuators

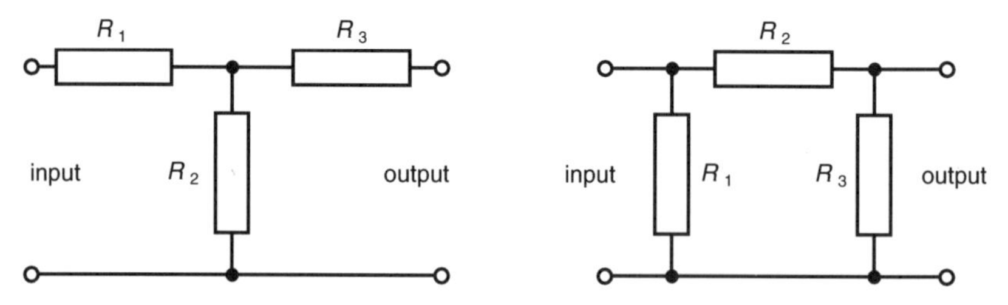

Figure 17.2 (a) A T-attenuator (b) a π-attenuator

The simplest, and most popular, forms of attenuators are the **T-attenuator** and the **π-attenuator** shown in Figure 17.2. In **symmetrical** versions of the attenuator, $R_1 = R_3$, and in **unsymmetrical** versions $R_1 \neq R_3$.

If the resistance between the input terminals is measured when the output is open-circuited, we obtain a resistance called R_{OC}. If the resistance between the input terminals is measured when the output is short-circuited, we obtain a resistance called R_{SC}. These values are referred to in the following.

For any attenuator of the type in Figure 17.2, there is always a value of load resistance which is equal to the resistance measured between the input terminals. In the case of a symmetrical attenuator, this is known as the **characteristic resistance**, R_0, of the attenuator.

For both forms of symmetrical attenuator, the characteristic resistance is

$$R_0 = \surd(R_{OC}R_{SC})$$

In the case of an *unsymmetrical attenuator*, the input resistance depends on which pair of terminals is taken as the input to the circuit; the input resistance which is equal to the load resistance is known as the **iterative resistance**.

17.5 Symmetrical T-attenuator

Referring to Figure 17.2(a)

$$R_1 = R_3$$

$$R_0 = \surd(R_{OC}R_{SC}) = \surd(R_1^2 + 2R_1R_2)$$

(as an exercise, the reader should verify the final equation). It can be shown that the *numerical attenuation*, N, produced by a symmetrical T-attenuator when a resistor of resistance R_O is connected between its output terminals is

$$N = \frac{R_1 + R_O}{R_O - R_1}$$

If we are provided with the values of N and R_0, the value of R_1 and R_2 are calculated from

$$R_1 = R_0(N - 1)/(N + 1)$$

$$R_2 = 2NR_0/(N^2 - 1)$$

The dB attenuation produced is

$$n = 20 \log_{10} N \text{ dB.}$$

Worked Example 17.5

(a) If a T-attenuator of the type in Figure 17.2(a) has values of $R_1 = R_3 = 300\ \Omega$, and $R_2 = 400\ \Omega$, and is terminated by a load equal to R_0, calculate (i) its characteristic resistance, (ii) the numerical and dB attenuation produced.

(b) A symmetrical T-attenuator has a characteristic resistance of 223.6 Ω, and provides an attenuation of 8.36 dB when terminated in a load equal to its characteristic resistance. Calculate the value of the components R_1 and R_2 in a circuit of the type in Figure 17.2(a).

Solution

(a) (i) The characteristic resistance of the attenuator is

$$R_0 = \sqrt{(R_1{}^2 + 2R_1R_2)} = \sqrt{(300^2 + (2 \times 300 \times 400))}$$
$$= 574.5\ \Omega$$

(ii) The numerical attenuation is given by

$$N = (R_1 + R_0)/(R_0 - R_1)$$
$$= (300 + 574.5)/(574.5 - 300) = 3.19$$

and the dB attenuation is

$$20 \log_{10} 3.19 = 10.08\ \text{dB}$$

(b) Since the attenuation produced by the filter is 8.36 dB, the numerical attenuation is

$$N = 10^{8.36/20} = 10^{0.418} = 2.618$$

The value of R_1 is given by

$$R_1 = R_0(N - 1)/(N + 1)$$
$$= 223.6(2.618 - 1)/(2.618 + 1) = 100\ \Omega$$

and

$$R_2 = 2NR_0/(N^2 - 1) = 2 \times 2.618 \times 223.6/(2.618^2 - 1)$$
$$= 200\ \Omega$$

17.6 Symmetrical π-attenuator

Referring to Figure 17.2(b), in the case of a symmetrical attenuator in which

$$R_1 = R_3$$

then

$$R_0 = \sqrt{(R_{OC}R_{SC})} = \sqrt{\left[\frac{R_1R_2 + R_1^2}{2R_1 + R_2} + \frac{R_1R_2}{R_1 + R_2}\right]}$$

(As an exercise, the reader should verify the equation for R_0.) It can be shown that the *numerical attenuation*, N, produced by a symmetrical π-attenuator when terminated in a resistance equal to R_0 is

$$N = \frac{R_1 + R_0}{R_1 - R_0}$$

and the corresponding dB attenuation is

$$n = 20 \log_{10} \text{N dB}.$$

Worked Example 17.6

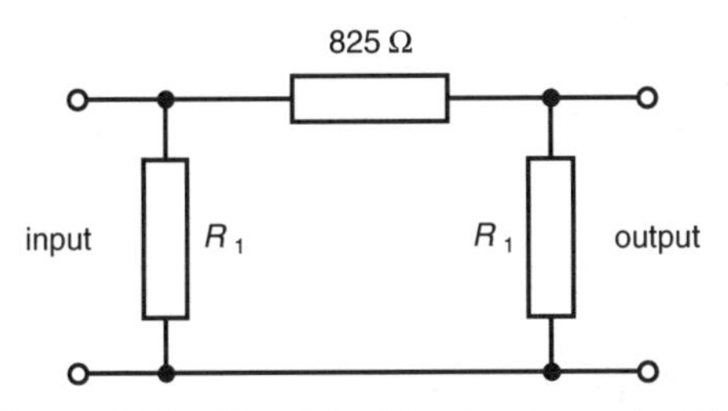

Figure 17.3 Circuit for Worked Example 17.6

The balanced π-attenuator in Figure 17.3 provides an attenuation of 10.06 dB when it is terminated in a resistance equal to its characteristic resistance of 574.5 Ω. Evaluate (a) the numerical attenuation produced by the attenuator, (b) the value of R_1.

Solution

(a) The numerical attenuation is

$$N = 10^{n/20} = 10^{10.06/20} = 10^{0.504} = 3.184$$

That is, an input of 1 V produces an output of

$$1\ \text{V}/3.184 = 0.314\ \text{V}$$

(b) We can manipulate the equation

$$N = \frac{R_1 + R_0}{R_1 - R_0}$$

to give

$$R_1 = R_0 \frac{N+1}{N-1} = 574.5 \times \frac{3.184 + 1}{3.184 - 1} = 1100\ \Omega$$

17.7 T–π and π–T conversion for resistive networks

For every resistive T-circuit of the type in Figure 17.4(a), there is an *equivalent* π-circuit of the type in Figure 17.4(b), and vice versa. The two circuits are related as follows:

T-to-π conversion

$$\begin{aligned} P_1 &= T_1 + T_3 + T_1 T_3 / T_2 \\ P_2 &= T_1 + T_2 + T_1 T_2 / T_3 \\ P_3 &= T_2 + T_3 + T_2 T_3 / T_1 \end{aligned}$$

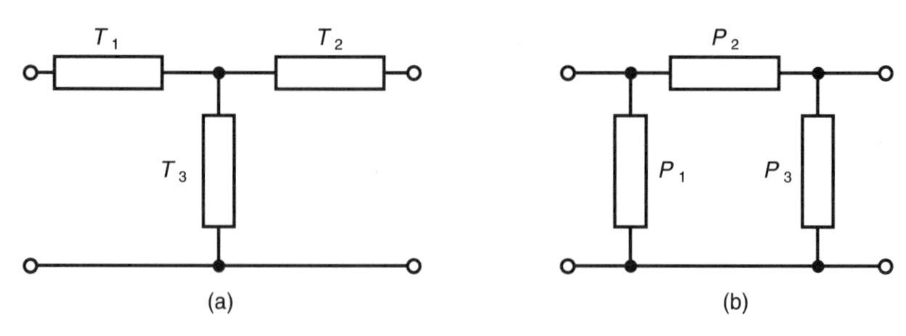

Figure 17.4 T–π and π–T conversion

π-to-T conversion

If $P = P_1 + P_2 + P_3$, then

$$\begin{aligned} T_1 &= P_1 P_2 / P \\ T_2 &= P_2 P_3 / P \\ T_3 &= P_3 P_1 / P. \end{aligned}$$

Worked Example 17.7

(a) A symmetrical T-attenuator has resistive arms of $T_1 = T_2 = 300\ \Omega$ and $T_3 = 400\ \Omega$. Determine the values in the equivalent π-attenuator.

(b) An unsymmetrical π-attenuator has resistive branches of $P_1 = 650\ \Omega$, $P_2 = 990\ \Omega$ and $P_3 = 110\ \Omega$. What are the resistive values in the equivalent T-attenuator?

Solution

(a) From the equation for T-to-π conversion we get

$$\begin{aligned} P_1 &= T_1 + T_3 + T_1 T_3 / T_2 \\ &= 300 + 400 + (300 \times 400)/300 = 1100\ \Omega \end{aligned}$$

$$\begin{aligned} P_2 &= T_1 + T_2 + T_1 T_2 / T_3 \\ &= 300 + 300 + (300 \times 300)/400 = 825\ \Omega \end{aligned}$$

$$\begin{aligned} P_3 &= T_2 + T_3 + T_2 T_3 / T_1 \\ &= 300 + 400 + (300 \times 400)/300 = 1100\ \Omega \end{aligned}$$

The reader will note that the initial values correspond to the symmetrical T-circuit in Worked Example 17.5, and the results correspond to the symmetrical π-attenuator in Worked Example 17.6. That is, the two circuits are identical in every respect. This is verified by the fact that the attenuators in those examples have the same value of characteristic resistance and attenuation.

(b) From the equation for π-to-T conversion

$$P = P_1 + P_2 + P_3 = 650 + 990 + 110 = 1750\ \Omega$$

hence

$$T_1 = P_1P_2/P = 650 \times 990/1750 = 367.7\ \Omega$$
$$T_2 = P_2P_3/P = 990 \times 110/1750 = 62.23\ \Omega$$
$$T_3 = P_1P_3/P = 650 \times 110/1750 = 40.86\ \Omega$$

17.8 Filter circuits

A **filter circuit** is one which will transmit signals within designated frequency ranges (the *pass-bands*), or suppress (attenuate) frequencies within other frequency ranges (the *stop-bands*).

The frequency (or frequencies) which separate the pass- and stop-bands are known as the **cut-off frequencies** or **corner frequencies**, f_c. The frequency at which cut-off is said to occur is where the magnitude of the output signal is 3 dB below that of the input signal, i.e. the frequency at which the output signal is 0.707 of the magnitude of the input signal.

Filters are classified according to the range of pass- and stop-bands in their frequency spectrum, the most popular characteristics being illustrated in Figure 17.5. These are

(a) the **low-pass filter**
(b) the **high-pass filter**
(c) the **band-pass filter**
(d) the **band-stop filter**.

Every filter contains one or more *reactive elements* such as a capacitor or an inductor in the circuit; the reader is warned that filters produce not only a change in the magnitude of the output signal, but also a change in the phase shift (which may be large) of the output signal when compared with the input signal.

A **passive filter** contains passive elements such as *R-L*, *R-C* or *L-C* elements, and **active filters** contain one or more op-amps (to give some gain) in addition to passive elements.

In the following calculations it will be assumed that the impedance of the load is very high, so that the current taken by the load is very small. That is, the load will not significantly alter the value of the output voltage obtained in the calculations.

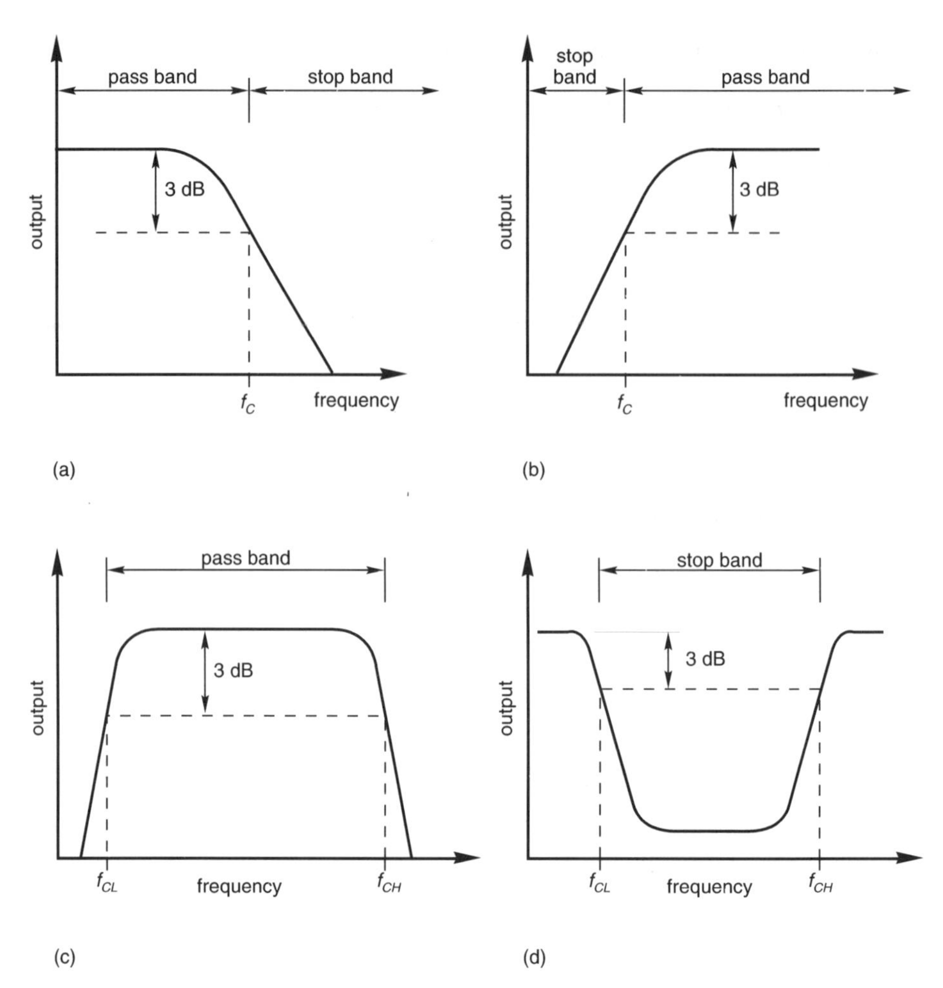

Figure 17.5 Filter circuit characteristics: (a) low-pass; (b) high-pass; (c) band-pass; (d) band-stop

17.9 Passive R-C low-pass filter

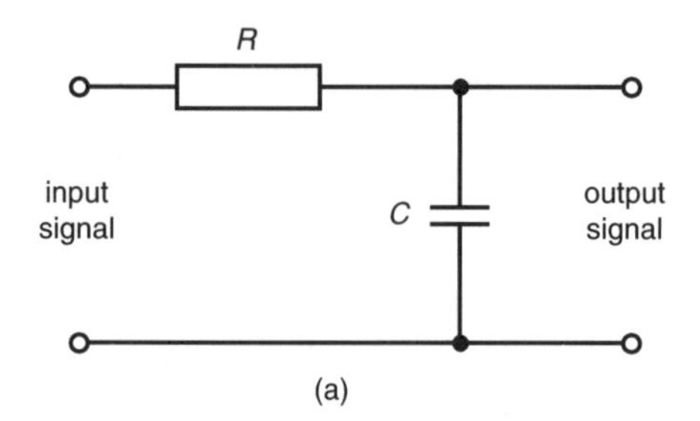

Figure 17.6 Passive R-C low-pass filter

The circuit of a typical passive *R-C* low-pass filter is shown in Figure 17.6. The corner frequency of the filter is

$$f_c = \mathbf{1/(2\pi RC)}\ \text{Hz}$$

At high frequencies, i.e. above the corner frequency, the reactance of the capacitor begins to short-circuit the load, and the output voltage rapidly reduces (see also Figure 17.5(a)).

Worked Example 17.8

In a passive *R-C* low-pass filter of the type in Figure 17.6, $R = 1\ \text{k}\Omega$ and $C = 0.1\ \mu\text{F}$. Determine the cut-off frequency of the filter. If the supply voltage is 1.5 V, calculate the output voltage at (a) 500 Hz, (b) the corner frequency and (c) 2.5 kHz. The impedance of the load can be assumed to be very high.

Solution

The cut-off frequency is

$$\begin{aligned} f_c &= 1/(2\pi RC) = 1/(2\pi \times 1000 \times 0.1 \times 10^{-6}) \\ &= 1591\ \text{Hz} \end{aligned}$$

Since the load impedance is very high, all the input current can be assumed to flow through *R* and *C*, and none of it through the load. The magnitude of the impedance between the input terminals is therefore

$$Z = \sqrt{(R^2 + X_C{}^2)}$$

where $X_C = 1/(2\pi fC)$. If V_S is the input voltage, the current flowing through the circuit is

$$I = V_S/Z$$

and from Ohm's law, the output voltage is

$$V_{\text{out}} = IX_C$$

(a) When $f = 500$ Hz

$$\begin{aligned} X_C &= 1/(2\pi fC) = 1/(2\pi \times 500 \times 0.1 \times 10^{-6}) \\ &= 3183\ \Omega \end{aligned}$$

and $Z = \sqrt{(R^2 + X_C{}^2)} = \sqrt{(1000^2 + 3183^2)} = 3336\ \Omega$

We see from the above that the output voltage is

$$V_{\text{out}} = X_C I = V_S X_C/Z = 1.5 \times 3183/3336 = 1.43\ \text{V}$$

(b) At the cut-off frequency of $f_c = 1591$ Hz

$$\begin{aligned} X_C &= 1/(2\pi f_c C) = 1/(2\pi \times 1591 \times 0.1 \times 10^{-6}) \\ &= 1000\ \Omega \end{aligned}$$

and $Z = \sqrt{(R^2 + X_C{}^2)} = \sqrt{(1000^2 + 1000^2)} = 1414\ \Omega$

We see from the above that the output voltage is

$$V_{out} = V_S X_C / Z = 1.5 \times 1000/1414 = 1.06 \text{ V}$$

From the information given in section 17.9, the output voltage at the corner frequency can be calculated as follows:

$$V_{out} = 0.707 V_{in} = 0.707 \times 1.5 = 1.06 \text{ V}$$

(c) At a frequency of 2.5 kHz, the reader should verify that

$$X_C = 636.6\ \Omega, Z = 1185\ \Omega \text{ and } V_{out} = 0.806 \text{ V}$$

The reader will see from the above results that the output voltage diminishes beyond the corner frequency.

17.10 Passive R-C high pass filter

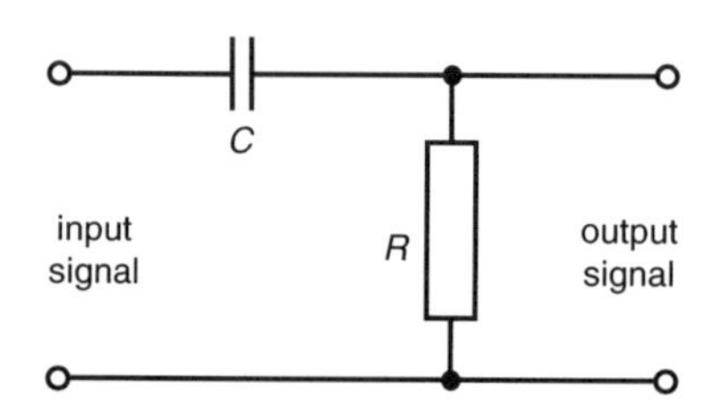

Figure 17.7 Passive R-C high-pass filter

A circuit diagram for this type of filter is shown in Figure 17.7. The cut-off frequency of the filter is

$$f_c = 1/(2\pi R_C) \text{ Hz}$$

and, once again, it will be assumed that the impedance of the load is sufficiently high so that it does not 'load' the filter circuit.

At low frequencies, i.e. frequencies below the cut-off frequency, the reactance of the capacitor is high enough to cause a significant reduction in output voltage. Above the corner frequency, the capacitive reactance is small enough not to produce a significant reduction in output voltage (see Figure 17.5(b)).

Worked Example 17.9

A passive R-C high-pass filter uses components of $R = 500\ \Omega$ and $C = 0.01\ \mu\text{F}$; what is the cut-off frequency of the filter?

If the load impedance is very high, and the input voltage is 2 V, what is the output voltage at (a) 20 kHz, (b) the cut-off frequency and (c) 44 kHz?

Solution

The cut-off frequency of the filter is

$$f_c = 1/(2\pi RC) = 1/(2\pi \times 500 \times 0.01 \times 10^{-6})$$
$$= 31830 \text{ Hz or } 31.83 \text{ kHz}$$

Since the load impedance is very high, the current drawn from the input circuit is

$$I = V_S/Z$$

where $Z = \sqrt{(R^2 + X_C^2)}$, and the output voltage is

$$V_{out} = IR = V_S R/Z$$

(a) At a frequency of $f = 20$ kHz

$$X_C = 1/(2\pi fC) = 1/(2\pi \times 20 \times 10^3 \times 0.01 \times 10^{-6}) = 796\ \Omega$$

and $Z = \sqrt{(R^2 + X_C^2)} = \sqrt{(500^2 + 796^2)} = 940\ \Omega$

We see from the above that the output voltage is

$$V_{out} = V_S R/Z = 2 \times 500/940 = 1.06 \text{ V}$$

(b) Using the method outlined above, the values at $f_c = 31.83$ kHz are

$$X_C = 500\ \Omega, Z = 707.1\ \Omega, V_{out} = 1.414 \text{ V}$$

Or, from earlier work, the output voltage at the cut-off frequency is

$$V_{out} = 0.7071 V_S = 1.414 \text{ V}$$

(c) At a frequency of 44 kHz, the corresponding values are

$$X_C = 361.7\ \Omega, Z = 617.1\ \Omega, V_{out} = 1.62 \text{ V}$$

That is the output voltage gradually rises from a low value below the cut-off frequency, to 0.7071 V_S at the corner frequency, and approaches the input voltage at higher frequencies.

17.11 Active R-C filters

Passive filter circuits are generally unable to give an overall voltage 'gain', and they cannot 'drive' a load. These factors are overcome if we use an **active filter**, which incorporate an op-amp.

In the following, we describe a low-pass and a high-pass active filter which, as will be shown in Worked Example 20.9 (Chapter 20), can be combined to act as an active *R-C* band-pass filter.

(a) *A phase-inverting active R-C low-pass filter*

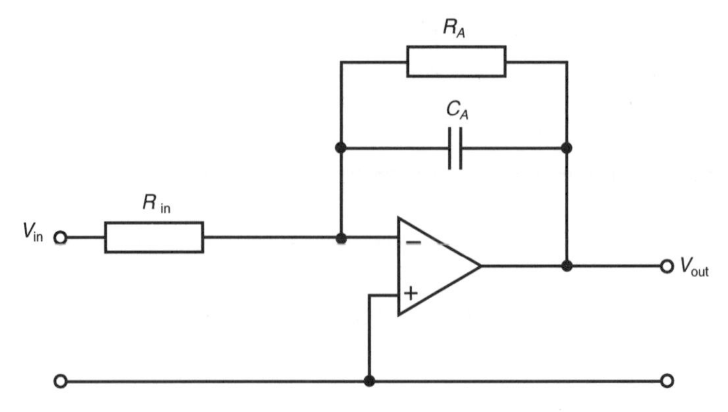

Figure 17.8 A phase-inverting active R-C low-pass filter

A phase-inverting active *R-C* low-pass filter is shown in Figure 17.8 (non-inverting configurations can also be constructed). The filter has a frequency response curve of the type in Figure 17.5(a), and its cut-off frequency is

$$f_c = 1/(2\pi R_A C_A) \text{ Hz}$$

and the magnitude of the gain at frequencies below the corner frequency is R_A/R_{in}. That is, R_A is involved in the cut-off frequency calculation, and R_{in} is involved in the low-frequency gain computation.

(b) *An active R-C high-pass filter*

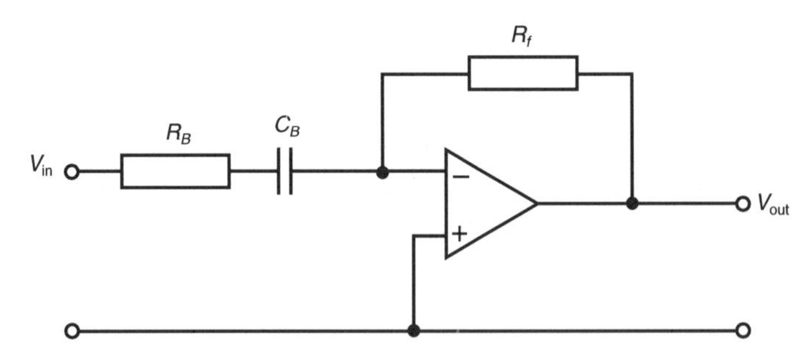

Figure 17.9 A phase-inverting active R-C high-pass filter

One form of *R-C* active high-pass filter is shown in Figure 17.9. The frequency response curve is shown in Figure 17.5(b), and its cut-off frequency is

$$f_c = 1/(2\pi R_B C_B) \text{ Hz}$$

and the magnitude of the gain at frequencies well above the corner frequency is R_F/R_B. That is, R_B is involved in the determination of the cut-off frequency, and R_F is used in the determination of the high-frequency gain.

Worked Example 17.10

(a) An active low-pass R-C filter of the type in Figure 17.8 is to have a cut-off frequency of 1 kHz, and uses a 0.01 μF capacitor in the feedback circuit. If the low-frequency gain is 10, determine the value of the other components in the circuit.

(b) Design an active high-pass R-C filter of the type in Figure 17.9, which has a cut-off frequency of 100 kHz and a high-frequency gain of 8. The circuit uses a 0.001 μF capacitor in the forward path.

Solution

(a) Referring to section (a), we see that the cut-off frequency of the low-pass filter is

$$f_c = 1/(2\pi R_A C_A)$$

or

$$R_A = 1/(2\pi f_c C_A) = 1/(2\pi \times 1000 \times 0.01 \times 10^{-6})$$
$$= 15920\ \Omega \text{ or } 15.92\ \text{k}\Omega$$

The magnitude of the low-frequency gain is given by

$$\text{magnitude of gain} = 10 = R_A/R_{\text{in}}$$

or

$$R_{\text{in}} = R_A/10 = 15.92/10 = 1.592\ \text{k}\Omega$$

(b) From section (b), we see that the cut-off frequency of the high-pass filter is

$$f_c = 1/(2\pi R_B C_B)$$

or

$$R_B = 1/(2\pi \times 100\,000 \times 0.001 \times 10^{-6}) = 1592\ \Omega$$

The required high-frequency gain magnitude is 8, hence

$$\text{magnitude of gain} = 8 = R_F/R_B$$

or

$$R_F = 8R_B = 8 \times 1582 = 12736\ \Omega \text{ or } 12.736\ \text{k}\Omega.$$

17.12 Computer listings

There are four Computer listings which cover the basic requirements of this chapter.

Listing 17.1 Conversion of numerical gain to decibels and decibels to a numerical gain

```
10 CLS ' ** CH 17-1 **
20 PRINT "CONVERSION OF NUMERICAL GAIN TO dB"
30 PRINT "AND VICE VERSA": PRINT
40 PRINT "Select ONE of the following:": PRINT
50 PRINT "1. Convert a power ratio into dB.": PRINT
60 PRINT "2. Convert a voltage ratio into dB.": PRINT
70 PRINT "3. Convert a current ratio into dB.": PRINT
80 PRINT "4. Convert a dB power ratio into a numerical ratio.": PRINT
90 PRINT "5. Convert a dB voltage ratio into a numerical ratio.": PRINT
100 PRINT "6. Convert a dB current ratio into a numerical ratio.": PRINT
110 INPUT "Enter your selection here (1 - 6): ", S
120 IF S < 1 OR S > 6 THEN GOTO 10
130 IF S = 2 THEN CLS : GOTO 300
140 IF S = 3 THEN CLS : GOTO 400
150 IF S = 4 THEN CLS : GOTO 600
160 IF S = 5 THEN CLS : GOTO 700
170 IF S = 6 THEN CLS : GOTO 800
200 CLS : PRINT "CONVERT A POWER RATIO INTO dB": PRINT
210 INPUT "Power input (W) = ", P1
220 INPUT "Power output (W) = ", P2: PRINT
230 G = P2 / P1
240 PRINT "Gain = "; 10 * LOG(G) / LOG(10); "dB"
250 END
300 PRINT "CONVERT A VOLTAGE RATIO INTO dB": PRINT
310 INPUT "Voltage input (V) = ", V1
320 INPUT "Voltage output (V) = ", V2: PRINT
330 G = V2 / V1: GOTO 500
400 PRINT "CONVERT A CURRENT RATIO INTO dB": PRINT
410 INPUT "Current input (A) = ", I1
420 INPUT "Current output (A) = ", I2: PRINT
440 G = I2 / I1
500 PRINT "Gain = "; 20 * LOG(G) / LOG(10); "dB"
550 END
600 PRINT "CONVERT dB POWER RATIO IN NUMERICAL RATIO."
610 PRINT : INPUT "dB power ratio = ", dB
620 G = dB / 10: GOTO 900
700 PRINT "CONVERT dB VOLTAGE RATIO IN NUMERICAL RATIO."
710 PRINT : INPUT "dB voltage ratio = ", dB
720 G = dB / 20: GOTO 900
800 PRINT "CONVERT dB CURRENT RATIO IN NUMERICAL RATIO."
810 PRINT : INPUT "dB current ratio = ", dB
820 G = dB / 20
900 PRINT : PRINT "Numerical ratio = "; 10 ^ G
910 END
```

Listing 17.1 deals with the conversion of power, voltage and current ratios into decibels, and vice versa. Since computer use natural logarithms, or logarithms to base e (e = 2.71828. . .), the result in natural logarithms is converted into logarithms to base 10 in lines 240 and 500 using the relationship

$$\log_{10} G = \log_e G / \log_e 10$$

The reverse process (antilogarithm) is carried out in line 900, where the numerical ratio is determined from the decibel value.

Listing 17.2 Attenuator calculations

```
10 '** CH 17-2 **
20 CLS : PRINT "ATTENUATOR CALCULATIONS": PRINT
30 PRINT "Select ONE of the following:": PRINT
40 PRINT "1. Symmetrical T attenuator design"
50 PRINT "      using component values.": PRINT
60 PRINT "2. Symmetrical T attenuator design using characteristic"
70 PRINT "      resistance and attenuation.": PRINT
80 PRINT "3. Symmetrical Pi attenuator design"
90 PRINT "      using component values.": PRINT
100 PRINT "4. Symmetrical Pi attenuator design using characteristic"
110 PRINT "      resistance and attenuation": PRINT
120 INPUT "Enter your selection here (1 - 4): ", S
130 IF S < 1 OR S > 4 THEN GOTO 10
140 IF S = 2 THEN CLS : GOTO 300
150 IF S = 3 THEN CLS : GOTO 500
160 IF S = 4 THEN CLS : GOTO 600
200 CLS : PRINT "SYMMETRICAL T ATTENUATOR DESIGN"
210 PRINT "USING COMPONENT VALUES": PRINT
220 INPUT "R1 (ohms) = ", R1
230 INPUT "R2 (ohms) = ", R2: PRINT
240 RO = SQR((R1 ^ 2) + (2 * R1 * R2))
250 N = (R1 + RO) / (RO - R1)
260 PRINT "Characteristic resistance (ohms) = "; RO
270 PRINT "Numerical attenuation = "; N
280 PRINT "dB attenuation = "; 20 * LOG(N) / LOG(10)
290 END
300 PRINT "SYMMETRICAL T ATTENUATOR DESIGN USING"
310 PRINT "dB ATTENUATION AND CHARACTERISTIC RESISTANCE."
320 PRINT : INPUT "dB attenuation = ", ndB
330 INPUT "Characteristic resistance (ohms) = ", RO
340 N = 10 ^ (ndB / 20): PRINT
350 PRINT "Numerical attenuation = "; N
360 R1 = RO * (N - 1) / (N + 1)
370 R2 = 2 * N * RO / ((N ^ 2) - 1)
380 PRINT "R1 (ohms) = "; R1
390 PRINT "R2 (ohms) = "; R2
```

```
400 END
500 PRINT "SYMMETRICAL PI ATTENUATOR DESIGN"
510 PRINT "USING COMPONENT VALUES": PRINT
520 INPUT "R1 (ohms) = ", R1
530 INPUT "R2 (ohms) = ", R2: PRINT
540 RO = SQR((R1 ^ 2) * R2 / ((2 * R1) + R2))
550 N = (RO + R1) / (R1 - RO)
560 PRINT "Characteristic resistance (ohms) = "; RO
570 PRINT "Numerical attenuation = "; N
580 PRINT "dB attenuation = "; 20 * LOG(N) / LOG(10)
590 END
600 PRINT "SYMMETRICAL PI ATTENUATOR DESIGN USING"
610 PRINT "dB ATTENUATION AND CHARACTERISTIC RESISTANCE."
620 PRINT : INPUT "dB attenuation = ", ndB
630 INPUT "Characteristic resistance (ohms) = ", RO
640 N = 10 ^ (ndB / 20): PRINT
650 PRINT "Numerical attenuation = "; N
660 R1 = RO * (N + 1) / (N - 1)
670 R2 = RO * (N ^ 2 - 1) / (2 * N)
680 PRINT "R1 (ohms) = "; R1
690 PRINT "R2 (ohms) = "; R2
700 END
```

Listing 17.2 performs a series of calculations on symmetrical attenuators. Options 1 and 3 evaluate the characteristic resistance, the numerical attenuation and the dB attenuation for a T- and π-circuit, respectively. Options 2 and 4 determine the values of resistors needed in a T- and π-circuit, respectively, together with the numerical attenuation when the characteristic resistance and dB attenuation are supplied. The T- and π-circuits which are referred to are those in Figure 17.2, in which $R_1 = R_3$.

Listing 17.3 T-π and π-T conversion for resistive networks

```
10 CLS '** CH 17-3 **
20 PRINT "T-PI AND PI-T CONVERSION FOR RESISTIVE NETWORKS.": PRINT
30 PRINT "Select ONE of the following:": PRINT
40 PRINT "1. T-PI conversion.": PRINT
50 PRINT "2. PI-T conversion.": PRINT
60 INPUT "Enter your selection here (1 or 2): ", S
70 IF S < 1 OR S > 2 THEN GOTO 10
80 IF S = 2 THEN GOTO 300
200 CLS : PRINT "T-PI CONVERSION": PRINT
210 INPUT "T1 (ohms) = ", T1
220 INPUT "T2 (ohms) = ", T2
230 INPUT "T3 (ohms) = ", T3: PRINT
240 PRINT "P1 (ohms) = "; T1 + T3 + (T1 * T3) / T2
```

```
250 PRINT "P2 (ohms) = "; T1 + T2 + (T1 * T2) / T3
260 PRINT "P3 (ohms) = "; T2 + T3 + (T2 * T3) / T1
270 END
300 CLS : PRINT "PI-T CONVERSION": PRINT
310 INPUT "P1 (ohms) = ", P1
320 INPUT "P2 (ohms) = ", P2
330 INPUT "P3 (ohms) = ", P3: PRINT
340 P = P1 + P2 + P3
350 PRINT "T1 (ohms) = "; P1 * P2 / P
360 PRINT "T2 (ohms) = "; P2 * P3 / P
370 PRINT "T3 (ohms) = "; P1 * P3 / P
380 END
```

The two routines in Listing 17.3 converts any resistive T-network into its equivalent π-network, and vice versa. The components referred to are those shown in Figure 17.4.

Listing 17.4 ***R*-*C*** filters

```
10 CLS '** CH 17-4 **
20 PRINT "R-C FILTERS": PRINT
30 PRINT "Select ONE of the following:": PRINT
40 PRINT "1. R-C filter cut-off frequency.": PRINT
50 PRINT "2. Low-pass filter frequency response.": PRINT
60 PRINT "3. High-pass filter frequency response.": PRINT
70 INPUT "Enter your selection here (1 - 3): ", S: PRINT
80 IF S < 1 OR S > 3 THEN GOTO 10
90 INPUT "R (ohms) = ", R
100 INPUT "C (microfarads) = ", C: PI = 3.14159
110 fc = 1 / (2 * PI * R * C * 10 ^ (-6)): PRINT
120 PRINT "Cut-off frequency = "; fc; "Hz": PRINT
130 IF S = 1 THEN END
140 INPUT "Minimum frequency (Hz) = ", fmin
150 IF fmin >= fc THEN PRINT "Error": GOTO 140
160 inc = (fc - fmin) / 5
170 INPUT "Input voltage (V) = ", V: PRINT
180 PRINT "Freq (Hz)"; TAB(15); "Output voltage"
190 FOR f = fmin TO (fmin + (10 * inc)) STEP inc
200     Xc = 1 / (2 * PI * f * C * 10 ^ (-6))
210     Z = SQR(R ^ 2 + Xc ^ 2)
220     I = V / Z
230     IF S = 2 THEN Vout = I * Xc ELSE Vout = I * R
240     PRINT f; TAB(15); Vout
250 NEXT f
260 END
```

The three options in Listing 17.4 refer to the passive *R-C* filters in Figures 17.6 and 17.7. Option 1 determines the cut-off frequency in Hz of the filters, and is part of the complete listing, rather than a separate section of the listing.

Options 2 and 3 respectively produce a table of eleven results which give an important part of the frequency response of *R-C* filters. The computer requests the value of R in ohms and the value of C in μF. In the case of options 2 and 3, you are also asked to supply the minimum frequency required; line 150 prevents a value being given which is less than the cut-off frequency of the filter. You also need to supply the input voltage to the filter.

The valuc of 'inc' calculated in line 160 of this listing, is the increment of frequency used in the calculation of the output voltage. The computer produces an output voltage table commencing at the minimum frequency, and increasing in increments up to a maximum frequency of

minimum frequency + (10 × frequency increment)

The table, of course, produces the output voltage at the cut-off frequency of the filter.

Exercises

17.1 What is the *power gain* in decibels of an attenuator if the input power and output power, respectively, are (a) 10 W and 0.8 W, (b) 100 mW and 700 mW?

17.2 If the power *attenuation* in a attenuator is 6.02 dB, and the output power is 20 mW, determine the input power.

17.3 A system has a power gain of 6.41 dB, and an input power is 8W. What is the power output?

17.4 A circuit has an input and output voltage, respectively, of (a) 0.8 V and 0.01 V, (b) 0.2 V and 3 V. Determine the dB 'gain' of the system.

17.5 A filter has an input current of 10 mA and an output current of 1.6 mA. What attenuation is provided by the filter?

17.6 The voltage gain of each stage of a two-stage network are 10 dB and −6 dB, respectively. If the input voltage is 2.1 V, determine the output voltage of (a) the first stage, (b) the second stage.

17.7 The first and third stages of a three-stage network have respective voltage gains of 20 dB and 6 dB. Determine the numerical gain of the second stage if the input and output voltage of the complete system are 1 V and 0.1 V.

17.8 A four-stage system has power gains of 10 dB, 8 dB, −15 dB and −6 dB. What is the overall numerical power gain of the system?
If the input power is 10 mW, calculate the output power.

17.9 A symmetrical T-attenuator (Figure 17.2(a)) uses $R_1 = R_3 = 200\ \Omega$ and $R_2 = 300\ \Omega$. Calculate the value of the characteristic resistance, the numerical attenuation and the dB attenuation of the attenuator when it is terminated in its characteristic resistance.

17.10 A symmetrical π-attenuator (see Figure 17.2(b)) uses $R_1 = R_3 = 400\ \Omega$ and $R_2 = 200\ \Omega$. Determine the characteristic resistance, the numerical attenuation and the dB attenuation of the attenuator when it is terminated by its characteristic resistance.

17.11 A symmetrical T-attenuator has an attenuation of 12 dB and a characteristic resistance of 300 Ω. What numerical attenuation is produced when the attenuator is terminated by its characteristic resistance, and what resistor values are used in the attenuator?

17.12 Determine the component values in a symmetrical π-attenuator which is equivalent to the attenuator in Exercise 17.9.

17.13 Calculate the component values in a symmetrical T-attenuator which is equivalent to the symmetrical π-attenuator in Exercise 17.10.

17.14 Determine the cut-off frequency of a passive *R-C* filter whose values are $R = 796\ \Omega$ and $C = 0.01\ \mu F$.

17.15 If the component values in Exercise 17.14 are used in a low-pass filter, determine the output voltage when the r.m.s. input voltage is 60 mV at frequencies of (a) 8 kHz, (b) the corner frequency and (c) 40 kHz.

17.16 If the component values in Exercise 17.14 are used in a high-pass filter, calculate the output voltage when the r.m.s. input voltage is 250 mV at a frequency of (a) 12 kHz, (b) the corner frequency, (c) 36 kHz.

Summary of important facts

An **attenuator** is a circuit which reduces the magnitude of the input power, voltage or current by a constant amount at all frequencies.

Attenuation is measured in **decibels** (dB). If P_1 and P_2 are the input power to and output power from an attenuator, the *power ratio* in decibels is

$$n = 10 \log_{10} \frac{P_2}{P_1} \text{ dB}$$

and the numerical power ratio is calculated from the decibel ratio as follows:

$$\frac{P_2}{P_1} = 10^{n/10}$$

If the input and output voltage associated with an attenuator are V_1 and V_2, respectively, then the attenuation in decibels is

$$n = 20 \log_{10} \frac{V_2}{V_1} \text{ dB}$$

and the numerical voltage ratio is calculated from the decibel ratio as follows:

$$\frac{V_2}{V_1} = 10^{n/20}$$

Similarly, for an input current and an output current of I_1 and I_2, the current ratio in decibels is

$$n = 20 \log_{10} \frac{I_2}{I_1} \text{ dB}$$

and the numerical current ratio is calculated from the decibel ratio as follows:

$$\frac{I_2}{I_1} = 10^{n/20}$$

Decibel ratios can be used to advantage where stages of an electronic system are cascaded, since the ratios can be *added together* (which is equivalent to multiplying the numerical ratios) to give the overall dB ratio. If n_1, n_2 and n_3 are the dB ratios of three cascaded stages, the overall dB ratio is

$$n = n_1 + n_2 + n_3 \text{ dB}$$

Basic attenuators can be constructed from the T- and π-sections in Figure 17.2. If $R_1 = R_3$ in either of these circuits, they are known as **symmetrical attenuators**. If R_1 is not equal to R_3, the circuit is known as an **unsymmetrical attenuator**.

A T-circuit can be converted into an equivalent π-circuit (see Figure 17.4), and vice versa, using the following equations.

T-to-π conversion

$$P_1 = T_1 + T_3 + T_1 T_3 / T_2$$
$$P_2 = T_1 + T_2 + T_1 T_2 / T_3$$
$$P_3 = T_2 + T_3 + T_2 T_3 / T_1$$

π-to-T conversion

If $P = P_1 + P_2 + P_3$, then

$$T_1 = P_1 P_2 / P$$
$$T_2 = P_2 P_3 / P$$
$$T_3 = P_3 P_1 / P$$

A **filter circuit** transmits frequencies within designated **pass-bands**, and suppresses or attenuates frequencies within its **stop-bands**.

Frequency response characteristics of typical filters are illustrated in Figure 17.5, and are

(a) **low-pass filter**, which passes low frequencies, and severely attenuates frequencies above its **cut-off frequency**
(b) **high-pass filter**, which severely attenuates frequencies below its cut-off frequency, and passes higher frequencies with very little loss
(c) **band-pass filter**, which severely attenuates frequencies which lie outside the pass-band of the filter
(d) **band-stop filter**, which severely attenuates frequencies which lie inside the **stop-band** of the filter.

Binary systems and computer logic

18.1 Introduction

In this chapter we look at basic numbering systems including binary (base 2), octal (base 8), decimal or denary (base 10), and hexadecimal (base 16), together with computer logic. By the end of this chapter the reader will be able to

- Understand binary, octal, decimal and hexadecimal numbering systems.
- Add, subtract, multiply and divide binary numbers.
- Deal with negative binary numbers.
- Design circuits using logic gates.
- Use software to convert a decimal number into binary and vice versa.

18.2 Basis of numbering systems

When we write down a number such as

$$5 \times 10^6$$

the value 5 is simply a **multiplying coefficient**, the value 10 is the **base** or **radix** of the number system, and 6 is the **power** or **exponent** to which the base has been raised.

The value of the multiplying coefficient in any position of a number can lie in the range zero to a value of (base – 1). For example in the decimal system (base 10), any digit can have a value between 0 and 9. In the binary system (base 2), each digit has a value of 0 or 1, and in the octal (base 8) system each digit can have a value in the range 0 to 7, etc.

An interesting fact is that any number raised to the power zero is equal to unity, i.e. $10^0 = 1$, $2^0 = 1$, $8^0 = 1$, $16^0 = 1$, etc.

A table showing a range of popular numbering systems is given in Table 18.1.

Table 18.1 Basic numbering systems

System	*Binary*	*Octal*	*Decimal*	*Hexadecimal*
Base	2	8	10	16
	$2^4\ 2^3\ 2^2\ 2^1\ 2^0$	$8^1\ 8^0$	$10^1\ 10^0$	$16^1\ 16^0$
	0	0	0	0
	1	1	1	1
	1 0	2	2	2
	1 1	3	3	3
	1 0 0	4	4	4
	1 0 1	5	5	5
	1 1 0	6	6	6
	1 1 1	7	7	7
	1 0 0 0	1 0	8	8
	1 0 0 1	1 1	9	9
	1 0 1 0	1 2	1 0	a
	1 0 1 1	1 3	1 1	b
	1 1 0 0	1 4	1 2	c
	1 1 0 1	1 5	1 3	d
	1 1 1 0	1 6	1 4	e
	1 1 1 1	1 7	1 5	f
	1 0 0 0 0	2 0	1 6	1 0
	1 0 0 0 1	2 1	1 7	1 1
	1 0 0 1 0	2 2	1 8	1 2
	1 0 0 1 1	2 3	1 9	1 3
	1 0 1 0 0	2 4	2 0	1 4
	1 0 1 0 1	2 5	2 1	1 5
	1 0 1 1 0	2 6	2 2	1 6
	1 0 1 1 1	2 7	2 3	1 7
	1 1 0 0 0	3 0	2 4	1 8
	1 1 0 0 1	3 1	2 5	1 9

Since each digit in the **hexadecimal system** can have a value in the range zero to 15 (decimal equivalent), we give alphabetical values to number which exceed the value of 9. That is, the first sixteen hexadecimal values are 0, 1, 2, 3, 4, 5, 6, 7, 8, 9, a, b, c, d, e and f.

18.3 Representing a number of any base

We can define the base of the numbering system we are working with by writing the base of the numbering system as a subscript to the number. For example, binary 11 is written 11_2, octal 11 is written 11_8, decimal 11 is written 11_{10}, and hexadecimal 11 is written 11_{16} (note the use of the decimal subscript).

Care should be taken when describing numbers, since 11_{10} is described as 'eleven', whereas binary 11 is described as 'one, one, binary'.

18.4 Bits and bytes

A single *b*inary dig*it* is known as a **bit**, and *eight consecutive bits*, i.e. 10101010 or 11001100, is known as a **byte**.

18.5 Converting an integer of any base into decimal

To convert an integer of any base into decimal, we simply write down the number as a series of powers of the base, and add the separate terms together, as shown in the following.

Worked Example 18.1

Convert 1101_2 and $3eb_{16}$ into decimal.

Solution

$$\begin{aligned} 1101_2 &= (1 \times 2^3) + (1 \times 2^2) + (0 \times 2^1) + (1 \times 2^0) \\ &= (8 + 4 + 0 + 1)_{10} = 13_{10} \end{aligned}$$

and

$$\begin{aligned} 3eb_{16} &= (3 \times 16^2) + (e \times 16^1) + (b \times 16^0) \\ &= (768 + 224 + 11)_{10} = 1003_{10}. \end{aligned}$$

18.6 Converting a decimal integer into another radix

In this case we divide the integer repeatedly by the new radix, successive remainders giving the required value (the final remainder being the most significant digit of the new number).

Worked Example 18.2

Convert 300_{10} into hexadecimal.

Solution

$$
\begin{array}{rl}
16\,\underline{)\,300} & \\
16\,\underline{)\;\;18} & \text{remainder } 12_{10} \text{ or } c_{16} \\
 & \qquad\text{(least significant digit)} \\
16\,\underline{)\;\;\;1} & \text{remainder } 2_{16} \\
0 & \text{remainder } 1_{16} \\
 & \qquad\text{(most significant digit)}
\end{array}
$$

That is $300_{10} \equiv 12c_{16}$.

18.7 Numbers having a fractional part

When converting a *non-decimal number into its decimal equivalent*, we use the method outlined in section 18.3 (see also Worked Example 18.3(a)).

When we convert a *decimal value into its non-decimal equivalent*, the integer part is dealt with as outlined in section 18.4. The fractional part is multiplied repeatedly by the radix of the system; the resulting integer part of the multiplication gives the required value (the fractional part of the multiplication is passed on to the next multiplication stage – see Worked Example 18.3(b)).

Worked Example 18.3

Convert (a) $79e.18_{16}$ into decimal, (b) 97.375_{10} into binary.

Solution

(a)
$$
\begin{aligned}
79e.15_{16} &= (7 \times 16^2) + (9 \times 16^1) + (14 \times 16^0) + (1/16) + (5/16^2) \\
&= (1792 + 144 + 14 + 0.0625 + 0.03125)_{10} \\
&= 1950.09375_{10}
\end{aligned}
$$

(b) As outlined above, we deal with the integer part and fractional part separately, as in the table below.

The integer part is initially divided repeatedly by 2 to give the integer part of the solution (1100001_2). Next, the fractional part is repeatedly multiplied by 2, the integer part of the calculation giving the fractional part ($.011_2$) of the solution, hence

$$97.375_{10} \equiv 1100001.011_2$$

Integer part		*Fractional part*	
$2\overline{)97}$			
$2\overline{)48}$	remainder 1	$0.375 \times 2 = 0$	.75
$2\overline{)24}$	remainder 0	$0.75 \times 2 = 1$	.5
$2\overline{)12}$	remainder 0	$0.5 \times 2 = 1$	(l.s.b.)
$2\overline{)6}$	remainder 0		
$2\overline{)3}$	remainder 0		
$2\overline{)1}$	remainder 1		
0	remainder 1 (m.s.b.)		

The reader should note that the **most significant bit** of the number is referred to as the *m.s.b.*, and the **least significant bit** is known as the *l.s.b.*

18.8 Pure binary addition and truth tables

If $a + b = c$

a is called the *augend*, *b* the *addend* and *c* the *sum*. If the sum exceeds the base of the numbering system, a **carry** is produced. That is

$0_2 + 0_2$ = a sum of 0_2 and no carry

$0_2 + 1_2$ = a sum of 1_2 and no carry

$1_2 + 0_2$ = a sum of 1_2 and no carry

$1_2 + 1_2$ = a sum of 0_2 and a carry of 1_2

Any carry produced by the above addition is the **carry-out** from the sum, which becomes the **carry-in** of the next higher addition.

The above process is known as **half-addition** because it only deals with the addition of the two binary digits, and does not deal with the carry-in bit from the previous addition. Any addition process which handles the augend, the addend and the carry-in is described as a **full-addition** process, and it produces a *sum* and a *carry-out*.

A **truth table** is a table showing all possible combinations of input and output values from a system. The truth table for full-addition is given in Table 18.2.

We see that, for a full-adder

1. when all inputs are '0', then sum = 0 and carry-out = 0
2. when only one input is '1', then sum = 1 and carry-out = 0
3. when two inputs are '1', then sum = 0 and carry-out = 1
4. when all inputs are '1', then sum = 1 and carry-out = 1

Table 18.2 Truth table for full-addition

Input signals			*Output signals*	
augend	*addend*	*carry-in*	*sum*	*carry-out*
0	0	0	0	0
1	0	0	1	0
0	1	0	1	0
1	1	0	0	1
0	0	1	1	0
1	0	1	0	1
0	1	1	0	1
1	1	1	1	1

We can use the above summary of the full-adder truth table in Worked Example 18.4 to show how two binary numbers are added together.

Worked Example 18.4

Add together the binary numbers 1011 and 1011.

Solution

The arithmetic procedure is as follows:

augend		1	0	1	1
addend		1	0	1	1
carry-in	1	0	1	1	0
	↖	↖	↖	↖	
carry-out	0	1	0	1	1
sum	1	0	1	1	0

The reader should remember that a full-addition *always* requires three input signals, namely an augend, an addend and a carry-in (which may be zero).

In the initial addition, both the augend and the addend are '1', and the carry-in is '0', so that the carry-out is '1' and the sum is '0'. The carry-out from this addition becomes the carry-in for the next higher addition, where all three input signals are '1'. Consequently, the sum of the second addition is '1' and the carry-out is '1'.

It is left as an exercise for the reader for the reader to check the solution for the remaining steps of the addition. In the final stage, the carry-out of '1' is transferred to yet another stage, which has an augend and an addend of '0', resulting in a sum of '1' and a carry-out of '0'. That is

$$1011_2 + 1011_2 = 10110_2$$

The reader will note that when we add two binary numbers of 'length' n bits, the result may be $(n + 1)$ bits long.

18.9 Negative binary numbers

When dealing with **positive numbers only**, we describe them as **unsigned binary numbers**. However, when we need to handle both *positive and negative numbers*, we must use **signed binary numbers**.

A computer or calculator can discover the 'sign' of a binary number by inspecting the most significant bit (m.s.b.) of the binary 'word'*, which is known as the **sign bit**.

If the m.s.b. of a 'signed' binary number is '0', then the number is positive, and the number is stored in pure binary. If the sign bit is '1', the number is *negative*, and it is stored in what is known as **binary complement notation**. There are two forms of complement notation, namely the **1's complement** and the **2's complement** (or **true complement**).

The 1's complement value of a binary number is obtained as follows:

Change the 0's into 1's, and the 1's into 0's throughout the number.

The 2's binary complement of as binary number is obtained by either of the following methods:

1. **Form the 1's complement of the number, and add '1' to the least significant bit of the number so formed.**
2. **Copy the number (commencing with the l.s.b.) up to and including the least significant '1'; thereafter, change all the 1's into 0's and 0's into 1's.**

For the purpose of uniformity, we will deal with computer words of length 8 bits (or 1 byte).

Worked Example 18.5

Form the 1's and 2's complement of the following *signed binary numbers* (i.e. the m.s.b. is a sign bit): 00000000, 00000001, 10101010 and 11111111.

Solution

Since we are dealing with signed binary values, we see that the first two values are positive and the second two are negative.

Applying the rules for the 1's complement form, we obtain the following results:

binary value	*1's complement value*
00000000	11111111
00000001	11111110
10101010	01010101
11111111	00000000

* The *word length* of a digital system is usually measured in *bytes*, and it may be any length from 1 byte (8 bits) up to 4 or more bytes.

Interestingly, since the first value is $+0_2$, we see that $-0_2 = 11111111_2$. Clearly, the final value given in the example is 11111111_2 or -0_2, so that

$$1\text{'s complement of } (-0_2) = 00000000_2 = +0_2$$

which agrees with what we know from basic algebra. Also, since the second value given in the example is $+1_2$, then in 1's complement form $-1_2 \equiv 1111110_2$.

Applying the rules for 2's complement numbers, we obtain the following results.

binary value	*2's complement value*
00000000	00000000
00000001	11111111
10101010	01010110
11111111	00000001

In this case we see that both the value of $+0_2$ and -0_2 are identical. The second binary number is $+1_2$; hence $-1_2 \equiv 11111111_2$. We also see that the final value is 11111111_2 or -1_2, so that $-(-1_2) = 00000001_2$; that is $-(-1_2) = +1_2$.

18.10 Binary subtraction

If $a - b = c$, a is the **minuend**, b the **subtrahend** and c the **difference**. We can re-write the equation in the following way:

$$a + (-b) = c$$

That is, to subtract a number, *we add its negative value to the minuend*, i.e. we add the binary complement of b to a. There are different rules for the 1's and 2's complement subtraction, and we explain the rules for the 2's complement below.

To subtract using the 2's complement, add the 2's complement of the subtrahend to the minuend.

If the sign bit of the sum is '0', the result is the true difference. If the sign bit of the sum is '1', the result is the 2's complement of the difference. Any overflow (NOT the sign bit) produced by the calculation is 'lost'.

Worked Example 18.6

Subtract 110100_2 (or 52_{10}) from 1011001_2 (or 89_{10}).

Solution

Initially we express 110100_2 as an 8-bit 'word', and then convert it into its 2's complement as follows:

$$\text{2's complement of } 00110100_2 = 11001100_2$$

Next we add the 2's complement to the 8-bit version of the minuend as follows:

minuend =	0101 1001
2's complement of subtrahend =	1100 1100
ADD	(1) 0010 0101

overflow (lost) ← (1)

The overflow bit (the 9th bit) is ignored. That is

$$0101\,1001_2 - 0011\,0100_2 = 0010\,0101_2$$

or $$101\,1001_2 - 110100_2 = 100101_2$$

This can be verified by the calculation

$$89_{10} - 52_{10} = 37_{10} \equiv 100101_2.$$

Worked Example 18.7

Subtract the 8-bit signed binary number 0001 0000 (or $+16_{10}$) from 1111 1110 (or -2_{10}).

Solution

The 2's complement of 0001 0000 is 1111 0000. The subtraction is performed as follows:

minuend =	1111 1110
2's complement of subtrahend =	1111 0000
ADD	(1) 1110 1110

overflow (lost) ← (1)

Once again, the overflow bit is ignored. Since the m.s.b. (the 8th bit) of the solution is '1', *the answer is negative*, and the result is given in 2's complement form. The *magnitude* of the final solution is

$$\text{2's complement of } 1110\,1110 = 0001\,0010_2 \text{ or } 18_{10}$$

Hence

$$1111\,1110_2 - 0001\,0000_2 = 1110\,1110_2 \equiv -18_{10}$$

This can be seen to be correct from the calculation

$$-2_{10} - (+16_{10}) = -18_{10}$$

18.11 Binary multiplication

If $a \times b = c$, a is the **multiplicand**, b the **multiplier** and c the **product**. *Hand multiplication* of binary numbers is generally similar to hand multiplication of decimal numbers, with the exception that we only need to multiply either by '1' or '0'.

In a computer or calculator, the multiplicand, the multiplier and the product are stored in *registers* in the machine, the product register being a special 'double length' register.

Worked Example 18.8

Multiply 1000.1_2 (or 8.5_{10}) by 110.1_2 (or 6.5_{10}).

Solution

As with decimal multiplication by hand, we initially 'remove' the radix points, and re-insert them later. In the following, 'mb' is the value of the multiplier bit.

multiplicand	1 0 0 0 1	
multiplier	1 1 0 1	
partial products	1 0 0 0 1	mb = 1
	0 0 0 0 0	mb = 0
	1 0 0 0 1	mb = 1
	1 0 0 0 1	mb = 1
sum of partial products	1 1 0 1 1 1 0 1	

The reader will see from the above why the product register length must be twice the length of the other registers. Since the multiplier has a length of 4 bits, there are four partial products in the process. Also, the total number of binary places in the multiplicand and the multiplier is two, the solution is

$$1000.1_2 \times 110.1_2 = 110111.01_2 \equiv 55.25_{10}.$$

18.12 Binary division

If $a \div b = c$, a is the **dividend**, b the **divisor** and c the **quotient**. Binary *hand division* is generally the same as for hand division of decimal numbers, with the exception that subtracting the divisor is carried out by *2's complement addition*.

The maximum 'length' of the quotient which can be stored by a calculator or computer depends on the length of the register in which the answer is stored.

Worked Example 18.9

Divide 111.1_2 (or 7.5_{10}) by 10.1_2 (or 2.5_{10}).

Solution

In this case, we 'remove' the binary points by 'moving' both binary points one place to the right to produce a simple integer ratio as follows:

$$\frac{111.1}{10.1} = \frac{1111 \div 2^0}{101 \div 2^0} = \frac{1111}{101}$$

The integer ratio gives the same answer. The process of subtracting the divisor from the dividend is carried out by 2's complement addition, as shown below. Each time we can divide by the divisor, we record a '1' in the quotient line, otherwise we record a '0' and bring down the next bit from the dividend.

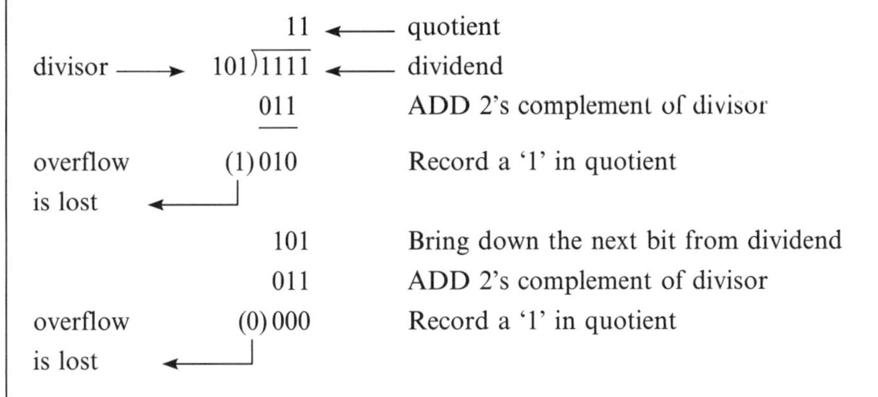

That is $111.1_2/10.1_2 = 11_2$, corresponding to $7.5_{10}/2.5_{10} = 3_{10}$.

18.13 Logic gates

The basic logic gates used in electronics are the **NOT, AND, OR, NAND, NOR** and **EXCLUSIVE-OR** (or **NOT-EQUIVALENT**) gates. Typical logic gate

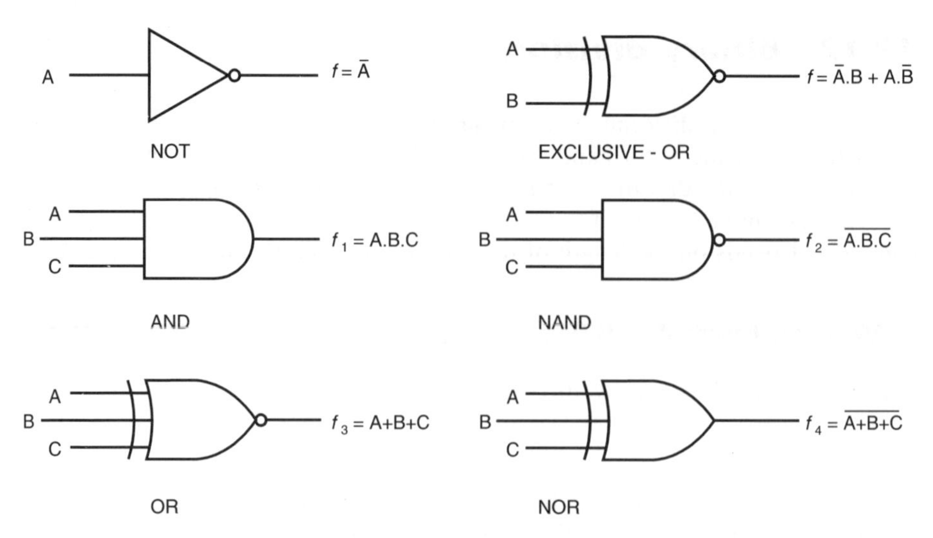

Figure 18.1 *International symbols for logic gates*

symbols are shown in Figure 18.1. The NOT gate has one input signal, the EXCLUSIVE-OR gate has two inputs, and the AND, NAND, OR and NOR gates can have many input signals (three are shown in the Figure). Truth tables for the gates are listed in Table 18.3.

Table 18.3 Truth table of basic gates

NOT *gate*		EXCLUSIVE-OR *gate*			AND (f_1), NAND (f_2), OR (f_3) and NOR (f_4)						
INPUT	OUTPUT	INPUTS		OUTPUT	INPUTS						
A	$\bar{A}$	A	B	f	A	B	C	f_1	f_2	f_3	f_4
0	1	0	0	0	0	0	0	0	1	0	1
1	0	1	0	1	1	0	0	0	1	1	0
		0	1	1	0	1	0	0	1	1	0
		1	1	0	1	1	0	0	1	1	0
					0	0	1	0	1	1	0
					1	0	1	0	1	1	0
					0	1	1	0	1	1	0
					1	1	1	1	0	1	0

Signals produced by computer systems and electronic logic gates are typically logic '0' (often corresponding to zero volts) and logic '1' (often corresponding to a voltage in the range +3 V to +5 V).

The relationship between the input signal (or input signals) to a gate and its output are given by the *truth table* of the gate, and those for a **NOT gate**, and an **EXCLUSIVE-OR** (or **NOT-EQUIVALENT**) gate are given in Table 18.3. The truth table for these gates are summarized as follows:

The output from a NOT gate is the logical opposite to the input signal (i.e. the output is NOT EQUAL to the input).

The output from an EXCLUSIVE-OR gate is '1' when signal A is NOT EQUIVALENT to signal B.

Additionally, the truth table for a 3-input **AND gate** (f_1), **NAND gate** (f_2), **OR gate** (f_3), and **NOR gate** (f_4) are also listed. The truth table for the AND, NAND, OR and NOR gates are summarized below.

The output from an AND gate is '1' only when ALL inputs are '1'.
The output from a NAND gate is '0' when ALL inputs are '1'.
The output from an OR gate is '1' when ANY input is '1'.
The output from a NOR gate is '0' when ANY input is '1'.

A NAND gate can be thought of as an AND gate followed by a NOT gate, as shown in Figure 18.2(a). A NOR gate can be thought of as an OR gate followed by a NOT gate, as illustrated in Figure 18.2(b).

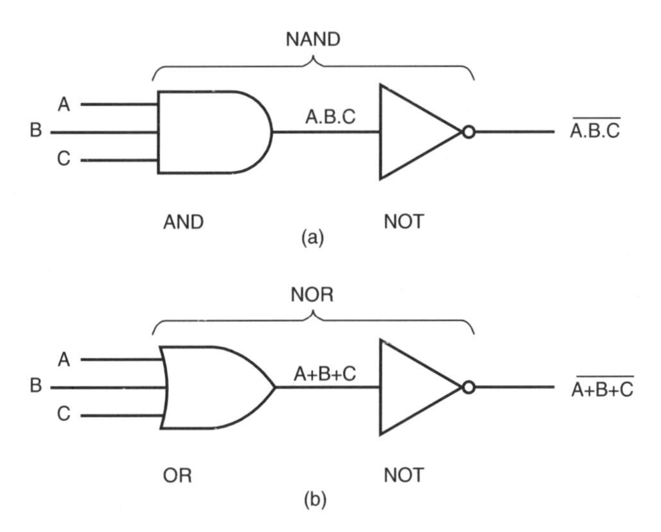

Figure 18.2 *Simple versions of: (a) NAND; and (b) NOR gates*

18.14 Logic circuit design from a Boolean equation

The mathematics of logic was first written down by the Reverend George Boole (who gave rise to the mathematics of Boolean algebra) in the nineteenth century. In the following, a '+' means the logical OR function, a '.' means the logical AND function, and a bar over the top of a variable or function, such as $\bar{A}$, means NOT A.

The basic theorems of logical algebra are as follows:

$A + 0 = A$	(1)	$A.0 = 0$	(2)
$A + 1 = 1$	(3)	$A.1 = A$	(4)
$A + A = A$	(5)	$A.A = A$	(6)
$A + \bar{A} = A$	(7)	$A.\bar{A} = 0$	(8)
$\bar{\bar{A}} = 1$	(9)		

We will refer to some of the numbered theorems in Worked Examples which follow.

There are also several laws referring to Boolean algebra, and the principal ones are given below.

The **commutative law**, which states that the order in which a variable appears is irrelevant, for example

$$A + B = B + A$$
$$A.B = B.A$$

The **associative law**, which states that the order in which identical functions are performed is irrelevant, such as

$$A + B + C = (A + B) + C = A + (B + C)$$
$$A.B.C = (A.B).C = A.(B.C)$$

The **distributive law**, which is expressed in two forms, namely the *product of sums form* and the *sum of products form* as follows:

$$A + (B.C.D. \ldots) = (A + B).(A + C).(A + D). \ldots$$
$$A.(B + C + D + \ldots) = A.B + A.C + A.D + \ldots$$

Worked Example 18.10

Signals from sensors A, B and C in a security system must give an indication of an intruder when the following logical equation is satisfied.

$$f = \bar{A}.B.C + A.\bar{B}.C + A.B.\bar{C}$$

where f is the output from the logic system. Simplify the expression for f, and draw a logic block diagram for the system.

Solution

The equation tells us that the alarm system is activated, i.e. $f = 1$, when any one or more of the three combinations on the right-hand side of the expression is satisfied. Initially, we can group the first and third combinations together as follows:

$$
\begin{aligned}
f &= A.B.C + A.B.\bar{C} + A.\bar{B}.C && \text{(commutative law)}\\
&= A.B.(C + \bar{C}) + A.\bar{B}.C && \text{(distributive law)}\\
&= A.B.1 + A.\bar{B}.C && \text{(theorem 7)}\\
&= A.B + A.\bar{B}.C && \text{(theorem 4)}\\
&= A.(B+\bar{B}.C) && \text{(2nd distributive law)}
\end{aligned}
$$

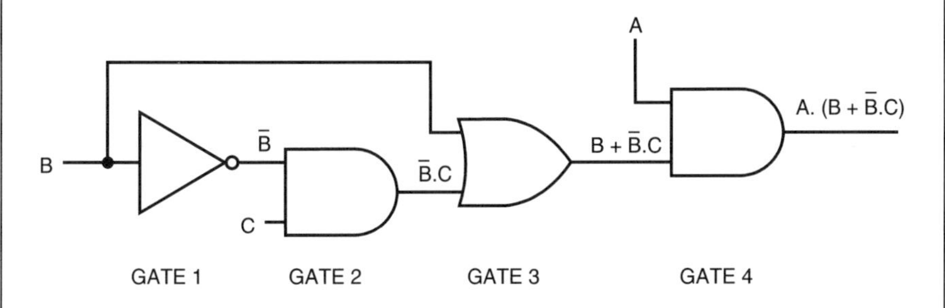

Figure 18.3 Block diagram for Worked Example 18.10

The resulting block diagram for this expression is given in Figure 18.3. The NOT gate (gate 1) produces $\bar{B}$, and this is ANDed in gate 2 with signal C to form the combination $\bar{B}.C$. In turn, this is ORed in gate 3 with signal B to give $(B + \bar{B}.C)$. Finally, in gate 4, the signal from gate 3 is ANDed with signal A to produce $A.(B+\bar{B}.C)$.

Table 18.4 Truth table for Worked Example 18.10

Line number	*Inputs* A	B	C	*Gate 1 output* $\bar{B}$	*Gate 2 output* $\bar{B}.C$	*Gate 3 output* $B+\bar{B}.C$	*Gate 4 output* $A.(B+\bar{B}.C)$
1	0	0	0	1	0	0	0
2	1	0	0	1	0	0	0
3	0	1	0	0	0	1	0
4	1	1	0	0	0	1	1
5	0	0	1	1	1	1	0
6	1	0	1	1	1	1	1
7	0	1	1	0	0	1	0
8	1	1	1	0	0	1	1

We now draw up the truth table for the above solution (see Table 18.4). Initially, we draw up a list of all the possible combinations of the input signals (see the input signal list for A, B and C). Next we look at the output from gate 1 in Figure 18.3, that is output $\bar{B}$, and put down the logical inverse of the signal in input column B. After this, we study the output of gate 2, which is the combination $\bar{B}.C$; this has a value of '0' either when $\bar{B} = 0$ or when $C = 0$. This is ORed with signal B in gate 3, so that the output from this gate is '1' whenever $B = 1$ or when $\bar{B}.C = 1$. The final output from the system is '1' in lines 4, 6 and 8; that is when any of the following combinations exists.

$$A.B.\bar{C} = 1$$

$$A.\bar{B}.C = 1$$

$$A.B.C = 1$$

That is

$$f = A.B.\bar{C} + A.\bar{B}.C + A.B.C$$

$$= A.B.C + A.B.\bar{C} + A.\bar{B}.C \qquad \text{(commutative law)}$$

The solution given above is not the only possible solution, because the following also work.

$$f = A.(C + B.\bar{C})$$

$$f = A.(B + C)$$

The reader should verify that the above equations satisfy the problem.

18.15 Computer listings

There are two items of BASIC language software associated with this chapter, and they are

Computer listing 18.1 – Convert a decimal value into binary.
Computer listing 18.2 – Convert a binary value into decimal.

Listing 18.1 Converting a decimal value to a binary value

```
10 CLS '**** CH18-1 ****
20 PRINT "CONVERTING A DECIMAL VALUE TO A BINARY VALUE"
30 PRINT "(TO 6 BINARY PLACES)": PRINT
40 INPUT "Decimal value = ", D
50 PRINT "INTEGER part (FIRST VALUE is the LEAST SIGNIFICANT BIT)"
60 WHOLE = FIX(D): FRACT = D - WHOLE
70 A = WHOLE / 2
80 B = INT(A)
90 IF B = A THEN PRINT "0" ELSE PRINT "1"
100 A = B / 2
110 IF A > .5 THEN GOTO 80
120 IF A = .5 THEN PRINT "1"
130 IF FRACT = 0 THEN END
140 PRINT "FRACTIONAL PART (to 6 binary places)."
150 PRINT ".";
160 N = 0 ' ** N is a 'counter' **
170 A = 2 * FRACT
```

```
180 IF A < 1 THEN PRINT "0";
190 IF A = 1 THEN PRINT "1": END
200 IF A > 1 THEN A = A - 1: PRINT "1";
210 A = 2 * A: N = N + 1
220 IF N = 6 AND A > 0 THEN PRINT " Binary fraction continues."
230 IF N = 6 THEN END
240 GOTO 180
```

In line 40 of Listing 18.1, you are asked to supply a decimal value (which may have a fractional part), which is converted into a binary value which has up to six binary fractional places. The method used here is the same as that used earlier in this chapter, so that the integer and fractional parts are dealt with separately.

In line 60, the instruction FIX(D) has the effect of separating the integer or WHOLE part from the number, and FRACT is the fractional part. Lines 70 to 130 convert the decimal integer value into its binary equivalent. Since the value is calculated as described earlier in this chapter, the integer result is printed in a single column on the left of the screen, with the least significant bit at the top of the screen, and the most significant bit at the bottom. The INT(A) instruction in line 80 returns the integer value which is either less than or equal to A.

Lines 160 to 240 use the method illustrated in this chapter to calculate the value of the fractional part up to six binary places long. If the binary fractional part continues beyond six places, line 220 causes the text 'Binary fraction continues' to be displayed.

If the number does not have a fractional part, the program will convert a decimal integer of up to 2 097 151 into binary. If the number has a fractional part, the magnitude of the integer part may be up to 131 071.

Listing 18.2 Converting a binary number to decimal

```
10 CLS '**** CH18-2 ****
20 PRINT "CONVERT A BINARY NUMBER TO DECIMAL": PRINT
30 PRINT "The number is limited to the length"
40 PRINT "          ABCD EFGH.IJKL"
50 PRINT "where A has the value 2^7, H has the value 2^0,"
60 PRINT "I has the value 1/2 and L has the value 1/16."
70 PRINT
80 INPUT "Value of A (0 or 1) = ", A: A = A * 2 ^ 7
90 INPUT "Value of B (0 or 1) = ", B: B = B * 2 ^ 6
100 INPUT "Value of C (0 or 1) = ", C: C = C * 2 ^ 5
110 INPUT "Value of D (0 or 1) = ", D: D = D * 2 ^ 4: PRINT
120 INPUT "Value of E (0 or 1) = ", E: E = E * 2 ^ 3
130 INPUT "Value of F (0 or 1) = ", F: F = F * 2 ^ 2
140 INPUT "Value of G (0 or 1) = ", G: G = G * 2 ^ 1
150 INPUT "Value of H (0 or 1) = ", H
160 PRINT "BINARY POINT HERE"
170 INPUT "Value of I (0 or 1) = ", I: I = I / 2
```

```
180 INPUT "Value of J (0 or 1) = ", J: J = J / 4
190 INPUT "Value of K (0 or 1) = ", K: K = K / 8
200 INPUT "Value of L (0 or 1) = ", L: L = L / 16
210 Vi = A + B + C + D + E + F + G + H
220 Vf = I + J + K + L: PRINT
230 PRINT "Decimal value = "; Vi + Vf
240 END
```

Listing 18.2 converts a binary value up to 1111 1111.1111 (or 255.9375_{10}) into decimal. The smallest binary value which can be converted is equivalent to 0.0625_{10}. The value V_i in line 210 is the integer part of the answer, and V_f is the fractional part.

The program asks you to supply each binary digit in turn, after which it determines the equivalent decimal value.

Exercises

18.1 Translate the following decimal numbers into their pure binary equivalents: (a) 5268, (b) 23.75, (c) 0.0125.

18.2 Convert the following into decimal: (a) 7892_9, (b) 2734_8, (c) 253_6, (d) 421_5.

18.3 Convert the following into pure binary: (a) 1/16, (b) 7/8, (c) 2/3, (d) 17/32.

18.4 Convert the following pure binary numbers into decimal: (a) 10111011, (b) 0.0111, (c) 10111.01.

18.5 Convert the following decimal numbers into values which have the following base or radix: (a) 989 into base 8, (b) 732 into radix 6, (c) 876 into radix 5, (d) 932 into base 3.

18.6 Convert the following into the numbering systems with the given base: (a) 857_9 into base 6, (b) 576_8 into base 9, (c) 222_5 into base 3.

18.7 Convert the following decimal numbers into pure binary numbers, and then add them together using binary arithmetic: (a) 4 + 7, (b) 16 + 20, (c) 17.5 + 12.75, (d) $5\frac{5}{16} + 2\frac{3}{8}$, (e) $7.5 + 5.3 + 5\frac{7}{8}$.

18.8 Convert the following decimal numbers into 8-bit pure binary numbers, and subtract them using 2's complement arithmetic: (a) 5 – 2, (b) 20 – 10, (c) 4 – 5, (d) 6.25 – 4.25.

18.9 Convert the following values into pure binary, and multiply them together: (a) 6×3, (b) 7.5×3.75, (c) 0.25×4.25.

18.10 Convert the following decimal numbers into pure binary, and divide them using binary arithmetic: (a) $6 \div 3$, (b) $7.5 \div 2.6$, (c) $1.125 \div 6$.

18.11 Using truth tables, verify that

$$\text{A} + \text{B.C} = (\text{A} + \text{B}).(\text{A} + \text{C})$$

$$\text{A}.(\text{B} + \text{C}) = \text{A.B} + \text{A.C}$$

18.12 Using Boolean algebra, prove that

$$A.\bar{\text{C}} + B + C.\bar{\text{B}} = A + B + C$$

18.13 Which of the following is equal to logic '0'?
(a) $A + 1$, (b) $A.\bar{A}$, (c) $A + A.\bar{A}$, (d) $\bar{A} + 1$,
(e) $(\bar{A} + \bar{B}).B.(A+\bar{A}.\bar{B})$

18.14 What is the simplest Boolean expression using simple AND, OR and NOT terms which satisfies Table 18.5?

Table 18.5 Truth table for Exercise 18.14

Inputs			*Output*
A	B	C	f
0	0	0	0
1	0	0	1
0	1	0	1
1	1	0	1
0	0	1	1
1	0	1	0
0	1	1	1
1	1	1	1

18.15 What input conditions at A and B in Figure 18.4 give an output of (a) $f = 0$, (b) $f = 1$?

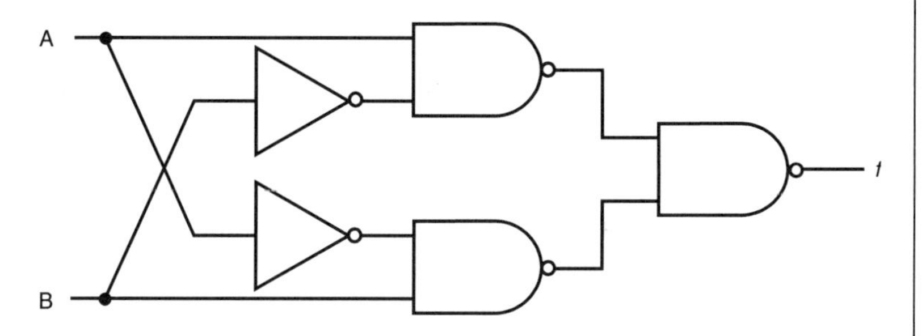

Figure 18.4 Block diagram for Exercise 18.15

18.16 One row is missing from table 18.6. Add the details of the missing line.

Table 18.6 Truth table for Exercise 18.16

Inputs		*Output*
A	B	$f = \bar{A} + B$
0	0	1
0	1	1
1	1	1

Summary of important facts

Any number can be written down in the following form

$$6 \times 10^8$$

where 6 is a **multiplying coefficient**, 10 is the **base** or **radix** of the numbering system, and 8 is the **power** or **exponent**.

Numbering systems in popular use are **binary** (base 2), **octal** (base 8), **decimal** or **denary** (base 10) and **hexadecimal** (base 16). When we need to show what the base of the numbering system we are using, we write it down as a *decimal subscript* to the number. For example 1010_2 is a binary number, 1010_{10} is a decimal number, and 1010_{16} is a hexadecimal number.

If $a + b = c, a$ is the **augend**, b the **addend** and c the **sum**. A **carry** produced by the addition is known as the **carry-out**, which becomes the **carry-in** for the next higher stage of the addition.

A *binary digit* is known as a **bit**. In an **unsigned binary number**, every value is positive. In a **signed binary number**, the most significant bit (m.s.b.) of the number is used as the **sign bit**. If the sign bit is '0', the number has a positive value, and is stored as a pure binary value. If the sign bit is '1', the number is negative, and is stored in **binary compliment form**. There are two forms of binary complement and are, respectively, the **2's complement** (or *true complement*) and **1's complement** (or *reduced complement* form).

If $a - b = c, a$ is the **minuend**, b the **subtrahend** and c the **difference**. Subtraction is carried out by *adding the 2's complement* of the subtrahend to the minuend, i.e. $c = a - b = a + (-b)$.

If $a \times b = c, a$ is the **multiplicand**, b the **multiplier** and c the **product**.

If $a \div b = c, a$ is the **dividend**, b the **divisor** and c the **quotient**.

The basic logic gates are the **NOT**, the **AND**, the **OR**, the **NAND**, the **NOR** and the **EXCLUSIVE-OR** (or **NOT-EQUIVALENCE**). A NOT gate has only one input signal, an EXCLUSIVE-OR gate has two input signals, and the other gates may have many inputs.

The operation of a gate is described in terms of a **truth table**, which lists the output from the gate for every possible combination of input signals. The input and output signals are either logic '0' or logic '1', corresponding to 'false' or 'true'. The truth table for the basic gates are summarised below.

The output from a NOT gate is the logical compliment or inverse of the input signal.
The output from an EXCLUSIVE-OR gate is '1' if the two input signals are not equivalent to one another, otherwise the output is '0'.
The output from an AND gate is '1' when ALL the input signals are '1', otherwise the output is '0'.
The output from a NAND gate is '0' when ALL the input signals are '1', otherwise the output is '1'.
The output from an OR gate is '1' when ANY input signal is '1', otherwise the output is '0'.
The output from a NOR gate is '0' when ANY input signals is '1', otherwise the output is '1'.

19 Complex numbers

19.1 Introduction

So far we have dealt with phasor quantities which involve 'in-phase' and 'quadrature' components using conventional mathematics and trigonometry. Mathematicians have devised a method of dealing with phasor quantities with relative ease, and it is known as complex numbers.

The reader should not confuse 'complex numbers' with 'complicated numbers', because complex numbers do not involve any more mathematics than we have dealt with already in this book. Anyone hoping to extend their knowledge of electrical and electronic engineering will find that the information in this chapter particularly useful. By the end of this chapter, the reader will be able to

- Understand what a complex number is.
- Appreciate and use 'operator j'.
- Manipulate complex numbers in rectangular and polar form.
- Change a rectangular form of a complex number into its polar version, and vice versa.
- Add, subtract, multiply and divide complex numbers.
- Understand and use the conjugate of a complex number.
- Apply complex numbers to a.c. circuit calculations.
- Use software to solve complex number calculations.

19.2 The operator j

Consider the right-angled triangle in Figure 19.1(a). Side B is at right-angles to side A, and we say that side B is in *quadrature* with side A. We show this fact simply by saying that side B can be expressed as jB, where j simply means 'at right angles to the reference direction'. It is therefore possible to write down the following expression for side C

$$\mathbf{C} = A + jB$$

The reader will note that $\mathbf{C}$ is printed in **bold**, which indicates to us that it is a **complex number**, so that part of its length lies in the horizontal or *reference direction*, and part of its length is in the perpendicular or *quadrature direction*.

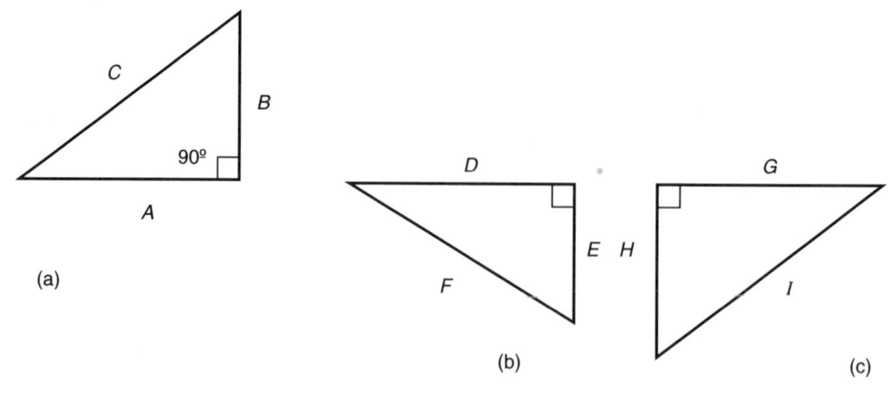

Figure 19.1 Introduction to the operator j

Alternatively, if the quadrature part is negative, as shown in Figure 19.1(b), we say that

$$\boldsymbol{F} = D - jE$$

and if both the horizontal part and the quadrature part are negative, as shown in Figure 19.1(c), we get

$$\boldsymbol{I} = -G - jH$$

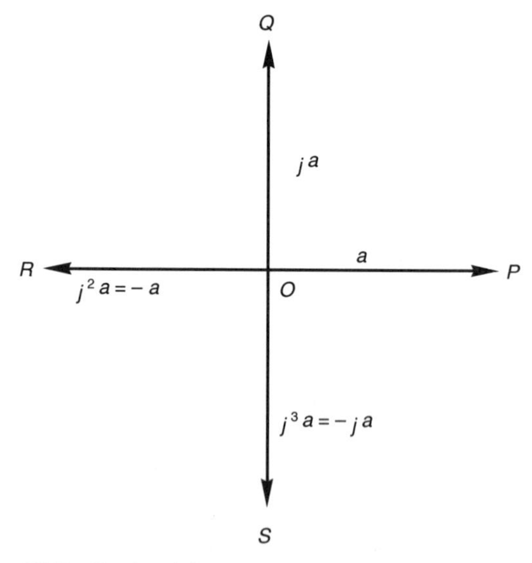

Figure 19.2 Further information about the operator j

Let us take another look at the process of rotating a line through 90°. Consider the line OP in Figure 19.2, which has length a. If we rotate this line through 90°, it becomes the line OQ, where

$$\boldsymbol{OQ} = ja$$

If we rotate OQ through another 90°, it arrives at OR, hence

$$\boldsymbol{OR} = ja \times j = j^2 a$$

(remember, 'j' simply implies the rotation of the line through 90°). However, since

$$\boldsymbol{OR} = -a$$

then

$$j^2 a = -a$$

that is

$$j^2 = -1$$

or

$$j = \sqrt{(-1)}$$

However, $\sqrt{(-1)}$ does not have a 'real' value because *any value* which is multiplied by itself gives a positive number. For this reason, j is described as an **imaginary operator**. However, so far as engineers are concerned, we can think of an 'imaginary' number in terms of the following expression:

imaginary number = imaginary operator (j) × real value

Once again, the process of multiplying a real value by the imaginary operator can be thought of as *rotating the real number in an anti-clockwise direction through 90°*.

Continuing around Figure 19.2, we see that

$$\boldsymbol{OS} = j \times \boldsymbol{OR} = j \times j^2 a = j^3 a$$

and

$$\boldsymbol{OP} = j \times \boldsymbol{OS} = j \times j^3 a = j^4 a = j^2 \times j^2 a$$
$$= (-1) \times (-1)a = a$$

It also follows that

$$\boldsymbol{OS} = j^3 a = j^2 \times ja = -ja.$$

19.3 The rectangular form and polar form of a complex number

Consider the triangle in Figure 19.3. The hypotenuse, r, can be expressed as the complex number

$$\boldsymbol{r} = a + jb$$

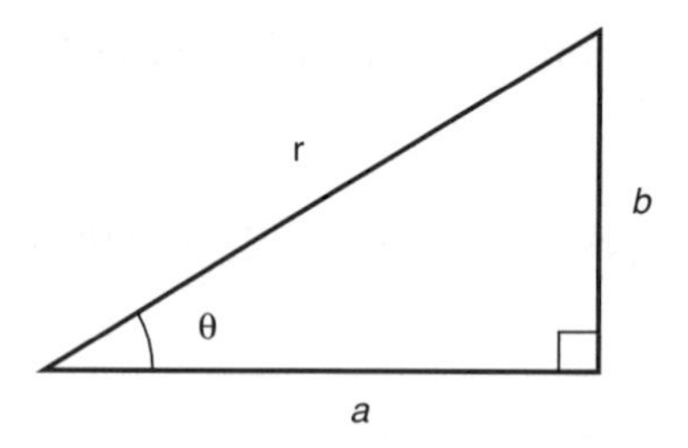

Figure 19.3 Rectangular and polar form of a complex number

This form of representation is known as the **rectangular form** or **Cartesian form** (after Descartes, the French mathematician) of the complex number.

The length a is known as the **real part** of the number, and jb as the **imaginary part**. Mathematicians replace 'j' by 'i' but, since i is used for current by engineers, we use 'j'.

It is also useful to express the complex value in terms of its magnitude or **modulus**, r, and its angle with respect to the reference direction, θ, which is known as the **argument**. The complex number $\mathbf{r}$ in Figure 19.3 can then be written in the form

$$\mathbf{r} = r\angle\theta$$

This form of representation is known as the **polar form** of the complex number. The reader should note that the expression '$\angle$' means 'at an angle of'.

RELATIONSHIP BETWEEN THE RECTANGULAR FORM AND POLAR FORM OF COMPLEX NUMBERS

The rectangular and polar form of a complex number are related by the following.

$$r = \sqrt{(a^2 + b^2)}$$
$$\theta = \arctan(b/a)$$

and

$$a = r\cos\theta$$
$$b = r\sin\theta$$

Worked Example 19.1

Convert the following rectangular complex values into their polar equivalents: (a) $2 + j3$, (b) $-3 + j4$, (c) $-4 - j2$, (d) $3 - j3$.

Solution

Each of the complex numbers can be represented on an **Argand diagram** (see Figure 19.4) – named after Jean Robert Argand.

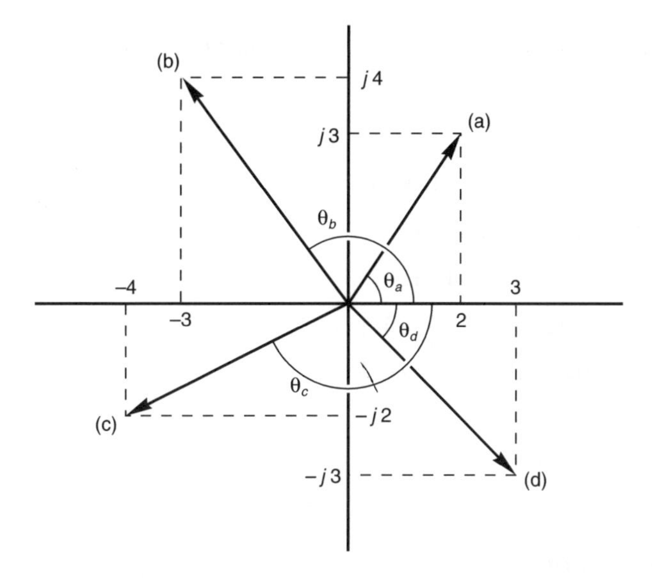

Figure 19.4 The Argand diagram for the complex numbers in Worked Example 19.1

(a) The magnitude of the value $(2 + j3)$ is

$$\sqrt{(2^2 + 3^2)} = 3.61$$

and its angle with respect to the reference direction is

$$\theta_a = \arctan(3/2) = 56.31^\circ$$

That is

$$2 + j3 \equiv 3.61\angle 56.31^\circ$$

(b) The magnitude of $(-3 + j4)$ is

$$\sqrt{((-3)^2 + 4^2)} = 5$$

Using the values given to determine the angle, a calculator will give its *principal value* which, since the angle is in the second quadrant, will be incorrect. The correct angle is

$$\begin{aligned} \theta_b &= \arctan(4/(-3)) + 180^\circ = -53.13^\circ + 180^\circ \\ &= 126.9^\circ \end{aligned}$$

hence

$$-3 + j4 \equiv 5\angle 126.9^\circ$$

(c) The magnitude of this value is

$$\sqrt{((-4)^2 + (-2)^2)} = 4.47$$

This complex number lies in the third quadrant (see Figure 19.4) and, since the tangent of the angle is positive, a calculator would indicate that the angle is in the first quadrant. The correct angle is

$$\theta_c = \arctan((-2)/(-4)) - 180° = 26.57° - 180°$$
$$= -153.43°$$

or

$$-4 - j2 \equiv 4.47\angle(-153.43°)$$

(d) In this case the magnitude is

$$\sqrt{(3^2 + (-3)^2)} = 4.24$$

and the phase angle is (see Figure 19.4)

$$\theta_c = \arctan((-3)/3) = -45°$$

or

$$3 - j3 \equiv 4.24\angle(-45°).$$

Worked Example 19.2

Convert the following polar complex values into their rectangular form: (a) $3\angle 60°$, (b) $5\angle 120°$, (c) $4\angle 200°$, (d) $6\angle 300°$.

Solution

The Argand diagram showing the complex values in this example is shown in Figure 19.5.

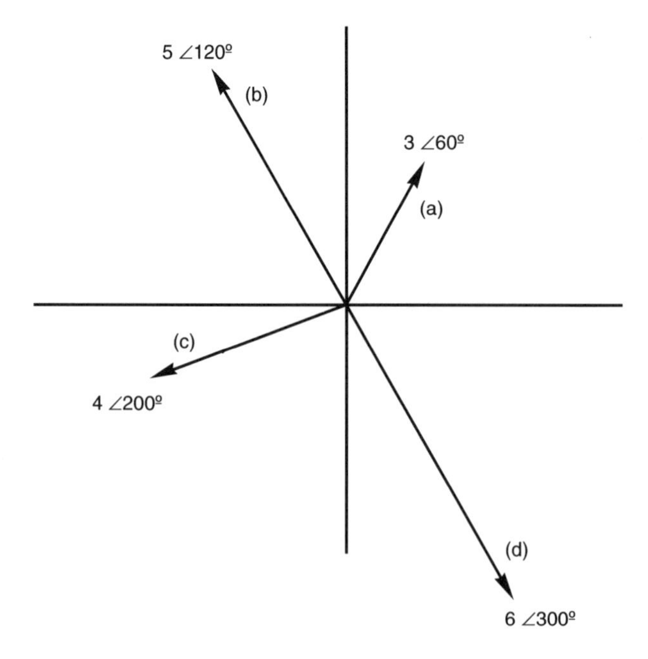

Figure 19.5 The Argand diagram for the complex numbers in Worked Example 19.2

(a) The real or 'inphase' part of $3\angle 60^\circ$ is

$$3\cos 60^\circ = 1.5$$

and the 'imaginary' or quadrature part is

$$3\sin 60^\circ = 2.6$$

hence

$$3\angle 60^\circ \equiv 1.5 + j2.6$$

(b) The real or 'inphase' part of $5\angle 120^\circ$is

$$5\cos 120^\circ = -2.5$$

and the imaginary part is

$$5\sin 120^\circ = 4.33$$

therefore

$$5\angle 120^\circ \equiv -2.5 + j4.33$$

(c) In this case the real or 'inphase' part of $4\angle 200^\circ$ is

$$4\cos 200^\circ = -3.76$$

and the quadrature part is

$$4\sin 200^\circ = -1.37$$

or

$$4\angle 200^\circ \equiv -3.76 - j1.37$$

The reader should note that

$$200^\circ \equiv 200^\circ - 360^\circ = -160^\circ$$

hence

$$4\angle 200^\circ \equiv 4\angle(-160^\circ) \equiv -3.76 - j1.37$$

(d) The real part of the final number is

$$6\cos 300^\circ = 3$$

and the quadrature part is

$$6\sin 300^\circ = -5.2$$

hence

$$6\angle 300^\circ \equiv 3 - j5.2$$

Also

$$300^\circ = 360^\circ - 300^\circ \equiv -60^\circ$$

therefore

$$6\angle 300^\circ \equiv 6\angle(-60^\circ) \equiv 3 - j5.2$$

19.4 The effect of a negative magnitude

If we have a complex number $\boldsymbol{a}$, then what is meant by $-\boldsymbol{a}$? In fact, the process of multiplying a number by (-1) is simply to rotate $\boldsymbol{a}$ through $180°$, as shown in Figure 19.6.

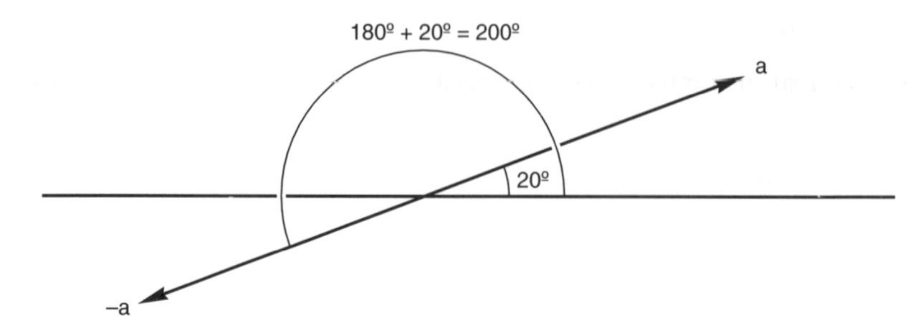

Figure 19.6 The effect of a negative magnitude

For example, if $\boldsymbol{a}$ has the value $4\angle 20°$, then

$$-\boldsymbol{a} = 4\angle(20° + 180°) = 4\angle 200° \equiv 4\angle(-160°).$$

Worked Example 19.3

Convert the following polar complex values into their equivalent rectangular complex values: (a) $-3\angle 60°$, (b) $-4\angle(-300°)$.

Solution

(a) We can write the complex number as follows:

$$-3\angle 60° = 3\angle(60 + 180)° = 3\angle 240°$$

The real part of this number is

$$3\cos 240° = -1.5$$

also

$$-3\cos 60° = -1.5$$

and the imaginary part is

$$3\sin 240° = -2.6$$

also

$$-3\sin 60° = -2.6$$

hence

$$-3\angle 60° \equiv -1.5 - j2.6$$

(b) In this case

$$-4\angle(-300)^\circ = 4\angle(-360 + 180)^\circ = 4\angle - 120^\circ$$

The real part of this number is

$$4\cos(-120^\circ) = -2$$

or

$$-4\cos(-300^\circ) = -2$$

and its imaginary part is

$$4\sin(-120^\circ) = -3.46$$

or

$$-4\sin(-300^\circ) = -3.46$$

hence

$$-4\angle(-300^\circ) \equiv -2 - j3.46$$

19.5 Addition of complex numbers

Before we can add complex numbers together, **each of them MUST be converted into their equivalent rectangular form**, and we proceed as follows:

1. add the real parts together to give the real part of the sum, and
2. add the imaginary parts together to give the imaginary part of the sum.

That is, if $\boldsymbol{A} = a + jb$ and $\boldsymbol{M} = c + jd$, then

$$\boldsymbol{A} + \boldsymbol{M} = (a + jb) + (c + jd) = (a + c) + j(b + d)$$

The sum of the real parts $(a + c)$ is the real part of the answer, and the sum of the imaginary parts $(b + d)$ is the imaginary part of the answer.

The reader should note that it is possible to add many complex numbers together by this method.

Complex numbers can also be *added together graphically* using the method described for adding phasors in Chapter 8.

Worked Example 19.4

Add the following complex values together:

(a) $(3 + j4) + (5 - j6)$
(b) $(-2 + j6) + 4\angle(-60^\circ)$
(c) $-3\angle 60^\circ + 5\angle 40^\circ$
(d) $(2 + j3) + (3 - j4) + 0.7071\angle 45^\circ$.

Solution

(a) Since both values in this part are given in rectangular complex form, they can be added together directly as follows:

$$\begin{aligned}(3+j4)+(5-j6) &= (3+5)+j(4-6)\\ &= 8-j2\end{aligned}$$

This is equivalent to $8.25\angle(-14^\circ)$ (the reader should verify this fact)

(b) The second value in this section is given in polar form, which must first be converted into rectangular form as follows:

$$\text{real part of } 4\angle(-60^\circ) = 4\cos(-60^\circ) = 2$$
$$\text{imaginary part of } 4\angle(-60^\circ) = 4\sin(-60^\circ) = -3.46$$

hence

$$4\angle(-60^\circ) \equiv 2-j3.46$$

The solution is therefore

$$\begin{aligned}(-2+j6)+4\angle(-60^\circ) &= (-2+j6)+(2-j3.46)\\ &= (-2+2)+j(6-3.46)\\ &= 0+j2.54 = j2.54\end{aligned}$$

That is, the 'real' part of the sum is zero, so that the solution is

$$j2.54 \equiv 2.54\angle 90^\circ$$

(c) In this case we must convert both values into their rectangular form. For the value $-3\angle 60^\circ$

$$\begin{aligned}\text{real part} &= -3\cos 60^\circ = -1.5\\ \text{imaginary part} &= -3\sin 60^\circ = -2.6\end{aligned}$$

and for the value $5\angle 40^\circ$ we have

$$\begin{aligned}\text{real part} &= 5\cos 40^\circ = 3.83\\ \text{imaginary part} &= 5\sin 40^\circ = 3.21\end{aligned}$$

The result is

$$\begin{aligned}-3\angle 60^\circ + 5\angle 40^\circ &\equiv (-1.5-j2.6)+(3.83+j3.21)\\ &= (-1.5+3.83)+j(-2.6+3.21))\\ &= 2.33+j0.61\end{aligned}$$

which is equivalent to $2.41\angle 14.67^\circ$

(d) In this case, we must convert the polar complex quantity $(0.7071\angle 45^\circ)$ into its rectangular equivalent, which is

$$\begin{aligned}\text{real part} &= 0.7071\cos 45^\circ = 0.5\\ \text{imaginary part} &= 0.7071\sin 45^\circ = 0.5\end{aligned}$$

hence

$$(2+j3)+(3-j4)+0.7071\angle 45^\circ = (2+3+0.5)+j(3-4+0.5)$$
$$= 5.5-j0.5$$

which is equivalent to $5.52\angle(-5.19^\circ)$.

19.6 Subtraction of complex numbers

As with the addition of complex numbers, **subtraction MUST be carried out using the rectangular complex value**, and the procedure is as follows:

1. the difference between the real parts of the numbers is the real part of the solution, and
2. the difference between the complex parts of the numbers gives the complex part of the solution.

That is, if $W = a + jb$ and $X = c + jb$, then

$$W - X = (a+jb) - (c+jd) = (a-c) + j(b-d)$$

The reader should note that we can only subtract one complex number from another at a time.

Complex numbers can be *subtracted from one another graphically* using the method described for subtracting phasors in Chapter 8.

Worked Example 19.5

Perform the following calculations:

(a) $(2-j2)-(3-j4)$
(b) $5\angle 70^\circ - (1.5-j4)$
(c) $-5\angle 40^\circ - 3\angle 120^\circ$.

Solution

(a) Since all the information is given in rectangular complex form we can proceed as follows:

$$(2-j2)-(3-j4) = (2-3)+j(-2+4)$$
$$= -1+j2$$

which is equivalent to the polar complex value $2.24\angle 116.6^\circ$.

(b) Initially, we must convert the polar complex value $5\angle 70^\circ$ into its rectangular form as follows:

$$\text{real part} = 5\cos 70^\circ = 1.71$$
$$\text{imaginary part} = 5\sin 70^\circ = 4.7$$

hence

$$\begin{aligned} 5\angle 70^\circ - (1.5 - j4) &\equiv (1.71 + j4.7) - (1.5 - j4) \\ &= (1.71 - 1.5) + j(4.7 + 4) = 0.21 + j8.7 \end{aligned}$$

which is equivalent to the polar value $8.703\angle 88.62^\circ$

(c) In this case we must convert both polar complex values into their rectangular form, as follows:
For $-5\angle 40^\circ$

$$\text{real part} = -5\cos 40^\circ = -3.83$$
$$\text{imaginary part} = -5\sin 40^\circ = -3.21$$

For $3\angle 120^\circ$

$$\text{real part} = 3\cos 120^\circ = -1.5$$
$$\text{imaginary part} = 3\sin 120^\circ = 2.6$$

hence

$$\begin{aligned} -5\angle 40^\circ - 3\angle 120^\circ &\equiv (-3.83 - j3.21) - (-1.5 + j2.6) \\ &= (-3.83 + 1.5) + j(-3.21 - j2.6) \\ &= -2.33 - j5.81 \end{aligned}$$

which is equivalent to the polar complex value $6.26\angle(-111.9^\circ)$.

19.7 Multiplication of complex numbers

Multiplication is best carried out using *polar complex values*; it can be performed using rectangular complex values, but is more tedious and error-prone. Both methods are described below.

Multiplication using polar values

If $\boldsymbol{L} = A\angle\phi_1$ and $\boldsymbol{M} = B\angle\phi_2$, the product of $\boldsymbol{L} \times \boldsymbol{M}$ is

$$\boldsymbol{LM} = AB\angle(\phi_1 + \phi_2)$$

For example, if $\boldsymbol{L} = 2\angle 45^\circ$ and $\boldsymbol{M} = 3\angle 30^\circ$, then

$$\boldsymbol{LM} = (2 \times 3)\angle(45^\circ + 30^\circ) = 6\angle 75^\circ$$

Using polar values, many complex numbers can be multiplied together at the same time.

Multiplication using rectangular values

If $W = a + jb$ and $X = c + jd$, the product of $W \times X$ is

$$WX = (a + jb)(c + jd) = ac + jbc + jad + j^2bd$$
$$= (ac - bd) + j(bc + ad)$$

For example, if $W = 2 + j3$ and $X = 4 + j5$, then

$$WX = (2 + j3)(4 + j5) = 8 + j12 + j10 + j^2 15$$
$$= (8 - 15) + j(12 + 10) = -7 + j22.$$

Worked Example 19.6

Perform the following multiplications:

(a) $-3\angle 40^\circ \times 2\angle 20^\circ$, (b) $4\angle 120^\circ \times (2 + j2)$, (c) $(-5 - j6) \times (10 + j8)$.

Solution

(a) Since both values are in polar form we proceed as follows:

$$-3\angle 40^\circ \times 2\angle 20^\circ = (-3 \times 2)\angle (40 + 20)^\circ = -6\angle 60^\circ$$
$$= 6\angle (60 - 180)^\circ = 6\angle (-120^\circ)$$

(b) In this case we shall convert the rectangular complex value into its polar form, as follows:

$$\text{modulus of } (2 + j2) = \sqrt{(2^2 + 2^2)} = \sqrt{8} = 2.828$$

and

$$\text{angle of } (2 + j2) = \arctan(2/2) = 45^\circ$$

hence

$$2 + j2 \equiv 2.828\angle 45^\circ$$

therefore

$$4\angle 120^\circ \times (2 + j2) = 4\angle 120^\circ \times 2.828\angle 45^\circ$$
$$= 11.314\angle (120 + 45)^\circ = 11.314\angle 165^\circ$$

which is equivalent to $(-10.93 + j2.93)$

(c) In this case we will perform the calculation directly using the rectangular values as follows:

$$(-5 - j6)(10 + j8) = -5(10 + j8) - j6(10 + j8)$$
$$= (-50 - j40) - j60 - j^2 48$$
$$= -50 - j40 - j60 + 48 = -2 - j100$$

which is equivalent to $100\angle(-91.15^\circ)$. The reader should verify the solution by converting the original rectangular complex values into their polar equivalents, and completing the calculation using polar values.

19.8 The conjugate of a complex number

The complex number $V = a + jb = r\angle\theta$ has a **conjugate** value V^*, which has the same magnitude as the original complex value, but the mathematical sign of the imaginary part is reversed. That is

$$V^* = a - jb = r\angle(-\theta)$$

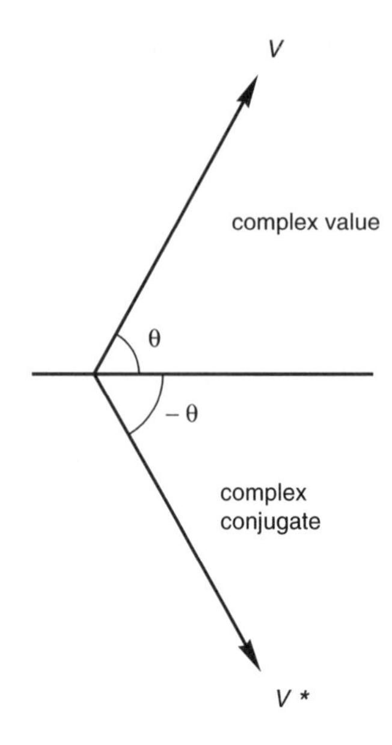

Figure 19.7 The conjugate of a complex number

This is illustrated in Figure 19.7. For example, if

$$V = 4 + j3 \equiv 5\angle 36.87^\circ$$

then

$$V^* = 4 - j3 \equiv 5\angle(-36.87^\circ)$$

A useful feature of a complex conjugate is that the product $V \times V^*$ is a real number which has no imaginary part (see section 19.9). For example, using $V = 5\angle 36.87^\circ$, then

$$V \times V^* = 5\angle 36.87^\circ \times 5\angle(-36.87^\circ) = 25\angle 0^\circ$$

and

$$\begin{aligned} V \times V^* &= (4 + j3)(4 - j3) = 16 - j12 + j12 - j^2 9 \\ &= 25 + j0. \end{aligned}$$

19.9 Division of complex numbers

Division of one complex number by another is best carried out using *polar complex values*. It may be performed using rectangular complex values, but is more tedious and error-prone. Both methods are described here.

Division using polar complex values

If $\boldsymbol{L} = A\angle\phi_1$ and $\boldsymbol{M} = B\angle\phi_2$, then

$$\frac{\boldsymbol{L}}{\boldsymbol{M}} = \frac{A\angle\phi_1}{B\angle\phi_2} = \frac{A}{B}\angle(\phi_1 - \phi_2)$$

For example, if $\boldsymbol{L} = 4\angle 80^\circ$ and $\boldsymbol{M} = 8\angle 40^\circ$, then

$$\frac{\boldsymbol{L}}{\boldsymbol{M}} = \frac{4\angle 80^\circ}{8\angle 40^\circ} = \frac{4}{8}\angle(80 - 40)^\circ = 0.5\angle 40^\circ$$

Division using rectangular complex values

If $\boldsymbol{W} = a + jb$ and $\boldsymbol{X} = c + jd$, then

$$\frac{\boldsymbol{W}}{\boldsymbol{X}} = \frac{a + jb}{c + jd}$$

To carry out the division, we must first reduce the denominator of the equation to a *real number* by multiplying both the denominator and the numerator by the conjugate of the denominator, as follows.

If $\boldsymbol{W} = 3 + j4$ and $\boldsymbol{X} = 5 + j6$, then

$$\frac{\boldsymbol{W}}{\boldsymbol{X}} = \frac{3 + j4}{5 + j6} = \frac{(3 + j4)(5 - j6)}{(5 + j6)(5 - j6)}$$

$$= \frac{15 + j20 - j18 - j^2 24}{25 + j30 - j30 - j^2 36}$$

$$= \frac{(15 + 24) + j(20 - 18)}{25 + 36}$$

$$= \frac{39 + j2}{61} = 0.639 + j0.0328$$

which is equivalent to $0.64\angle 2.94^\circ$. The reader should verify the result using the polar equivalents of $\boldsymbol{W}$ and $\boldsymbol{X}$.

Worked Example 19.7

Perform the following calculations, leaving the result in its polar form:

(a) $4\angle 120°/(-16\angle(-150°))$, (b) $(-2+j3)/5\angle(-20°)$,
(c) $(4+j5)/(-2-j3)$.

Solution

(a) Since the values are in polar form, we can proceed directly as follows:

$$\frac{4\angle 120°}{-16\angle(-150°)} = \frac{4}{-16}\angle(120-(-150))° = -0.25\angle 270°$$
$$= 0.25\angle(270-180)° = 0.25\angle 90°$$

(b) In this case we will convert the numerator into its polar complex equivalent as follows:

$$\text{modulus} = \surd((-2)^2+3^2) = 3.61$$

Since the complex value of the numerator lies in the second quadrant, and assuming that the reader is using a calculator to convert the value, then

$$\text{angle} = \arctan(3/(-2)) + 180° = -56.31° + 180°$$
$$= 123.69°$$

hence

$$\frac{-2+j3}{5\angle(-20°)} = \frac{3.61\angle 123.69°}{5\angle(-20°)} = 0.722\angle(123.69-(-20))$$
$$= 0.722\angle 143.69°$$

(c) Since the numerator and the denominator are both in rectangular complex form, we will proceed directly with them as follows (this method is troublesome and error-prone, but we illustrate it here).

$$\frac{4+j5}{-2-j3} = \frac{(4+j5)(-2+j3)}{(-2-j3)(-2+j3)}$$
$$= \frac{-8-j10+j12+j^2 15}{(-2)^2+3^2}$$
$$= \frac{(-8-15)+j(-10+12)}{13} = \frac{-23+j2}{13}$$
$$= -1.769+j0.154$$

The modulus of the number is

$$\surd((-1.769)^2+0.154^2) = 1.78$$

and since the angle lies in the second quadrant

$$\text{angle} = \arctan(0.154/(-1.769)) + 180^\circ$$
$$= -5^\circ + 180^\circ = 175^\circ$$

That is, the polar solution to the problem is $1.78\angle 175^\circ$.

19.10 Impedance in complex numbers

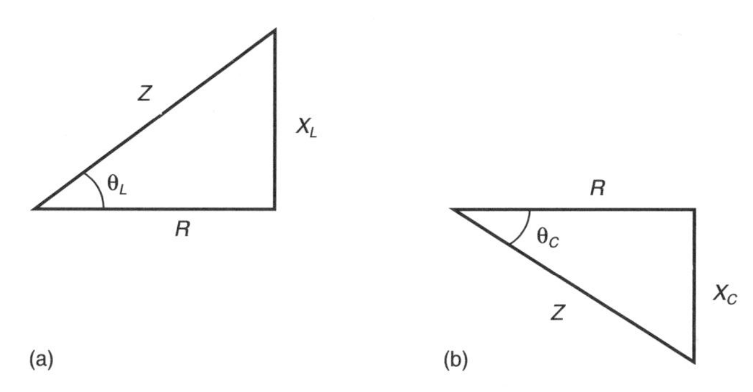

Figure 19.8 Impedance triangle for: (a) an R-L series circuit; (b) an R-C series circuit

The impedance triangle for (a) a series *R-L* circuit and (b) a series *R-C* circuit is shown in Figure 19.8. Using our knowledge of complex numbers we can write down the complex impedance for each type of circuit as follows:

For the series R-L circuit

$$\boldsymbol{Z} = R + jX_L \equiv Z_L \angle \phi_L$$

where

$$Z_L = \text{modulus of the circuit impedance}$$
$$= \sqrt{(R^2 + X_L^2)}$$

and

$$\phi_L = \arctan(X_L/R)$$

For the series R-C circuit

$$\boldsymbol{Z} = R - jX_C \equiv Z_C \angle \phi_C$$

where

$$Z_C = \text{modulus of the circuit impedance}$$
$$= \sqrt{(R^2 + (-X_C)^2)}$$

and

$$\phi_C = \arctan(-X_C/R)$$

(a) Impedances in series

If n impedances are connected in series, the net complex impedance of the circuit is

$$Z_T = Z_1 + Z_2 + Z_3 + \ldots + Z_n.$$

(b) Impedances in parallel

If n branches containing complex impedances are connected in parallel, the reciprocal of the total impedance of the circuit is

$$\frac{1}{Z_T} = \frac{1}{Z_1} + \frac{1}{Z_2} + \frac{1}{Z_3} + \ldots + \frac{1}{Z_n}$$

In the special case of two parallel-connected impedances, the total impedance of the circuit is

$$Z_T = \frac{Z_1 Z_2}{Z_1 + Z_2}.$$

Worked Example 19.8

Determine the complex value of the effective impedance of a resistor of 10 Ω in series with an inductive reactance of 25 Ω.

Solution

From the equations for the series circuit

$$Z_T = R + jX_L = 10 + j25\ \Omega$$

which is equivalent to an impedance of $26.93\angle 68.2°\ \Omega$.

Worked Example 19.9

What is the net impedance of a resistor of 1 kΩ in series with a capacitance of reactance 1.5 kΩ?

Solution

$$Z_T = R - jX_C = 10 - j1.5\ \text{k}\Omega$$

which is equivalent to $1.8\angle(-56.3)°\ \Omega$.

Worked Example 19.10

Determine the resistance in a series *R-L-C* circuit if the net impedance of the circuit is $1.58\angle(-18.43°)$ kΩ, the inductive reactance is 2 kΩ and the capacitive reactance is 2.5 kΩ.

Solution

The equation for the impedance of the circuit is

$$Z_T = R + j(X_L - X_C)$$

or

$$1.58\angle(-18.43°) = R + j(2 - 2.5) = R - j0.5 \text{ k}\Omega$$

That is

$$R = 1.58\angle(-18.43°) + j0.5 \text{ k}\Omega$$

In order to perform the subtraction, we must convert the value $1.58\angle(-18.43°)$ into its rectangular complex value, as follows:

$$\text{real component } = 1.58\cos(-18.43°) = 1.5 \text{ k}\Omega$$
$$\text{imaginary component } = 1.58\sin(-18.43°) = -0.5 \text{ k}\Omega$$

hence

$$R = (1.5 - j0.5) + j0.5 = 1.5 \text{ k}\Omega.$$

Worked Example 19.11

A two-branch parallel a.c. circuit consists of an inductive impedance of $10\angle 60°$ Ω in parallel with a capacitor of reactance 20 Ω. Evaluate the effective complex impedance of the circuit.

Solution

The impedance of the capacitor is

$$Z_2 = R_C - jX_C = 0 - j20 = -j20\ \Omega \equiv 20\angle(-90)°\ \Omega$$

The effective impedance of the circuit is

$$Z_T = \frac{Z_1 Z_2}{Z_1 + Z_2} = \frac{10\angle 60° \times 20\angle(-90°)}{10\angle 60° + (-j20)}$$
$$= \frac{200\angle(60-90)°}{(5 + j8.66) - j20} = \frac{200\angle(-30°)}{5 - j11.34}$$
$$= \frac{200\angle(-30°)}{12.4\angle(-66.2°)} = \frac{200}{12.4}\angle(-30 + 66.2)°$$
$$= 16.1\angle 36.2°\ \Omega$$

That is, the circuit has a net inductive reactance. It is left as an exercise for the reader to verify the accuracy of the polar-to-rectangular and the rectangular-to-polar conversions.

19.11 Ohm's law for a.c. circuits

To apply Ohm's law calculations to a.c. circuits, we merely use complex values throughout, as shown in the following calculations.

Worked Example 19.12

A resistance of 10 Ω and an inductive reactance of 5 Ω are connected in series to a 12 V sinusoidal supply. Determine (a) the complex impedance of the circuit, (b) the complex value of the current in the circuit, (c) the power factor of the circuit, and (d) the magnitude of the voltage across each element in the circuit.

Solution

(a) The impedance of the circuit is

$$\boldsymbol{Z} = R + jX_L = 10 + j5 \equiv 11.18\angle 26.57^\circ\ \Omega$$

Throughout this example, it is left as an exercise for the reader to make the necessary polar-to-rectangular (and vice versa) complex conversions.

(b) From Ohm's law

$$\boldsymbol{I} = \frac{\boldsymbol{V_S}}{\boldsymbol{Z}}$$

For our own convenience, we will assume that the voltage is in the reference direction, hence

$$\boldsymbol{I} = 12\angle 0^\circ / 11.18\angle 26.57^\circ = 1.073\angle(-26.57^\circ)\ \text{A}$$

That is, the current has a magnitude of 1.073 A, and lags behind $\boldsymbol{V_S}$ by 26.57°.

(c) The power factor of the circuit is

$$\cos(-26.57^\circ) = 0.894\ \text{(lagging)}$$

(d) The magnitude of voltage across each element in the circuit is equal to *the product of the magnitude of the current in the circuit, and the magnitude of the impedance of the element*

For the resistance

$$V_R = IR = 1.073 \times 10 = 10.73\ \text{V}$$

and for the inductive reactance

$$V_L = IX_L = 1.073 \times 5 = 5.365\ \text{V}$$

At this point we can check the result as follows:

$$\sqrt{(V_R^2 + V_L^2)} = \sqrt{(10.73^2 + 5.365^2)} = 12\ \text{V} = V_S$$

Worked Example 19.13

An impedance of $10\angle 45°\ \Omega$ is connected in parallel with an impedance of $(8 - j12)\ \Omega$, the combination being connected to a 100 V sinusoidal a.c. supply. Determine (a) the complex impedance of the circuit, (b) the current in each branch, (c) the total current drawn from the supply and (d) the power factor of the circuit.

Solution

(a) When determining the complex impedance of the parallel circuit, we must both multiply and add complex values. We therefore convert both impedance values as follows: (the reader should verify the conversions):

$$10\angle 45° \equiv 7.071 + j7.071\ \Omega$$
$$8 - j12 \equiv 14.42\angle(-56.31°)\ \Omega$$

The impedance of the parallel circuit is

$$\begin{aligned} \boldsymbol{Z_T} &= \frac{\boldsymbol{Z_1 Z_2}}{\boldsymbol{Z_1} + \boldsymbol{Z_2}} = \frac{10\angle 45° \times 14.42\angle(-56.31°)}{(7.071 + j7.071) + (8 - j12)} \\ &= \frac{144.2\angle(-11.31°)}{15.071 - j4.929} = \frac{144.2\angle(-11.31°)}{15.86\angle(-18.1°)} \\ &= 9.09\angle 6.79° \equiv 9.03 + j1.07\ \Omega \end{aligned}$$

That is the circuit has a net inductive reactance

(b) Since the voltage across each branch is $100\angle 0°$V, the current in each branch is

$$\boldsymbol{I_1} = \frac{\boldsymbol{V_S}}{\boldsymbol{Z_1}} = \frac{100\angle 0°}{10\angle(45°)} = 10\angle -45° \equiv 7.07 - j7.07\ \text{A}$$
$$\begin{aligned} \boldsymbol{I_2} &= \frac{\boldsymbol{V_S}}{\boldsymbol{Z_2}} = \frac{100\angle 0°}{14.42\angle(-56.31°)} \\ &= 6.93\angle 56.31° \equiv 3.84 + j5.77\ \text{A} \end{aligned}$$

(c) The total current drawn from the supply is

$$\begin{aligned} \boldsymbol{I} &= \boldsymbol{I_1} + \boldsymbol{I_2} = (7.07 - j7.07) + (3.84 + j5.77) \\ &= 10.91 - j1.3 \equiv 10.99/(-6.8°)\ \text{A} \end{aligned}$$

That is, 10.99 A is drawn from the supply lagging by 6.8° behind the supply voltage.

(d) The power factor of the complete circuit is

$$\cos(-6.8°) = 0.993\ \text{(lagging)}.$$

19.12 Three-phase calculations

A knowledge of complex numbers is invaluable when dealing with three-phase calculations, as illustrated by Worked Examples 19.14 and 19.15.

Worked Example 19.14

The voltage between the red line and neutral of a three-phase, star-connected system is $250\angle 0°$V, and that between the yellow line and neutral is $250\angle(-120°)$V. Determine the complex expression for the voltage of the red line with respect to the yellow line (i.e. the red-to-yellow line voltage).

Solution

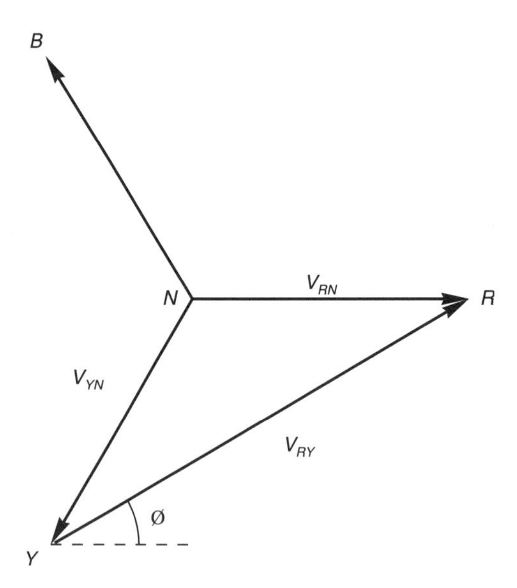

Figure 19.9 Phasor diagram for Worked Example 19.14

The phasor diagram showing $\boldsymbol{V_{R_N}}$ and $\boldsymbol{V_{YN}}$ is shown in Figure 19.9. The voltage of the red line with respect to the yellow line is

$$\boldsymbol{V_{RY}} = \boldsymbol{V_{RN}} - \boldsymbol{V_{YN}} = 250\angle 0° - 250\angle(-120°)$$

Initially, we must convert both voltages into their rectangular complex form as follows:

For $\boldsymbol{V_{RN}}$

$$\text{horizontal component} = 250\cos 0° = 250 \text{ V}$$
$$\text{vertical component} = 250\sin 0° = 0 \text{ V}$$

That is $\boldsymbol{V_{RN}} = 250\angle 0° \equiv 250 + j0$ V

For V_{YN}

$$\text{horizontal component} = 250\cos(-120°) = -125 \text{ V}$$
$$\text{vertical component} = 250\sin(-120°) = -216.5 \text{ V}$$

That is $V_{YN} = 250\angle(-120°) \equiv -125 - j216.5$ V
hence

$$\begin{aligned} V_{RY} &= V_{RN} - V_{YN} \\ &= (250 + j0) - (-125 - j216.5) = 375 + j216.5 \text{ V} \\ &\equiv 433\angle 30° \text{ V} \end{aligned}$$

That is V_{RY} has a value of 433 V at a phase angle of 30° with respect to the reference direction.

Worked Example 19.15

An unbalanced star-connected, four-wire load consumes the following currents:

$$I_R = 100\angle 0° \text{ A}, I_Y = 150\angle(-140°) \text{ A}, I_B = 80\angle 130° \text{ A}$$

Obtain a complex expression for the current in the neutral wire.

Solution

As an exercise, the reader should draw the phasor diagram accurately to scale and obtain his/her own result, which can be compared with the one obtained here.

The neutral wire current is given by

$$I_N = I_R + I_Y + I_B$$

where all the values are complex. Since we are adding the phasors, we must convert the values provided into their rectangular complex form.

For I_R

$$\text{horizontal component} = 100\cos 0° = 100 \text{ A}$$
$$\text{vertical component} = 100\sin 0° = 0 \text{ A}$$

That is $100\angle 0° \equiv 100 + j0$ A

For I_Y

$$\text{horizontal component} = 150\cos(-140°) = -114.9 \text{ A}$$
$$\text{vertical component} = 150\sin(-140°) = -96.42 \text{ A}$$

That is $150\angle(-140°) \equiv -114.9 - j96.42$ A

For I_B

$$\text{horizontal component} = 80\cos 130° = -51.42 \text{ A}$$
$$\text{vertical component} = 80\sin 130° = 61.28 \text{ A}$$

That is $80\angle 130° \equiv -51.42 + j61.28$ A

The current in the neutral wire is therefore

$$\begin{aligned} I_N &= I_R + I_Y + I_B \\ &= (100 + j0) + (-114.9 - j96.42) + (-51.42 + j61.28) \\ &= 66.32 - j35.14 \equiv 75.1\angle(-152°) \text{ A} \end{aligned}$$

That is, 75.1 A flows in the neutral wire at an angle of $-152°$ with respect to the reference direction (or with respect to I_R).

Computer listings

Listing 19.1 Rectangular to polar conversion

```
10 CLS '*** CH19-1 ***
20 DIM D(6, 4): PI = 3.14159
30 PRINT "RECTANGULAR TO POLAR CONVERSION": PRINT
40 FOR R = 0 TO 6: FOR C = 0 TO 4: D(R, C) = 0: NEXT C: NEXT R
50 INPUT "Number of values to be converted (Max. 5) = ", No: PRINT
60 IF No < 1 OR No > 5 THEN GOTO 10
70 FOR N = 1 TO No
80     PRINT TAB(8); "For value"; N
90     INPUT "Horizontal (real) component = ", D(N, 1)
100     INPUT "Vertical (quadrature) component = ", D(N, 2)
110     D(N, 3) = SQR(D(N, 1) ^ 2 + D(N, 2) ^ 2)
120     IF D(N, 1) = 0 AND D(N, 2) > 0 THEN PHI = PI / 2: GOTO 160
130     IF D(N, 1) = 0 AND D(N, 2) < 0 THEN PHI = -PI / 2: GOTO 160
140     IF D(N, 1) < 0 AND D(N, 2) = 0 THEN PHI = PI: GOTO 160
150     PHI = ATN(D(N, 2) / D(N, 1))
160     D(N, 4) = PHI * 180 / PI
170     IF D(N, 1) < 0 AND D(N, 2) > 0 THEN D(N, 4) = D(N, 4) + 180
180     IF D(N, 1) < 0 AND D(N, 2) < 0 THEN D(N, 4) = D(N, 4) - 180
190 NEXT N
200 CLS
210 FOR N = 1 TO No
220     PRINT TAB(8); "For value"; N
230     PRINT TAB(3); "Inphase component = "; D(N, 1); ": Quadrature component = ";
D(N, 2)
240     PRINT TAB(3); "Modulus = "; D(N, 3); ": Angle (deg.) = "; D(N, 4)
250 NEXT N
260 END
```

Computer listing 19.1 deals with the conversion of up to five complex values in rectangular form into polar form. This makes the values particularly suitable for use in multiplication and division. The magnitude of each value is calculated in line 110. Lines 120 to 140 deal with angles of 90°, −90° and 180°, respectively, otherwise the principal angle is calculated in line 150. The angle is converted into its value in degrees in line 160, and lines 170 and 180 handle angles which lie in the 2nd and 3rd quadrants, respectively.

Listing 19.1 therefore gives a magnitude (which is positive), together with an angle in the range 0 to +180° or 0 to −180°. The screen printing routine in lines 210 to 250 gives both the rectangular and polar value of each complex number.

Listing 19.2 Multiplication of complex numbers

```
10 CLS '*** CH 19-2 ***
20 PRINT "MULTIPLICATION OF COMPLEX NUMBERS": PRINT
30 PI = 3.14159: DIM D(6, 4)
40 FOR R = 0 TO 6: FOR C = 0 TO 4: D(R, C) = 0: NEXT C: NEXT R
50 INPUT "Number of values to be multiplied (Max. 5) = ", No: PRINT
60 IF No < 2 OR No > 5 THEN GOTO 10
70 D(0, 3) = 1: D(0, 4) = 0 '** Holding locations of modulus and angle **
80 FOR N = 1 TO No
90     PRINT TAB(8); "For value "; N
100      INPUT "Modulus = ", D(N, 3)
110      INPUT "Phase angle (deg.) = ", D(N, 4): D(N, 4) = D(N, 4) * PI / 180
120      ' ** Correction if magnitude is negative **
130      IF D(N, 3) < 0 THEN D(N, 4) = D(N, 4) + PI: D(N, 3) = D(N, 3) * (-1)
140      D(0, 3) = D(0, 3) * D(N, 3) ' ** Store product magnitude in D(0, 3) **
150      D(0, 4) = D(0, 4) + D(N, 4) ' ** Store product angle in D(0, 4) **
160 NEXT N
170 '** Correction if angle is >+180 deg or <-180 deg **
180 IF D(0, 4) > 2 * PI THEN D(0, 4) = D(0, 4) - 2 * PI * INT(D(0, 4) / (2 * PI))
190 IF D(0, 4) > PI THEN D(0, 4) = D(0, 4) - 2 * PI
200 IF D(0, 4) < -2 * PI THEN D(0, 4) = D(0, 4) + 2 * PI * INT(ABS(D(0, 4)) / (2 * PI))
210 IF D(0, 4) < -PI THEN D(0, 4) = D(0, 4) + 2 * PI
220 CLS
230 FOR N = 1 TO No
240     PRINT TAB(8); "For value "; N
250     PRINT "Modulus = "; D(N, 3); ": Angle (deg.) ="; D(N, 4) * 180 / PI
260 NEXT N
270 PRINT
280 PRINT "Modulus of product = "; D(0, 3)
290 PRINT "Angle of product (deg.) = "; D(0, 4) * 180 / PI
300 END
```

Listing 19.2 handles the multiplication of up to five complex numbers; *the data MUST be supplied in polar complex form.* If the product gives an angle which is either greater than +180° or is less than −180°, lines 180 to 210 convert the angle of the product into one of these two ranges. The screen printing routine (lines 230 to 290) prints all the values supplied, together with the product.

Listing 19.3 Division of complex numbers by one another

```
10 CLS ' *** CH19-3 ***
20 PRINT "DIVISION OF ONE COMPLEX NUMBER BY ANOTHER": PRINT
30 DIM D(6, 4)
40 FOR R = 0 TO 6: FOR C = 0 TO 4: D(R, C) = 0: NEXT C: NEXT R
50 PRINT
60 INPUT "Modulus of numerator = ", D(1, 3)
70 INPUT "Phase angle (deg.) of numerator = ", D(1, 4): PRINT
80 INPUT "Modulus of denominator = ", D(2, 3)
90 INPUT "Phase angle (deg.) of denominator = ", D(2, 4)
100 IF D(2, 3) = 0 THEN CLS : PRINT "Error: Denominator is ZERO": GOTO 40
110 D(3, 3) = D(1, 3) / D(2, 3) ' ** Calculate magnitude **
120 D(3, 4) = D(1, 4) - D(2, 4)' ** Calculate phase angle **
130 ' ** Make correction if magnitude is negative **
140 IF D(3, 3) < 0 THEN D(3, 3) = -1 * D(3, 3): D(3, 4) = D(3, 4) + 180
150 ' ** Make correction if angle > +180 or < -180.
160 IF D(3, 4) > 360 THEN D(3, 4) = D(3, 4) - 360 * INT(D(3, 4) / 360)
170 IF D(3, 4) > 180 THEN D(3, 4) = D(3, 4) - 360
180 IF D(3, 4) < -360 THEN D(3, 4) = D(3, 4) + 360 * INT(ABS(D(3, 4)) / 360)
190 IF D(3, 4) < -180 THEN D(3, 4) = D(3, 4) + 360
200 PRINT
210 PRINT "Modulus of result = "; D(3, 3)
220 PRINT "Phase angle of result = "; D(3, 4)
230 END
```

The program in Computer listing 19.3 deals with the division of one polar complex number by another, and line 100 handles the case where the modulus of the denominator is zero. Line 130 converts a negative magnitude into its equivalent positive magnitude, and lines 160 to 190 deal with the case where the angle of the result is either greater than +180° or less than −180°.

So far nothing has been said about a program for the conversion of polar complex values into their equivalent rectangular form, or of the addition and subtraction of complex numbers. These have, in fact, already been dealt with in Computer listing 8.2, but was given in the guise of the addition and subtraction of phasors.

Exercises

19.1 Combine Listings 8.2, 19.1, 19.2 and 19.3 into one menu-driven listing, so that any form of complex number calculation can be handled.

19.2 Convert the following rectangular complex values into their equivalent polar complex value: (a) $-200+j150$, (b) $20-j60$, (c) $10+j4$, (d) $-5-j6$.

19.3 Convert the following polar complex values into their equivalent rectangular complex form: (a) $3\angle 50°$, (b) $-3\angle 50°$, (c) $2\angle 200°$, (d) $-2\angle 120°$, (e) $4\angle(-60°)$, (f) $6\angle 500°$, (g) $-6\angle(-500°)$.

19.4 Add the following complex values, and give the result both in rectangular and polar form:
(a) $(2+j3)+(-4-j5)$
(b) $(500-j600)+(-600-j300)$
(c) $(5+j6)+(5-j4)+(-8+j4)$.

19.5 Add the following complex values, giving the result in both rectangular and polar complex form:
(a) $2\angle 60° + (-3+j2)$
(b) $-4\angle 60° + (5+j6)$
(c) $6\angle 500° + (-8\angle(-400°)) + (5+j6)$.

19.6 Perform the following subtractions, giving the result both in polar and rectangular complex form:
(a) $(2-j3)-(-4-j5)$
(b) $(4-j5)-5\angle 220°$
(c) $-6\angle 40° - 4\angle(-120°)$.

19.7 When three complex numbers have been added together, their sum is $6\angle(-130°)$. If two of the complex numbers are $(4+j5)$ and $2\angle 120°$, determine the value of the third number.

19.8 Multiply the following complex numbers, leaving the result both in polar and rectangular complex form:
(a) $10\angle 50° \times 8\angle 60°$
(b) $-4\angle 10° \times 9\angle(-70°) \times 16\angle 190°$.

19.9 Multiply the following complex values, leaving the result both in polar and rectangular complex form:
(a) $(2+j3)\times(5+j6)$
(b) $(-4-j5)\times(8+j2)$
(c) $5\angle 20° \times (-2-j4)$
(d) $9\angle 100° \times (-2-j4) \times 5\angle 20°$.

19.10 Solve the following, giving the answer in both polar and rectangular complex form:
(a) $10\angle 70° \angle 6\angle(-30°)$, (b) $-5\angle 90°/(-2+j3)$, (c) $(-5-j4)/(3+j4)$.

19.11 Solve the following, giving the solution in both polar and rectangular complex form:
(a) $((2-j3)+(3+j6))/2\angle 80°$
(b) $(3\angle 40° - 4\angle 90°)/((5+j6)+(3-j4))$
(c) $4\angle 100° \times (1+j3)/(2\angle 70° - 4\angle 20°)$.

19.12 In the following, **A**, **B**, **C**, and **D** are complex values. If $A = 5 - j6, B = -2\angle 80°, C = -5 + j2$, determine D if $A = B \times D/C$.

19.13 The following complex voltages are connected in series with one another.

$$V_1 = 6 + j7V,\ V_2 = 8 - j5\ \text{V},\ V_3 = -9 - j8V,\ V_4 = -3 - j2\ \text{V}$$

Determine the resultant voltage both in rectangular and polar complex form.

19.14 Two circuits X and Y are connected in parallel with one another, in which circuit X draws a current of $2 + j3$ A. If the total current drawn by the parallel circuit is $8.25\angle(-14.04°)$ A, determine the current drawn by circuit Y.

19.15 If a current of $(5 + j6)$ A flows into a circuit of impedance $10\angle 30°\ \Omega$, calculate the voltage applied to the circuit.

19.16 A voltage of $240\angle 120°$ V is applied to a circuit of impedance $(10 + j5.77)\ \Omega$. Determine a complex expression for the current flowing into the circuit.

19.17 A resistanceless coil of inductance 0.1 H and a resistor of 500 Ω are connected (a) in series, (b) in parallel to a 10 V, 1 kHz a.c. supply. What is (i) the complex impedance of the circuit, (ii) the current in the circuit?

19.18 The p.d. across a series circuit is $(25 + j40)$ V. If the circuit comprises a coil of resistance 12.5 Ω and inductance 0.04 H, and the angular frequency of the supply is 500 rad/s, determine the complex expression for the circuit current in polar and rectangular form.

19.19 The p.d. across a circuit is $(3000 + j600)$ V, and the current through it is $(10 - j5)$ A. Determine the power, the reactive VA and the apparent power consumed. State whether the reactive VA is lagging or leading.

Summary of important facts

A **complex number** is one having *magnitude and direction*. It can be expressed in **rectangular complex form** or **Cartesian form** as follows:

$$Z = R + jX$$

where Z is the complex value (printed in **bold**), R is the **real** ('reference' or horizontal component) of the number, and X is the **imaginary**, 'quadrature' or vertical component. (**Note**: the word 'imaginary' merely means that it is at right-angles to the reference direction.)

The quantity 'j' is an **imaginary operator**, which can either be thought of as an operator which 'rotates' the imaginary part through 90° in an anti-clockwise direction, or as the value

$$j = \sqrt{(-1)}$$

Also

$$j^2 = -1$$
$$j^3 = j^2 \times j = -j$$
$$j^4 = j^2 \times j^2 = -(-1) = 1$$

Since operating on a number by 'j' corresponds to rotating the value through 90°, then

$$j \equiv 1\angle 90^\circ$$
$$j^2 \equiv 1\angle 180^\circ$$
$$j^3 \equiv 1\angle 270^\circ$$
$$j^4 \equiv 1\angle 360^\circ \equiv 1\angle 0^\circ$$

Mathematicians use the letter 'i' as the imaginary operator but, since 'i' is used to represent current in electrical engineering, we use the letter 'j'.

A complex number can be represented in its **polar form** as follows:

$$\boldsymbol{Z} = M\angle\phi$$

where M is the magnitude or **modulus** of the complex value, and ϕ is the angle or **argument** of the complex number with respect to the reference direction.

The rectangular and polar forms of complex number are related by the following equations:

$$\boldsymbol{M} = \sqrt{(\boldsymbol{R}^2 + \boldsymbol{X}^2)}$$
$$\boldsymbol{\phi} = \textbf{arctan}\,(\boldsymbol{X}/\boldsymbol{R})$$
$$\boldsymbol{R} = \boldsymbol{M}\,\textbf{cos}\,\boldsymbol{\phi}$$
$$\boldsymbol{X} = \boldsymbol{M}\,\textbf{sin}\,\boldsymbol{\phi}$$

If $\boldsymbol{Z} = M\angle\phi$ is a complex value, then

$$-\boldsymbol{Z} = M\angle(\phi \pm 180^\circ)$$

The **conjugate** of the complex number $\boldsymbol{Z} = M\angle\phi = R + jX$ is

$$\boldsymbol{Z}^* = M\angle(-\phi) = R - jX$$

The product of a complex number and its conjugate is a real number, and has no imaginary part. If $\boldsymbol{Z} = M\angle\phi = R + jX$, then

$$\boldsymbol{ZZ}^* = M^2 = R^2 + X^2$$

Complex number must be **added** using the *rectangular form* of the numbers. Many complex numbers may be added together. If $\boldsymbol{Z_1} = R_1 + jX_1$, $\boldsymbol{Z_2} = R_2 + jX_2$, and $\boldsymbol{Z_3} = R_3 + jX_3$, then

$$\begin{aligned}\boldsymbol{Z_1} + \boldsymbol{Z_2} + \boldsymbol{Z_3} &= (R_1 + jX_1) + (R_2 + jX_2) + (R_3 + jX_3) \\ &= (R_1 + R_2 + R_3) + j(X_1 + X_2 + X_3)\end{aligned}$$

Complex numbers must be **subtracted** using the *rectangular form* of the numbers. If $\boldsymbol{Z_1} = R_1 + jX_1$ and $\boldsymbol{Z_2} = R_2 + jX_2$, then

$$\boldsymbol{Z_1} - \boldsymbol{Z_2} = (R_1 + jX_1) - (R_2 + jX_2) = (R_1 - R_2) + j(X_1 - X_2)$$

Multiplication of complex numbers can be carried out using either polar values or rectangular values, but *multiplication using polar values is recommended because it is quicker and is less error-prone.*

If $\boldsymbol{I} = I_1\angle\phi_1 \equiv A + jB$, and $\boldsymbol{Z} = Z_1\angle\phi_2 \equiv C + jD$ then, using polar values, we get

$$\boldsymbol{V} = I_Z = I_1\angle\phi_1 \times Z_1\angle\phi_2 = I_1Z_1\angle(\phi_1 + \phi_2)$$

and using rectangular complex values, we get

$$\begin{aligned} \boldsymbol{V} = \boldsymbol{IZ} &= (A + jB)(C + jD) \\ &= AC + jBC + jAD + j^2BD \\ &= (AC - BD) + j(BC + AD) \end{aligned}$$

Division of one complex number can be carried out using either polar or rectangular complex values, but *division using polar values is recommended because it is quicker and is less error-prone.*

If $\boldsymbol{V} = V_1\angle\phi_1 \equiv A + jB$, and $\boldsymbol{Z} = Z_1\angle\phi_2 \equiv C + jD$ then, using polar values, we get

$$\boldsymbol{I} = \frac{\boldsymbol{V}}{\boldsymbol{Z}} = \frac{V_1\angle\phi_1}{Z_1\angle\phi_2} = \frac{V_1}{Z_1}\angle(\phi_1 - \phi_2)$$

Using rectangular complex values we get

$$\begin{aligned} \boldsymbol{I} = \frac{\boldsymbol{V}}{\boldsymbol{Z}} &= \frac{A + jB}{C + jD} = \frac{(A + jB)(C - jD)}{(C + jD)(C - jD)} \\ &= \frac{(AC + BD) + j(BC - AD)}{C^2 + D^2} \end{aligned}$$

The process of multiplying the denominator by its complex conjugate is known as **rationalising the denominator**, and has the effect of reducing the final denominator to a 'real' number.

Computer solution of electronic and electrical circuits

20.1 Introduction

Computers are an integral part of the education of electronic and electrical engineers, to the extent that many BASIC language programs for the solution of many types of problem have been included in this book. In this chapter, we look at the solution of a range of examples using one of the popular software packages, namely **SPICE** (**S**imulation **P**rogram with **I**ntegrated **C**ircuit **E**mphasis). We shall be using a version known as *PSpice*, of which a *low-cost evaluation version* is widely available to students. This has the full range of the facilities of SPICE itself (although it deals with smaller systems), and is particularly suited to student use. It is widely available from shareware sources (information can be obtained from any good computer magazine), and the evaluation version with full documentation is available from the address at the foot of the page.*

By the end of this chapter, the reader will be able to

- Write and understand SPICE programs.
- Solve problems concerning electronic and electrical engineering.

20.2 The concept of SPICE

In SPICE, we describe each circuit to the computer in an **input file** or **source file**, in which each circuit element or operation is specified in one program line. When the input file is complete, it is passed to a **software analyser**, which checks the file for errors. If the file is free from errors, SPICE carries out the required analysis and supplies the results. Depending on the type of analysis requested, the results may be given in numerical form, or in graphical form, or both.

*The evaluation version of PSpice and its documentation is available from ARS Microsystems, Herriard Business Centre, Alton Road, Basingstoke, RG25 2PN.

SPICE allows the use of comments in the input file, so that when it is read at a later date it can readily be understood.

SPICE is so versatile that any type of problem can be solved, ranging from simple d.c. circuits, through a.c. circuits and transients, up to the most complex integrated circuit problem.

20.3 A simple introductory example

Initially we will look at the simple series-parallel d.c. circuit in Figure 20.1.

Worked Example 20.1

Using PSpice determine, in Figure 20.1, the voltage at each node, the current in the circuit, and the total power consumed.

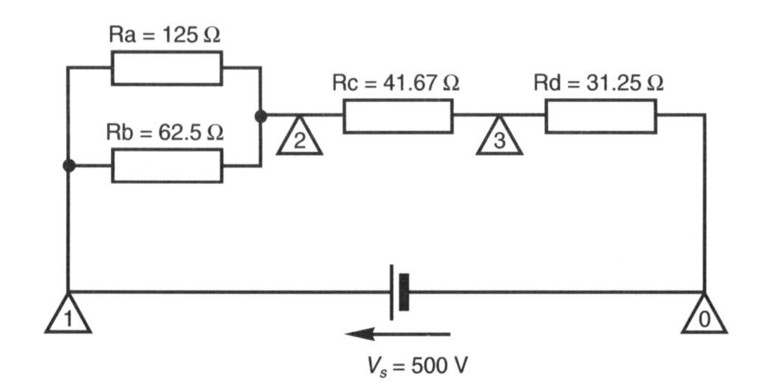

Figure 20.1 Circuit for Worked Example 20.1

Solution

When writing a SPICE file, the following rules should be followed.

1. Draw the circuit diagram and give it a NAME, which can be used in the TITLE LINE (see below for details).
2. LABEL every element in the circuit, and write its VALUE on the circuit. Both the label and the value are used in the input file.
3. Mark each node with a NODE NUMBER (which must either be a *positive number*, or *zero*), the REFERENCE NODE is node zero (0) – *do not* use capital O when entering this value in the input file. Each node *must have* (a) at least two connections to it, and (b) a d.c. path (i.e. a resistive path) to the reference node.
4. Decide what type of ANALYSIS is to be performed on the circuit. If an ANALYSIS type is not specified in the input file, SPICE automatically performs a SMALL SIGNAL BIAS analysis (see details below).

5. Create a SPICE INPUT FILE as follows:
   (a) The first line or TITLE LINE should contain a helpful title; alternatively, it should be left blank.
   (b) Write down a series of *element lines, comment lines, blank lines* and *control lines* (the meaning of these is explained later). The order in which they appear is immaterial.
6. Terminate the input file with a '.END' line (the '.' before END is important, so do not leave it out).

Table 20.1 PSpice input file for Worked Example 20.1

```
** Input file for Worked Example 20.1 **

* Any line commencing with a '*' is a COMMENT LINE
* Voltage source description.
* positive node
*    | negative node
*    |    | voltage
*    |    |    | anything after a ';' is a COMMENT
*    |    |    |    |
Vs   1    0   500   ;500 V d.c. source

* Circuit resistances
Ra   1   2   125
Rb   1   2   62.5ohm    ;'ohm' is optional
Rc   2   3   41.67
Rd   3   0   31.25ohm

.OPTIONS NOPAGE
.END     ;END of the input file
```

The input file for the circuit is given in Table 20.1 in which the first three steps outlined above have been completed. In our case, we simply write

'** Input file for Worked Example 20.1 **'

in the TITLE LINE; SPICE ignores anything in this line, so *do not write any data in this line which is required to solve the example*. The next line is left blank merely to improve the presentation of the file; SPICE ignores blank lines.

The next line in the file is a COMMENT LINE, which commences with a '*'. When SPICE encounters a comment line, it ignores it. The next six lines are comment lines, which serve to give useful information about the make-up of the ELEMENT LINE which describes the voltage source Vs. Each element line commences with an alphabetical character, and the 'V' in the Vs line tells SPICE that it is dealing with an independent voltage source. Lines describing resistors begin with an R, lines describing inductors begin with an L, capacitor lines begin with a C, etc.

The data in line Vs tells SPICE that node 1 is positive with respect to node 0 by 500 V. This line is terminated by an *in-line comment*, which follows a ';' in the line; SPICE ignores in-line comments.

Next we define the resistors in the circuit as follows: Resistor Ra is described as being connected between nodes 1 and 2, and has a resistance of 125 units of resistance; since the value has no decimal multiplier, SPICE assumes that the unit is the basic SI unit for that element, namely the ohm. The line for Rb uses the value of 62.5ohm (**notice**: there is no space between 62.5 and 'ohm'); the word 'ohm' is optional, and may be omitted (see also section 20.5). Resistors Rc and Rd are then described in much the same way.

The penultimate line is a '.OPTIONS' line (pronounced 'dot options'), which states which options SPICE must use. This line is, itself, optional, and may be omitted. The '.' before the word 'OPTIONS' is important, and must not be omitted because it tells SPICE that it is dealing with a CONTROL LINE, which is used to control the operation of SPICE. The NOPAGE option reduces the number of sheets of paper produced at print-out time, and is useful in saving the time required to produce the print-out and economises on the use of paper. The program is terminated by a '.END' line.

Table 20.2 PSpice small signal bias solution for Worked Example 20.1

```
****    SMALL SIGNAL BIAS SOLUTION     TEMPERATURE =  27.000 DEG C

 NODE  VOLTAGE    NODE  VOLTAGE    NODE  VOLTAGE

(    1) 500.0000 (    2) 318.1900 (    3) 136.3600

   VOLTAGE SOURCE CURRENTS
   NAME        CURRENT

   Vs        -4.364E+00
   TOTAL POWER DISSIPATION   2.18E+03  WATTS
```

When we run the source file for this example, SPICE produces the SMALL SIGNAL BIAS SOLUTION in Table 20.2. Unless told otherwise, the analysis is produced on the assumption that the temperature is 27°C; this value of temperature can be altered within the file, as shown in Worked Example 20.3 (section 20.6). The SMALL SIGNAL BIAS SOLUTION calculates the voltage at each node (relative to node 0), together with the *current flowing into the positive node* of each d.c. source, and the total d.c. power consumed.

Some of the data in Table 20.2 is presented in scientific notation, so that some care should be taken when interpreting the results. The output file tells us that the current *flowing into* the positive terminal of Vs is −4.364 A; that is, 4.364 A *flows out* of the positive terminal of Vs. We also see that the power consumed by the circuit is 2.18E+03 = 2.18×10^3 = 2180 W.

This example has, in fact, largely been solved earlier in the book, and the reader should refer to Worked Example 3.5 in Chapter 3.

The SPICE computer file which contains the circuit description has the file extension '.CIR', and the file containing the results has the same file name as the .CIR file, but has the extension '.OUT'. An advantage of the '.OUT' file is that it may be imported into almost any word processor so that it may be studied and selected portions printed. It is not necessary to have SPICE installed in the computer in order to study a '.OUT' file.

20.4 Solving a d.c. network

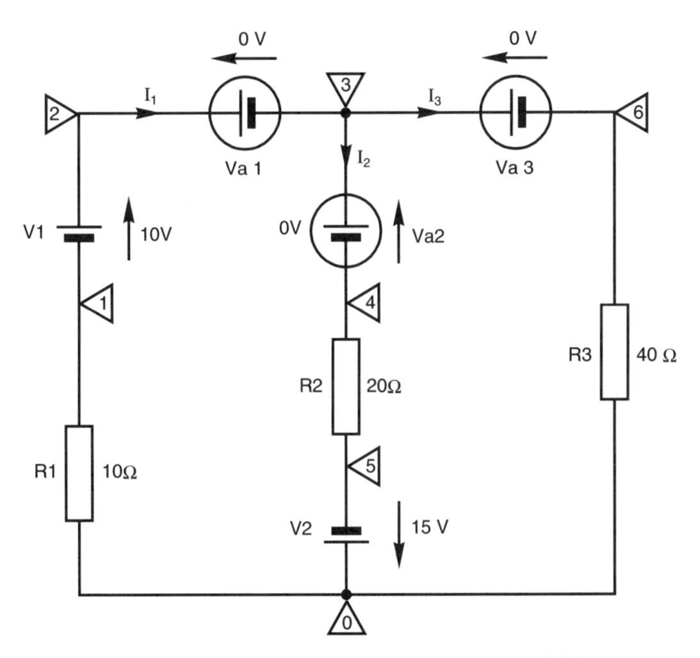

Figure 20.2 Circuit for Worked Example 20.2

We now use SPICE to solve the network in Figure 20.2. In this case we use SPICE 'ammeters' to measure the current flowing in each branch of the circuit, and these are shown as Va1, Va2 and Va3, respectively, to measure I_1, I_2 and I_3. The reader will recall that SPICE calculates the current flowing *into the positive node of a supply source*. If we connect a zero-voltage source, i.e. a **null source**, where we need to have an ammeter, then SPICE reports the value of the current in that branch. In this way we can use a null source as an ammeter.

Worked Example 20.2

Use SPICE to determine the current in each branch of Figure 20.2.

Solution

The circuit comprises two meshes containing e.m.fs V1 and V2, together with resistors R1, R2 and R3. SPICE ammeters Va1, Va2 and Va3 have been inserted in the three branches; it is important to note that the 'positive' direction of current in the branch flows into the positive node of each 'ammeter'.

Table 20.3 Input file for Worked Example 20.2

```
** Input file for Worked Example 20.2 **

* Voltage sources
V1   2   1   DC   10V   ;'DC' and 'V' optional
V2   0   5   15

* SPICE ammeters
Va1   2   3   0V   ;'V' optional
Va2   3   4   DC   0   ;'DC' optional
Va3   3   6   0

* Resistors
R1   1   0   10ohm   ;'ohm' optional
R2   4   5   20
R3   6   0   40

.OPTIONS   NOPAGE
.END
```

The input file for the circuit is shown in Table 20.3. Initially we define the two d.c. sources V1 and V2; the order in which circuit elements are defined is immaterial and we could, alternatively, define the voltage sources at a later stage in the file. The reader will note that we have added two optional items of information in the V1 line, and these are the letter 'DC' and 'V' (after the voltage value). Both of these are optional, and SPICE ignores them, but both are useful to the reader in terms of the information they convey. Care should be taken when giving such data, and for details the reader should refer to the SPICE manual. Section 20.5 shows the list of letters which can be used in association with SPICE as unit symbols.

The input file defines V1 as a 10 V d.c. source connected between nodes 2 and 1, with node 2 (the first numbered node) being positive with respect to node 1. Source V2 is a 15 V d.c. source which is connected between node 0 and node 5, with node 0 being positive with respect to node 5. Alternatively, we can define V2 as

```
V2   5   0   -15
```

The SPICE ammeters Va1, Va2 and Va3 are next defined, followed by the definition of the resistors (the reader will note that the dimension 'ohm' is optional, and can be omitted). Once again we use a '.OPTIONS NOPAGE' line, and the file is terminated by a '.END' line.

The results are listed in Table 20.4, in which the d.c. voltage at each node (relative to node 0) is given, together with the *current flowing into the positive node of each voltage source*. The current in Va1, Va2 and Va3 is the 'positive' direction of current in the appropriate branch, and the values given by SPICE should be compared with the values obtained in Worked Example 4.2 in Chapter 4.

Table 20.4 Small signal bias solution for Worked Example 20.2

```
****    SMALL SIGNAL BIAS SOLUTION      TEMPERATURE =  27.000 DEG C

NODE  VOLTAGE    NODE  VOLTAGE    NODE  VOLTAGE    NODE  VOLTAGE

(  1)  -8.5714 (  2)   1.4286 (  3)   1.4286 (  4)   1.4286
(  5) -15.0000 (  6)   1.4286

   VOLTAGE SOURCE CURRENTS
   NAME        CURRENT

   V1        -8.571E-01
   V2        -8.214E-01
   Va1        8.571E-01
   Va2        8.214E-01
   Va3        3.571E-02
   TOTAL POWER DISSIPATION  2.09E+01  WATTS
```

20.5 SI multiples in SPICE

Table 20.5 SPICE multiples

SPICE suffix	*Multiple*	*Metric prefix*
T	10^{12}	tera-
G	10^{9}	giga-
MEG	10^{6}	mega-
K	10^{3}	kilo-
M	10^{-3}	milli-
U	10^{-6}	micro-
N	10^{-9}	nano-
P	10^{-12}	pico-
F	10^{-15}	femto-

The value of a voltage, current, resistance, capacitance, etc. in SPICE may have any one of the metric multiple symbols given in Table 20.5 associated with its value. We can also use a standard scientific notation value as follows:

$$1260 \equiv 1.26\text{K} \equiv 0.00126\text{MEG} \equiv 1.26\text{E3}$$

The reader should note that, in SPICE, M means *milli-* or 10^{-3}, and MEG means *mega-* or 10^6. When defining a value, no space is left between the value and the multiple, as in the example above.

SPICE only recognises the suffixes in Table 20.5, and it ignores other characters. For example, the value of the 10 Ω resistor R_1 in Table 20.3 can either be given simply as 10, or as 10ohm (there is no space between the 10 and the 'ohm'); the reason is that the letter 'o' in 'ohm' is not included in Table 20.5, and is ignored. Similarly, in the 10V dimension for V1 in Table 20.3, the letter 'V' is ignored since 'V' does not appear in Table 20.5.

A WORD OF WARNING. Do be careful when dimensioning capacitance, because F means *femto-* or 10^{-15}. That is 10UF (10 μF) is allowable, as is 15FF (15 femtofarad), but 19F means 19×10^{-15} (it *does not* mean 19 farad).

20.6 Resistance–temperature coefficient

Worked Example 20.3

A resistor with a linear resistance–temperature coefficient of 0.004 per deg. C referred to 0°C, has a resistance of 4.63 Ω at 0°C. Use SPICE to determine its resistance at 20°C and 50°C.

Solution

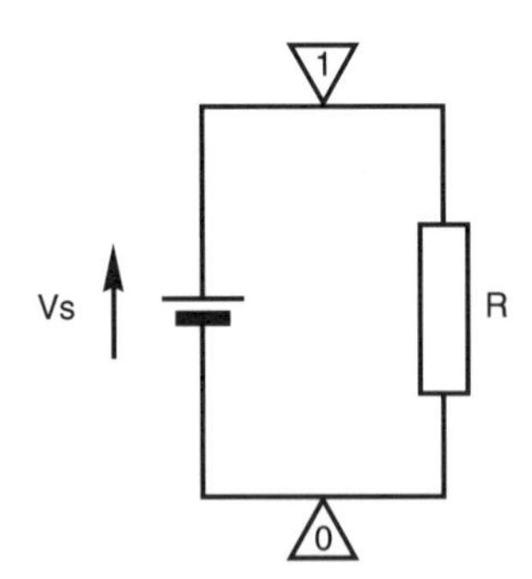

Figure 20.3 Circuit for Worked Example 20.3

The circuit used is the simple one in Figure 20.3, and the input file for the circuit is in Table 20.6.

The file lists the supply voltage Vs (although we have given it a value of 1 V, it could have any value in this case), and the resistor R, which has a value of 4.63 Ω at the *nominal temperature* at which SPICE calculates its results. This

Table 20.6 Input file for Worked Example 20.3

```
** Input file for Worked Example 20.3 **

* Supply source
Vs   1   0   1V

* Resistance with linear resistance-temperature coefficient
*        Linear resistance-temperature coefficient
*                          |
R   1   0   4.63ohms   TC = 0.004   ;4.63 ohms at TNOM

.OPTIONS   NOPAGE   TNOM = 0C   ;'C' optional

.TEMP   20C   50C   ;calculate result at stated temperature
.END
```

is normally 27°C, but we can alter this temperature in the '.OPTIONS' line by including a 'TNOM =' statement. We have set the nominal temperature at 0°C by means of the TNOM = 0C (the C is optional, and can be omitted). That is, our resistor has a resistance of 4.63 Ω at 0°C.

In this case the line describing R is modified to include the linear resistance–temperature coefficient referred to the nominal temperature by including the statement TC = 0.004. This describes the resistor as having a resistance–temperature coefficient of 0.004 per deg. C.

Next, we insert a '.TEMP' command line, in which we list the temperature values at which the answers are calculated. Since we have not specified an *analysis type* in the file, SPICE simply performs a small-signal bias solution at the TEMPERATURE ADJUSTED VALUES. In this case we are not interested in the small-signal bias solutions, and in Table 20.7 we list the temperature-adjusted resistance values at 20°C and 50°C. Once again, the solutions are in scientific notation.

Table 20.7 Temperature-adjusted values for Worked Example 20.3

```
****   TEMPERATURE-ADJUSTED VALUES   TEMPERATURE =   20.000 DEG C

**** RESISTORS
NAME      VALUE
R       5.000E+00

****   TEMPERATURE-ADJUSTED VALUES   TEMPERATURE =   50.000 DEG C

**** RESISTORS
NAME      VALUE
R       5.556E+00
```

The values in Table 20.7 should be compared with those in Worked Example 1.5 in Chapter 1.

20.7 A series R-L-C a.c. circuit

Worked Example 20.4

Using SPICE, determine the voltage across each element shown in full line in the circuit in Figure 20.4 at frequencies of 45, 50 and 55 Hz, together with the voltage applied to the circuit.

(Note: The 1 TΩ resistor, Rshunt, is not part of the series circuit, and the reason for its use is described below.)

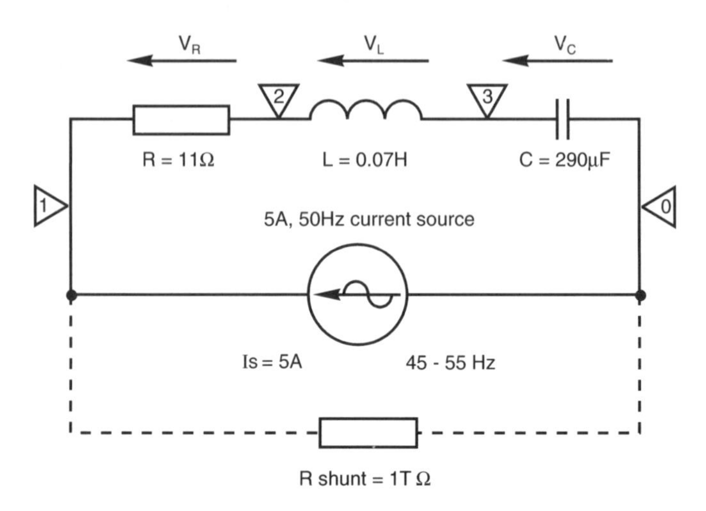

Figure 20.4 Circuit for Worked Example 20.4

Solution

The input file for the circuit is given in Table 20.8. In this case the circuit is energised by a 5 A *current source*, which has a frequency ranging from 45 to 55 Hz. The current source is described in the 'Is' element line, and the current flows *through the source* from the first numbered node to the second numbered node. That is the current flows *inside the source* from node 0 to node 1, i.e. the current leaves the source at node 1.

Since it is an alternating current source, we *must include* an 'AC' statement in the 'Is' line. This also implies that we must have a '.AC' command line in the file to describe features of the AC source (see below). Finally, we enter the magnitude of the current, which is 5 A; the 'A' is optional and can be omitted (there is no gap between the 5 and the 'A').

Next we describe the circuit elements in the usual way. The reader is reminded that 290UF corresponds to 290 μF or 290×10^{-6} F.

It has already been pointed out that each node must have a d.c. connection to node 0. Since the current source, Is, has an infinite internal resistance then, effectively, node 1 does not have a d.c. path to node 0. We

Table 20.8 Input file for Worked Example 20.4

```
 ** Input file for Worked Example 20.4 **

* Current source
*   Current flows from the circuit into this node
*     | Current flows from this node into the circuit
*     |  |
Is  0   1   AC   5A   ;'AC' must be stated

* Circuit elements
R   1   2   11ohms
L   2   3   0.07H
C   3   0   290UF

* Element Rshunt is necessary for SPICE in this case
Rshunt   3   0   1Tohm

.OPTIONS NOPAGE

*AC analysis
*   LINear sweep
*      | Number of points in sweep
*      |    | Start frequency (Hz)
*      |    |    | End frequency (Hz)
*      |    |    |    |
.AC  LIN   3   45   55

.PRINT   AC   VM(1,2)   VP(1,2)   VM(2,3)   VP(2,3)
.PRINT   AC   VM(3,0)   VP(3,0)   VM(1,0)   VP(1,0)

.END
```

therefore connect a resistance, which we call R_{shunt}, between node 1 and node 0. So long as its value is high, the value itself is immaterial; we have selected a value of $1\ T\Omega = 10^{12}\ \Omega$.

The '.AC' command line tells SPICE how to carry out the AC analysis on the circuit. The information in the Is line is as follows:

- LIN – this causes the frequency to be 'swept' LINearly from the starting frequency to the final frequency.
- 3 – this is the number of points in the sweep at which calculations are to be made.
- 45 – this is the starting frequency (in Hz) of the sweep.
- 55 – the final frequency of the sweep.

That is, the '.AC' line causes SPICE to calculate values at 45, 50 and 55 Hz.

Next, we have two '.PRINT' lines which cause SPICE to PRINT (that is, the results are made available in the OUTPUT FILE) the Magnitude and the Phase angle of voltages as follows:

VM means print the Magnitude of the Voltage between the stated nodes, and
VP means print the Phase angle of the voltage between the stated nodes relative to the reference direction.

Hence

VM (1,2) means print the Magnitude of the Voltage of node 1 relative to node 2, and
VP (1,2) means print the Phase angle of the voltage between nodes 1 and 2 relative to the reference direction.

Since the current, Is, is the reference quantity, the above statements cause the magnitude of phase angle of the voltage across R (i.e. V_R) relative to the supply current to be calculated and printed at each value of frequency. Similarly, VM(2,3) and VP(2,3) cause the magnitude of phase angle of V_L to be calculated, and VM(3,0) and VP(3,0) calculates the magnitude and phase of V_C. The corresponding output file is in Table 20.9.

Table 20.9 a.c. analysis for Worked Example 20.4

```
****    AC ANALYSIS                    TEMPERATURE =  27.000 DEG C

 FREQ        VM(1,2)     VP(1,2)     VM(2,3)     VP(2,3)

  4.500E+01  5.500E+01  7.402E-15  9.896E+01  9.000E+01
  5.000E+01  5.500E+01  7.402E-15  1.100E+02  9.000E+01
  5.500E+01  5.500E+01  1.480E-14  1.210E+02  9.000E+01

 FREQ        VM(3,0)     VP(3,0)     VM(1,0)     VP(1,0)

  4.500E+01  6.098E+01 -9.001E+01  6.683E+01  3.463E+01
  5.000E+01  5.488E+01 -9.000E+01  7.783E+01  4.504E+01
  5.500E+01  4.989E+01 -9.000E+01  8.986E+01  5.226E+01
```

The reader should compare the results at 50 Hz with those for Worked Example 9.4 in Chapter 9.

20.8 A parallel a.c. circuit

Worked Example 20.5

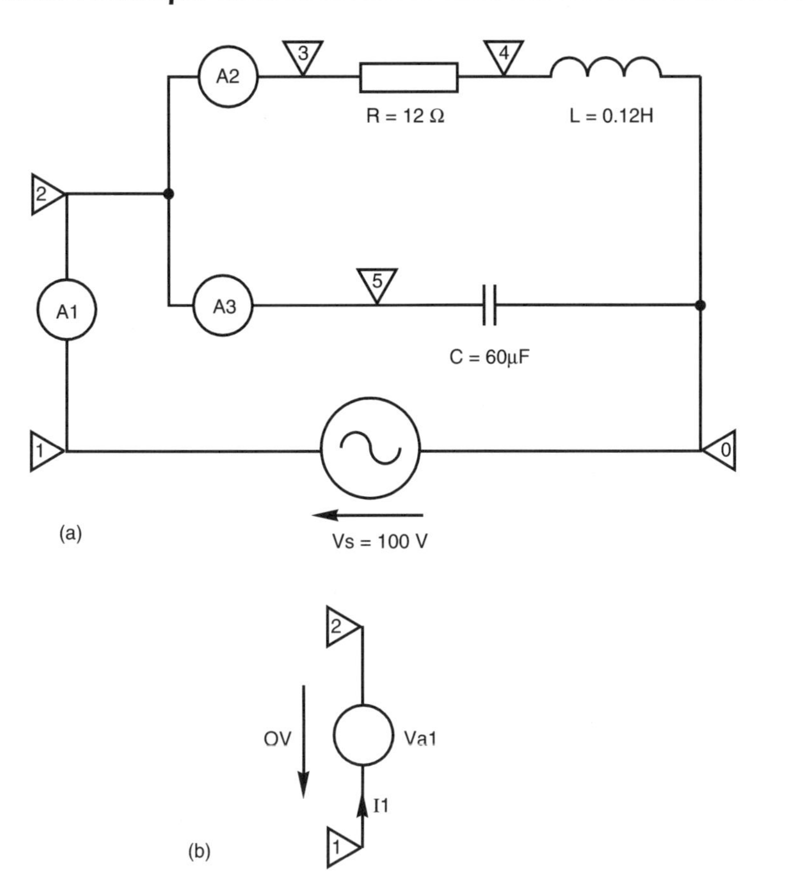

Figure 20.5 Circuit for Worked Example 20.5

For the parallel circuit in Figure 20.5(a), determine the magnitude of the current indicated by each of the three ammeters in the circuit, and the phase angle of the total current drawn by the circuit at frequencies of 47.1, 52.1, 57.1, 62.1, 67.1 Hz.

Solution

Once again we employ the concept of SPICE 'ammeters'. As we have mentioned earlier, SPICE reports the current *flowing into the positive node* of a voltage source. If we use a voltage source of zero volts (a null source), as shown in Figure 20.5(b), with the current entering the 'positive' terminal of the null source, we can use this as an ammeter; in the case of Figure 20.5(b),

Va1 replaces ammeter A1 in the circuit, with node 1 of the source being 'positive' with respect to node 2. The input file is given in Table 20.10, in which we use voltage sources Va1, Va2 and Va3 to replace ammeters A1, A2 and A3, respectively. The reader will note that the 'positive' node of each ammeter is given before its 'negative' node.

Table 20.10 Input file for Worked Example 20.5

```
** Input file for Worked Example 20.5 **

* Voltage source
Vs   1   0   AC   100V

* Circuit elements
R   3   4   12ohm
L   4   0   0.12H
C   5   0   60UF

* Ammeters
Va1   1   2   0V
Va2   2   3   0V
Va3   2   5   0V

.OPTIONS   NOPAGE

.AC     LIN   5   47.1   67.1
.PRINT   AC   IM(Va1)   IP(Va1)   IM(Va2)   IM(Va3)
.END
```

The '.AC' line specifies a LINear scan of five points, commencing at 47.1 Hz, and finishing at 67.1 Hz, as specified in the question.

The '.PRINT' line tells SPICE that the results are to be taken from the AC analysis, and says that we need the following results:

IM(Va1) – magnitude of the current in Va1
IP(Va1) – phase angle of the current in Va1
IM(Va2) – magnitude of the current in Va2
IM(Va3) – magnitude of the current in Va3.

The results are listed in Table 20.11. The reader should compare the results for the frequency of 57.1 Hz with those for Worked Example 11.4, where it was shown that the circuit was resonant at a frequency of 57.1 Hz.

The reader will note that SPICE reports that the phase angle of the current drawn from the supply is only -0.26° (see IP(Va1)) which, for all practical purposes, represents resonance.

Table 20.11 Output file for Worked Example 20.5

```
****     AC ANALYSIS                    TEMPERATURE =   27.000 DEG C

 FREQ         IM(Va1)      IP(Va1)      IM(Va2)      IM(Va3)

  4.710E+01   1.138E+00  -4.135E+01   2.668E+00   1.776E+00
  5.210E+01   7.991E-01  -2.712E+01   2.435E+00   1.964E+00
  5.710E+01   6.008E-01  -2.561E-01   2.237E+00   2.153E+00
  6.210E+01   6.143E-01   3.327E+01   2.069E+00   2.341E+00
  6.710E+01   7.940E-01   5.601E+01   1.923E+00   2.530E+00
```

20.9 A simple transistor amplifier

Worked Example 20.6

Use SPICE to simulate the transistor amplifier circuit in Figure 20.6(a), and determine the current gain and voltage gain of the circuit.

The parameters of the transistor are $h_{ie} = 1.5$ kΩ (the input resistance of the transistor), and $h_{fe} = 100$ (the common-emitter current gain of the transistor).

Solution

Initially, we convert the circuit in Figure 20.6(a) into the small-signal equivalent circuit in diagram (b).

In the latter diagram, the transistor is replaced by its simplified equivalent circuit contained within broken lines, comprising its input resistance parameter h_{ie}, and its current gain parameter h_{fe}. Also included (for the benefit of SPICE) within the broken lines is a SPICE ammeter VaBase, which measures the base current; the reason for this is described below. Since SPICE deals with resistance values (R values) and not h parameters, h_{ie} is replaced by Rie in the input file.

We arrive at the a.c. equivalent circuit in Figure 20.6(b) as follows: The battery, V_{CC}, is represented as an a.c. short-circuit, so that the 'top' end of Rb and Rc are connected to ground. Also, the reactance of each capacitor at the operating frequency (1 kHz) is assumed to be very small, so that they are replaced by short-circuits in the a.c. equivalent circuit. Finally, the transistor is replaced by the simple equivalent circuit shown in a broken line, in which $h_{ie} = \text{Rie} = 1.5$ kΩ represents the resistance between the base and emitter of the transistor, and the collector current is $h_{fe}I_b = 100I_b$.

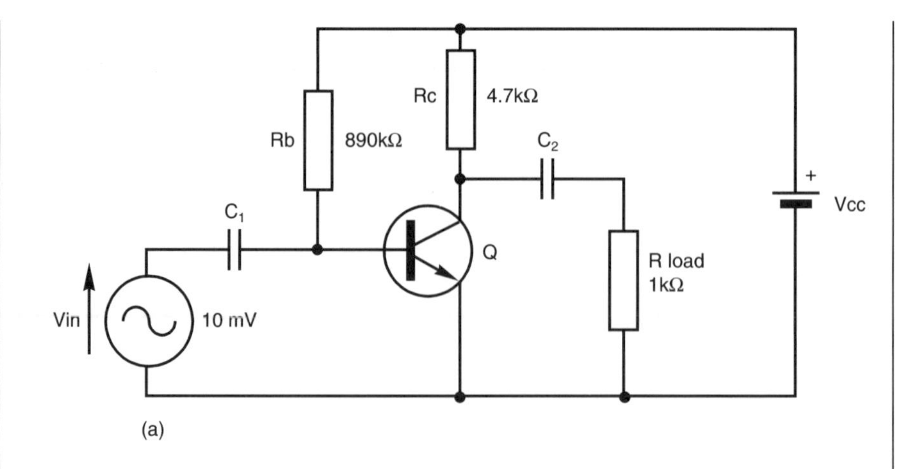

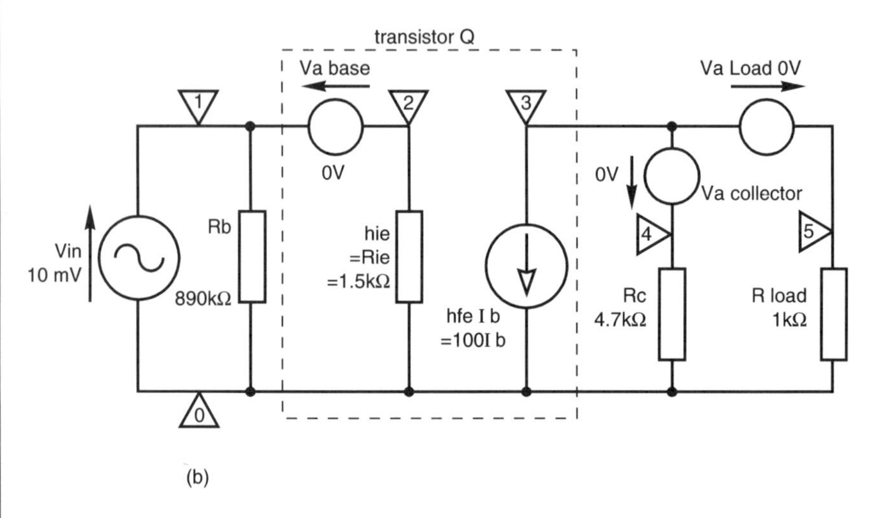

Figure 20.6 Solution of Worked Example 20.6 (a) Fixed bias, common emitter circuit (b) and equivalent circuit

Two other 'ammeters' have been inserted in diagram (b), and these are VaCollector (a.c. current in R_c) and VaLoad (a.c. current in the load).

The input file for the circuit is given in Table 20.12. New items in the file are the current-controlled current source, Fbjt, and **transfer function analysis**, '.TF', which are described below.

SPICE can simulate any one of four different **controlled sources**, as follows:

E device – a voltage-controlled voltage source
F device – a current-controlled current source
G device – a voltage-controlled current source
H device – a current-controlled voltage source

Table 20.12 Input file for Worked Example 20.6

```
 ** Input file for Worked Example 20.6 **

* Input a.c. signal
Vin   1   0   AC   10MV   ; 10 mV a.c. signal

* Base circuit
VaBase    1   2   0V       ; base current ammeter
Rb        2   0   890Kohm ; base bias resistor
Rie       2   0   1.5Kohm ; transistor input resistance

* Transistor Q
* Symbol for current-controlled current source
* | Current flows into this output node
* |    | Current flows out of this output node
* |    |   | Source of controlling current (base current)
* |    |   |    |     Current gain
* |    |   |    |       |
Fbjt  3   0   VaBase  100

* Collector and output circuit
VaCollector    4   3   0V       ; transisitor collector current
Rc             4   0   4.7Kohm ; collector resistor
VaLoad         5   4   0V       ; current in Rload
Rload          5   0   1Kohm    ; load resistor

.AC   LIN   1   1KHz   1KHz   ; calculate at 1 kHz

.PRINT  AC  IM(VaBase)  IM(VaCollector)  IM(VaLoad)
.PRINT  AC  VM(1)   VP(1)   VM(5)   VP(5)

*Transfer Function analysis
*|  Output variable
*|     |  Input source
*|     |     |
.TF  V(5)  Vin

.OPTIONS NOPAGE
.END
```

Typically uses of these devices are:

E device – operational amplifier
F device – bipolar junction transistor
G device – field-effect transistor
H device – separately excited d.c. generator.

We simulate transistor Q in the Fbjt line of the input file, and we must supply four items of information as follows: Firstly, the node where the current enters the source, secondly, where the current leaves the output of the F device, followed by the *name of the voltage source* in which the controlling current (the base current) flows and, finally, the current gain of the transistor. These are, respectively, nodes 3 and 0, VaBase, and 100.

The remaining elements in the circuit are then entered in the input file in the manner described earlier.

The '.AC' line differs from that given earlier in this chapter, insomuch that we only specify one frequency, namely 1 kHz; that is, the 'start' and 'end' frequencies are both 1 kHz. This enables SPICE to calculate values at one 'spot' frequency.

The first '.PRINT' line provides the magnitude of the transistor base current, the magnitude of the collector current, and the magnitude of the current in the load, Rload. The second '.PRINT' line gives the magnitude and phase angle of the input voltage and of the voltage across the load.

Finally, the input file contains a '.TF', or Transfer Function analysis. This determines

1. the voltage gain between the input voltage (Vin) and the specified output variable (V(5)),
2. the value of the input resistance as 'seen' by Vin (which is Rb in parallel with Rie),
3. the value of the output resistance between the output node and ground (Rc in parallel with Rload).

The relevant part of the output file is given in Table 20.13. Initially, SPICE completes the TRANSFER FUNCTION analysis, and shows that the overall voltage gain is -55.06; that is, the a.c. output voltage is 55.06 times greater than Vin, and the negative sign tells us that the phase shift imparted by the amplifier is 180°. Next we see that the input resistance is 1.497 kΩ, which corresponds to Rb = 890 kΩ in parallel with Rie = 1.5 kΩ. Finally, the TRANSFER FUNCTION analysis shows that the output resistance of the stage is 824.6 Ω (Rc in parallel with Rload).

Table 20.13 Output file for Worked Example 20.6

```
****   SMALL-SIGNAL CHARACTERISTICS

   V(5)/Vin = -5.506E+01
   INPUT RESISTANCE AT Vin =  1.497E+03
   OUTPUT RESISTANCE AT V(5) =  8.246E+02

 FREQ        IM(VamBase) IM(VamCollector)      IM(VamLoad)

1.000E+03     6.678E-06   6.678E-04              5.506E-04

 FREQ        VM(1)        VP(1)        VM(5)       VP(5)

1.000E+03     1.000E-02   0.000E+00   5.506E-01   1.800E+02
```

In the following we will perform a simple analysis, and compare the results with the SPICE solutions (which are shown in brackets).

$$\text{Base current (IM(VaBase))} = V_{in}/h_{ie} = 10\ \text{mV}/1.5\ \text{k}\Omega$$
$$= 6.667 \times 10^{-6}\ \text{A}\ (6.678 \times 10^{-6})$$
$$\text{Collector current (VaCollector)} = h_{fe}I_b$$
$$= 100 \times 6.667 \times 10^{-6}$$
$$= 0.6667 \times 10^{-3}\ \text{A}\ (6.678 \times 10^{-4})$$
$$\text{Load current (IM(VaLoad))} = \text{Ic} \times \text{Rc}/(\text{Rc} + \text{Rload})$$
$$= 0.55 \times 10^{-3}\ \text{A}\ (5.506 \times 10^{-4})$$
$$\text{a.c. output voltage } (V(5)) = \text{Iload} \times \text{Rload}$$
$$= 0.55 \times 10^{-3} \times 10^{3}$$
$$= 0.55\ \text{V}\ (5.501 \times 10^{-1})$$

The final line in Table 20.13 shows that the output voltage is 180° out of phase with the input voltage (see VP(5)).

20.10 An inverting summing amplifier

Worked Example 20.7

Determine the output voltage of the ideal operational amplifier circuit in Figure 20.7.

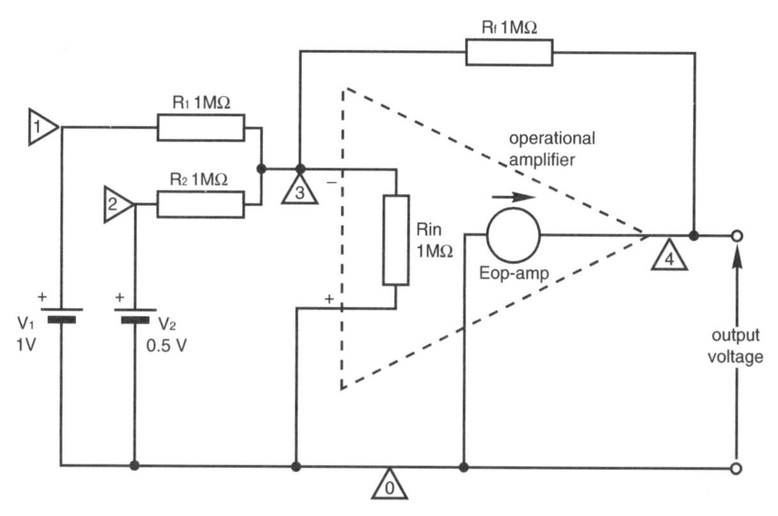

Figure 20.7 Circuit for Worked Example 20.7

Solution

Table 20.14 Input file for Worked Example 20.7

```
** Input file for Worked Example 20.7 **

* Voltage sources
V1   1   0   1V
V2   2   0   0.5V

* Resistors
R1   1   3   1MEGohm
R2   2   3   1MEGohm
Rf   3   4   1MEGohm

* Op-amp
*   (+) Output node
*        | (-) Output node
*        |    | (+) Controlling node
*        |    |    | (-) Controlling node
*        |    |    |    | Voltage gain
*        |    |    |    |      |
Eopamp 4    0    0    3     100000
Rin      3    0    10MEGohm

.OPTIONS NOPAGE
.END
```

The input file for the circuit is given in Table 20.14 and, initially, we define the two input sources V1 and V2, together with the two input resistors R1 and R2, and the feedback resistor Rf.

The op-amp can be regarded as a voltage-controlled voltage source, so that an ideal op-amp can be described as an E source (see also Worked Example 20.6). This type of source does not take into account some of the aspects of a practical op-amp such as saturation, maximum slew rate, etc., but is adequate for the purpose in hand. Practical op-amps can be handled those in the SPICE library.

The ideal op-amp is described in the Eopamp line. The first two nodes are the output nodes (nodes 4 and 0), which are followed by the non-inverting input node (node 0) and the inverting input node (node 3); finally we give the voltage gain (100 000) of the amplifier. As a touch of realism, we include the differential input resistance, Rin = 10MΩ, between the inverting and non-inverting input nodes.

The corresponding output file is in Table 20.15. The reader will see that the voltage at node 3 is only 15 μV, which is extremely small; it shows that, in this type of amplifier, the inverting input terminal is a *virtual earth* point.

Table 20.15 reports that the output voltage is −1.5 V; the result should be compared with those obtained in Worked Example 16.2.

Table 20.15 Output file for Worked Example 20.7

```
****    SMALL SIGNAL BIAS SOLUTION     TEMPERATURE =  27.000 DEG C

NODE  VOLTAGE    NODE  VOLTAGE    NODE  VOLTAGE    NODE  VOLTAGE


(  1)   1.0000 (   2)    .5000   (   3) 15.00E-06 (   4)  -1.5000
```

20.11 A non-inverting amplifier

Worked Example 20.8

Use SPICE to determine the output voltage from the non-inverting op-amp circuit in Figure 20.8.

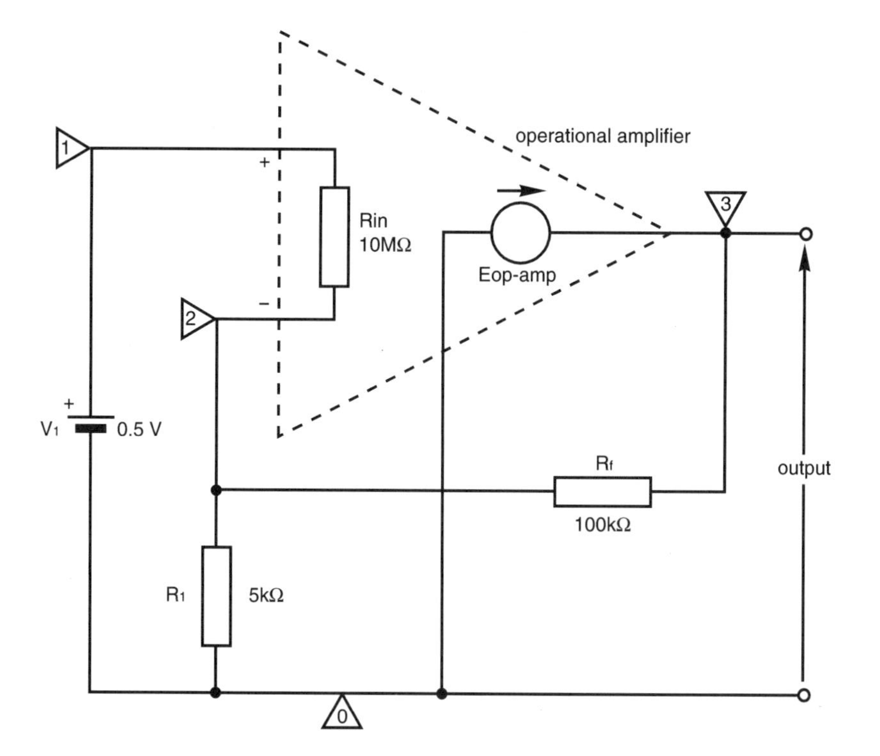

Figure 20.8 Circuit for Worked Example 20.8

Solution

Table 20.16 Input file for Worked Example 20.8

```
 ** Input file for Worked Example 20.8 **

* Signal voltage
V1   1   0   0.5V

* Resistors
R1   2   0     5Kohm
Rf   3   2   100Kohm

* Op-amp
*    (+) Output node
*        | (-) Output node
*        |    | (+) Input node
*        |    |    | (-) Input node
*        |    |    |    |   Voltage gain
*        |    |    |    |      |
Eopamp 3    0    1    2    100000
Rin      1    2    10MEGohm

.OPTIONS NOPAGE

.END
```

The input file for the circuit is given in Table 20.16, the op-amp being identical to that in Worked Example 20.7. Since this is a non-inverting amplifier, the input signal is applied to the non-inverting input of the op-amp (node 1). The theory of the circuit is given in Chapter 16; the relevant section of the output file for the circuit is listed in Table 20.17.

Table 20.17 Output file for Worked Example 20.8

```
****    SMALL SIGNAL BIAS SOLUTION      TEMPERATURE =   27.000 DEG C

NODE   VOLTAGE    NODE   VOLTAGE    NODE   VOLTAGE

(   1)    .5000   (   2)    .4999   (   3)   10.4980
```

The output voltage predicted by the theory in Chapter 16 is

$$V1 \times Rf/(Rf + R1) = 21 \times V1 = 21 \times 0.5 = 10.5 \text{ V}$$

which is very close to the value predicted by SPICE. The reader should compare the results in Table 20.17 with the solution of Worked Example 16.3 in Chapter 16.

20.12 Low-pass, high-pass and band-pass filters

In this section we will see how the frequency response curve showing the output magnitude of several filters changes with frequency.

Worked Example 20.9

Design a passive *R-C* low-pass filter with a corner frequency of 1 kHz, a passive *R-C* high-pass filter with a corner frequency of 10 kHz, and a simple active band-pass filter with a lower corner frequency of 1 kHz and an upper corner frequency of 100 kHz.

Solution

The analysis of these filters was described in Chapter 18, and we will apply the theory here.

The *low-pass filter* comprises the R_1C_1 circuit in Figure 20.9, for which the corner frequency is

$$w_{C(LP)} = \frac{1}{R_1C_1} = \frac{1}{1.592 \times 10^3 \times 0.1 \times 10^{-6}}$$
$$= 6281.4 \text{ rad/s or } 1 \text{ kHz}$$

The corner frequency of the *high-pass filter* is

$$w_{C(HP)} = \frac{1}{R_2C_2} = \frac{1}{1.592 \times 10^3 \times 0.01 \times 10^{-6}}$$
$$= 62\,814 \text{ rad/s or } 10 \text{ kHz}$$

To produce a simple *active band-pass filter*, we have cascaded an active low-pass filter with an active high-pass filter. The corner frequency of the low-pass filter is

$$w_{C1} = \frac{1}{R_4C_4} = \frac{1}{1.592 \times 10^3 \times 0.1 \times 10^{-6}} = 6281.4 \text{ rad/s or } 1 \text{ kHz}$$

and the corner frequency of the active high-pass filter is

$$w_{C2} = \frac{1}{R_6C_6} = \frac{1}{1.592 \times 10^3 \times 0.001 \times 10^{-6}} = 628\,140 \text{ rad/s or } 100 \text{ kHz}$$

The d.c. gain of the first stage is $R_4/R_5 = 10$, and the high frequency gain of the second stage is $R_7/R_6 = 10$, have been selected to enable the mid-band gain of the filter to reach zero dB (see the results in Table 20.19 and the graphs below).

The input file for the complete circuit is given in Table 20.18. The majority of the components and analyses in this file have been described earlier, and we will look in detail at the new items in the file.

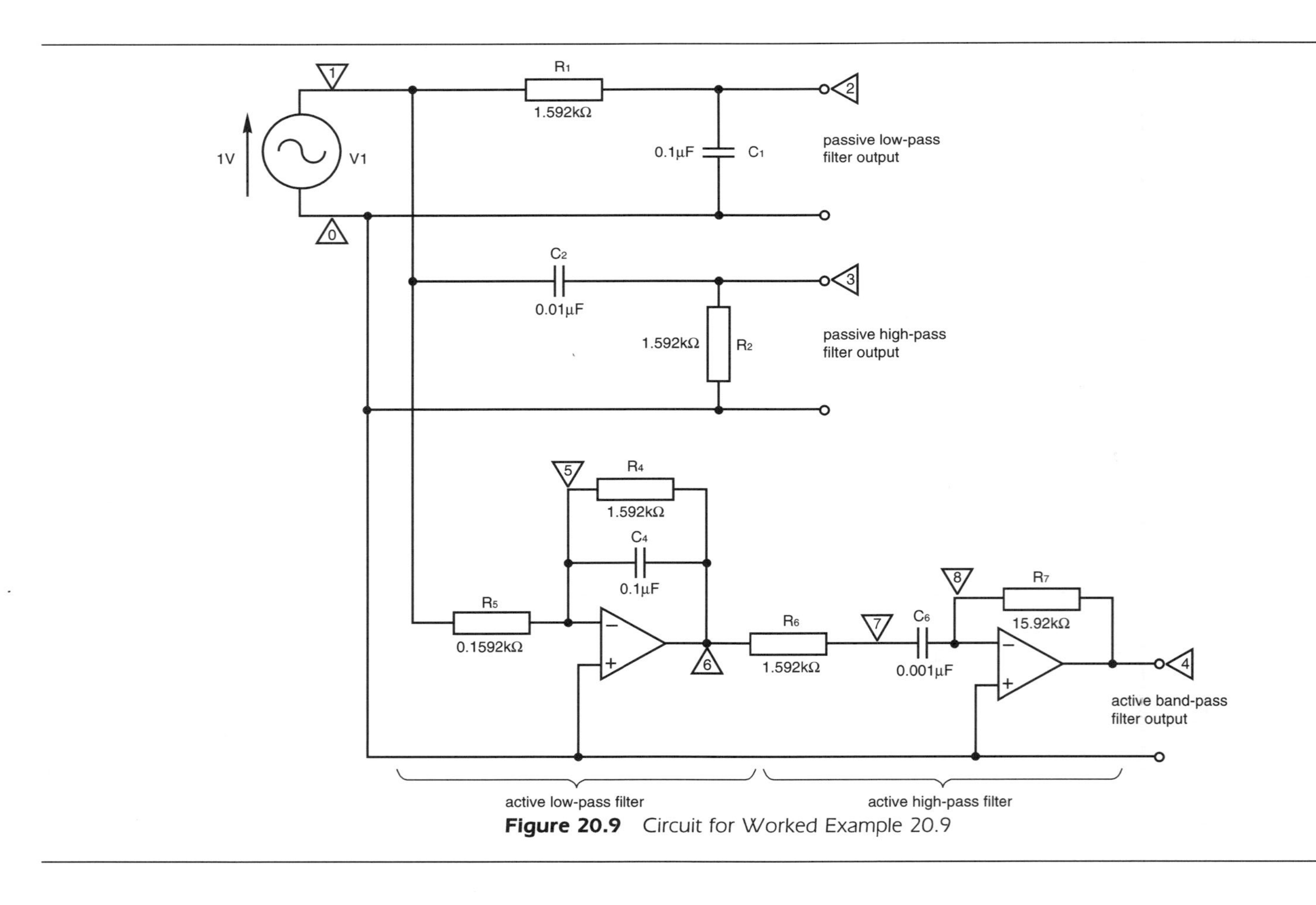

Figure 20.9 Circuit for Worked Example 20.9

Table 20.18 Input file for Worked Example 20.9

```
 ** Input file for Worked Example 20.9 **

* Input signal
V1      1   0   AC   1

* Passive low-pass filter
R1      1   2   1.592Kohm
C1      2   0   0.1UF

* Passive high-pass filter
R2      3   0   1.592Kohm
C2      1   3   0.01UF

* Active band-pass filter
R5      1   5   0.1592Kohm         ; low-pass section
R4      5   6   1.592Kohm          ; low-pass section
C4      5   6   0.1UF              ; low-pass section
Eopamp1  6   0   0   5   100000 ; low-pass section
C6      7   8   0.001UF             ; high-pass section
R6      6   7   1.592Kohm           ; high-pass section
R7      8   4   15.92Kohm           ; high-pass section
Eopamp2  4   0   0   8   100000   ; high-pass section

*AC analysis
*|  Sweep frequency by decades
*|       | 5 points per decade
*|       |    | commence at 100 Hz
*|       |    |     | Finish at 1 MHz
*|       |    |     |     |
.AC    DEC   5    100    1MEG

*PLOT results from the .AC analysis
* |      |  Plot the voltage gain at in dB at
* |      |  node(2)  node(3) node(4)
* |      |     |         |        |
* |      |     |         |        |
.PLOT   AC   VDB(2)   VDB(3)  VDB(4)

*PRINT results from the .AC analysis at nodes 2, 3 and 4
*  |     |
.PRINT  AC   VDB(2)   VDB(3)   VDB(4)

.OPTIONS   NOPAGE

.END
```

In particular, since we are looking into the frequency response of filters, we would like to sweep the frequency ranges logarithmically. We therefore say, in the '.AC' analysis line, that the frequency is to be swept in DECades, calculating 5 values at 5 points per decade. The lowest frequency is to be 100 Hz (1 decade lower than the lowest corner frequency), and the highest frequency is to be 1 MHz (2 decades above the highest corner frequency).

Next we look at the method used to PLOT the results. The '.PLOT' line causes SPICE to plot the specified results on the screen (and, if necessary, on a printer) using standard characters such as '+', '*' and '='. In our case we ask SPICE to use the results from the '.AC' analysis, showing the *decibel gain* at node 2 (the passive low-pass filter), node 3 (the passive high-pass filter) and node 4 (the active band-pass filter). To do this we merely say that we need the results for VDB(2), VDB(3) and VDB(4), respectively.

Table 20.19 a.c. analysis for Worked Example 20.9

FREQ	VDB(2)	VDB(3)	VDB(4)
1.000E+02	-4.324E-02	-4.000E+01	-2.004E+01
1.585E+02	-1.078E-01	-3.600E+01	-1.611E+01
2.512E+02	-2.659E-01	-3.200E+01	-1.226E+01
3.981E+02	-6.393E-01	-2.800E+01	-8.638E+00
6.310E+02	-1.456E+00	-2.401E+01	-5.455E+00
1.000E+03	-3.012E+00	-2.004E+01	-3.010E+00
1.585E+03	-5.457E+00	-1.611E+01	-1.456E+00
2.512E+03	-8.641E+00	-1.226E+01	-6.416E-01
3.981E+03	-1.227E+01	-8.637E+00	-2.727E-01
6.310E+03	-1.611E+01	-5.454E+00	-1.251E-01
1.000E+04	-2.005E+01	-3.009E+00	-8.662E-02
1.585E+04	-2.402E+01	-1.455E+00	-1.252E-01
2.512E+04	-2.801E+01	-6.386E-01	-2.730E-01
3.981E+04	-3.201E+01	-2.656E-01	-6.423E-01
6.310E+04	-3.600E+01	-1.077E-01	-1.458E+00
1.000E+05	-4.000E+01	-4.319E-02	-3.013E+00
1.585E+05	-4.400E+01	-1.725E-02	-5.458E+00
2.512E+05	-4.800E+01	-6.874E-03	-8.642E+00
3.981E+05	-5.200E+01	-2.738E-03	-1.227E+01
6.310E+05	-5.600E+01	-1.090E-03	-1.611E+01
1.000E+06	-6.000E+01	-4.342E-04	-2.005E+01

In addition we ask SPICE to PRINT the same set of results. The results from this analysis are listed in Table 20.19. Since the results are calculated over five frequency decades, there are $(5 \times 4) + 1 = 21$ results. The attention of the reader is directed to the fact that gain of the low-pass filter (VDB(2)) is -3 dB at 1 kHz, and the gain of the high-pass filter (VDB(3)) is -3 dB at 10 kHz. The gain of the active filter has been adjusted so that its mid-band gain is zero dB (or thereabouts), and the gain of the filter is -3 dB at the two

Table 20.20 Printer output for the '.PLOT' line in Worked Example 20.9

```
 LEGEND:

*: VDB(2)
+: VDB(3)
=: VDB(4)

 FREQ      VDB(2)

(*)---------- -8.0000E+01 -6.0000E+01 -4.0000E+01 -2.0000E+01  0.0000E+00
(+)---------- -4.0000E+01 -3.0000E+01 -2.0000E+01 -1.0000E+01  0.0000E+00
(=)---------- -3.0000E+01 -2.0000E+01 -1.0000E+01  0.0000E+00  1.0000E+01

                      - - - - - - - - - - - - - - - - - - - - - - - - - -
 1.000E+02 -4.324E-02 +           =            .           .            *
 1.585E+02 -1.078E-01 .   +          =         .           .            *
 2.512E+02 -2.659E-01 .        +  .         =  .           .            *
 3.981E+02 -6.393E-01 .           .  +         . =         .            *
 6.310E+02 -1.456E+00 .           .       +    .    =                  *.
 1.000E+03 -3.012E+00 .           .            +       =   .          * .
 1.585E+03 -5.457E+00 .           .            .    +     =.        *   .
 2.512E+03 -8.641E+00 .           .            .        + =.      *     .
 3.981E+03 -1.227E+01 .           .            .           = +  *       .
 6.310E+03 -1.611E+01 .           .            .           =  *  +      .
 1.000E+04 -2.005E+01 .           .            .           X        +   .
 1.585E+04 -2.402E+01 .           .            .        *  =          + .
 2.512E+04 -2.801E+01 .           .            .      *    =           +.
 3.981E+04 -3.201E+01 .           .            .    *     =.            +
 6.310E+04 -3.600E+01 .           .            .  *      = .            +
 1.000E+05 -4.000E+01 .           .            *       =   .            +
 1.585E+05 -4.400E+01 .           .         *  .    =      .            +
 2.512E+05 -4.800E+01 .           .       *    . =         .            +
 3.981E+05 -5.200E+01 .           .    *    =  .           .            +
 6.310E+05 -5.600E+01 .           .  * =       .           .            +
 1.000E+06 -6.000E+01 .           X            .           .            +
```

corner frequencies of 1 kHz and 100 kHz, giving an effective bandwidth of $(100 - 1) = 99$ kHz.

The printer output corresponding to the '.PLOT' line is shown in Table 20.20. Since we did not specify the scale to be used by the printer, SPICE makes its own decision about the scale to be used for each output result. It is for this reason that the 'graph' in Table 20.20 looks a little unusual. The graph marked with '*' represents the low-pass filter output, that marked with '+' is the high-pass filter, and the band-pass filter output is marked with '='. Where any two graphs cross at a frequency where a calculation is made, the corresponding point is marked with a capital 'X'.

The two columns to the left of Table 20.10 correspond to the first two columns in Table 20.19.

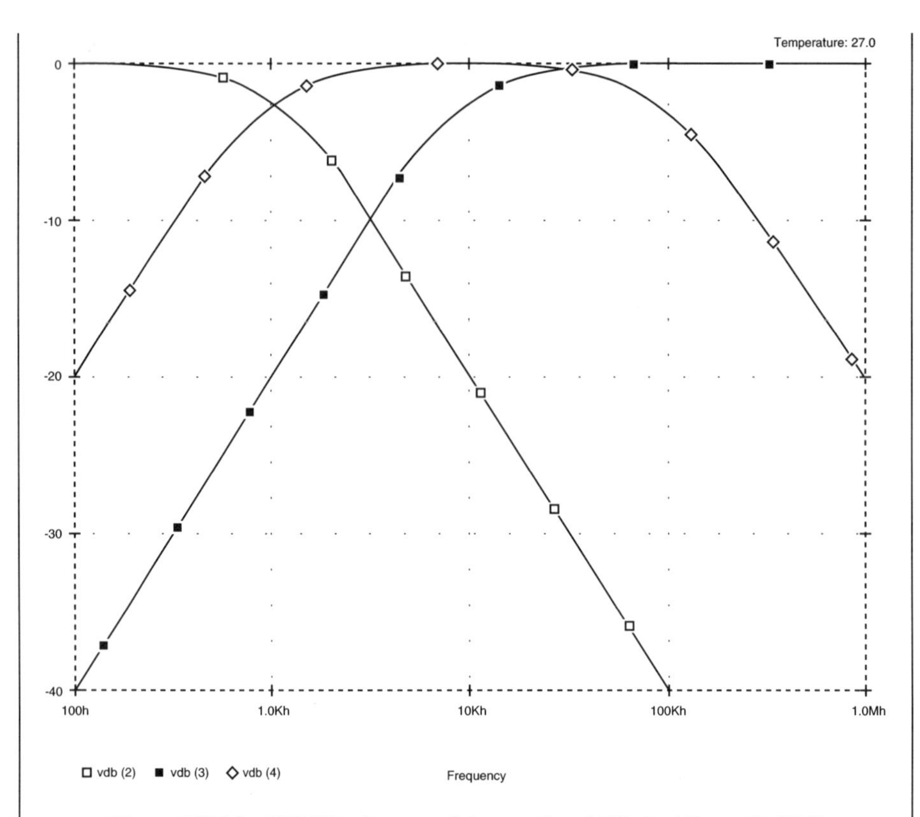

Figure 20.10 *PROBE print-out of the results of Worked Example 20.9*

An optional method of printing the results is by the use of a plotter or printer which can be used in a graphics mode. PSpice has an optional **graphics post-processor** called PROBE, which enables the computer system to be used as a **software oscilloscope**. The results produced by this example are plotted using PROBE and a 24–pin dot matrix printer in Figure 20.10, and the improvement in quality over Table 20.20 is quite marked. Once again, the frequency results of the low-pass filter (VDB(2)), the high-pass filter (VDB(3)), and the band-pass filter (VDB(4)) are plotted; the scales show the gain in dB plotted to a base of frequency on a logarithmic scale.

Exercises

20.1 Use SPICE to solve Exercise 2.12 in Chapter 2. The reader should insert a SPICE 'ammeter' in each branch to determine the current in that branch.

20.2 Using the circuit in Exercise 4.1 in Chapter 4, with the negative pole of the 10 V source as reference, determine the voltage at node B and at node A. What is the voltage of node B with respect to node A?

Insert a SPICE 'ammeter' in series with each voltage source to determine the current in each branch.

20.3 Using SPICE, solve Exercise 9.5 in Chapter 9.

20.4 A coil of inductance 31.8 mH and resistance 10 Ω is connected in parallel with a 159 μF capacitor to a 50 Hz supply. Using SPICE, verify that the circuit resonates with the supply frequency. What current flows in the capacitor if the circuit is energised by a 20 V, 50 Hz supply, and what is the dynamic impedance of the circuit? (See also Exercise 11.12 and 11.13 of Chapter 11.)

20.5 If, in the circuit of Worked Example 20.6 in this chapter, Vin = 100 mV, Rb = 850 kΩ, Rc = 4 kΩ, Rload = 1.5 kΩ, hie = 2 kΩ and hfe = 80, determine the a.c. input resistance of the stage, its a.c. output resistance, the stage gain, and the a.c. voltage developed across Rload.

20.6 If each of the operational amplifiers in Figure 20.11 has an differential input resistance of infinity and a gain of (a) 10^5, (b) 500, determine the value of V1 and V2.

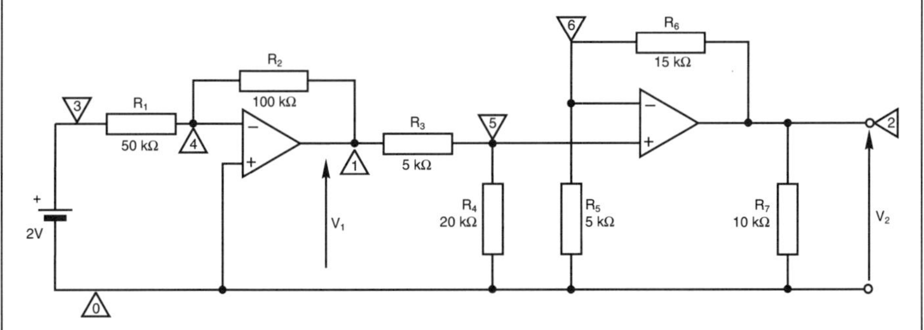

Figure 20.11 Circuit for Exercise 20.6

20.7 Write a SPICE input file to plot the overall gain (in dB) and phase shift of the circuit in Figure 20.12 (this is a *phase lead filter circuit* used in a control system) to a base of frequency (on a logarithmic scale) over the frequency range 100 Hz to 100 kHz. Estimate the maximum phase shift and the frequency at which it occurs.

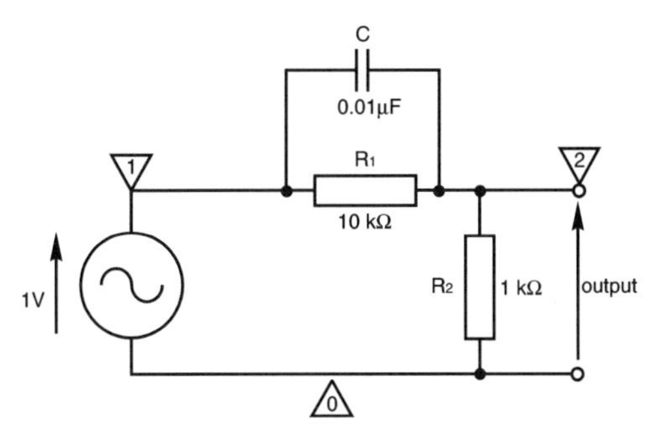

Figure 20.12 Circuit for Exercise 20.7

Summary of important facts

A range of software packages are available for the solution of electronic and electrical circuits, one of the most important being **SPICE** (**S**imulation **P**rogram with **I**ntegrated **C**ircuit **E**mphasis). Several versions of SPICE are available, one of the most accessible being PSpice. The *evaluation version* (or student version) of PSpice was used to solve the problems in this chapter. The range of problems which can be solved is practically unlimited, the primary restriction being the imagination of the user.

Each circuit is described in an **input file** or **source file**, which is passed to a **software analyser** for checking and analysing. If the input file is free of errors, the results (together with the input file) are collected together in an **output file**. The output file can be viewed on the computer monitor or printed out as a hard copy.

Most versions of SPICE have a **graphics post-processor** (the version in PSpice is called PROBE), which allows the computer system to be used as a **software oscilloscope**. A high-quality print-out of graphical results can be obtained in this way.

SPICE contains a large **library of subcircuits**, which includes many semiconductors and integrated circuits. Elements from the library can be called into use, allowing the design of complex systems to be carried out.

SPICE enables the user to design and analyse either analogue or digital systems.

Answers to exercises

Chapter 1

1.1 (a) 830; (b) 79 800; (c) 3 600 000.
1.2 (a) 9 700 000; (b) 82 000; (c) 0.000 058; (d) 8.17×10^{-11}.
1.3 (a) 1.8×10^{-3}; (b) 0.18.
1.4 $R_E - (R_1 + R_2)$; (b) $D^2/2W$; (c) $(F/\Phi) - S_1$; (d) μ/μ_0.
1.5 (a) $1/(\frac{1}{r_1} - \frac{4\pi e}{c})$; (b) $\sqrt{(Q/(4\pi Ee))}$; (c) $(V_1 - h_i I_1)/V_2$; (d) $1/(\omega_0^2 L - 1/C_1)$.
1.6 (a) 0.208 Ω; (b) 0.826 Ω.
1.7 1.87 mm.
1.8 (a) $39.65 \times 10^{-4}(°C)^{-1}$; (b) $33.09(°C)^{-1}$.
1.9 (a) 0.003; (b) 0.00287.
1.10 0.004.

Chapter 2

2.1 0.221 Ω.
2.2 5600 Ω.
2.3 410.5 Ω.
2.4 4.1 Ω.
2.5 6 kΩ.
2.6 12 Ω.
2.7 30 kΩ.
2.8 1.571 Ω.
2.9 (a) 0.225 S; (b) 4.444 Ω; (c) 2.25 Ω, (d) 1 A, 1.25 A.
2.10 25 Ω; 20 V, 30 V, 40 V, 50 V.
2.11 1.667 Ω.
2.12 2 A in 1 Ω and 4 Ω; 4 A in 0.5 Ω; 0.8 A in 10 Ω; 1.6 A in 2 Ω; 1.6 A in 2 Ω, 3 Ω and 5 Ω; 8 V across 10 Ω and 5 Ω; 2 V across 0.5 Ω; 4.8 V across 3 Ω; 3.2 V across 2 Ω; 2 V across 1 Ω; 8 V across 4 Ω.
2.13 33.12 Ω; 0.662 A in 8 Ω; 0.691 A in 1 Ω; 0.461 A in 10 Ω; 0.23 A in 20 Ω; 1.353 A in 70 Ω; 1.666 A in 40 Ω and 20 Ω; total current = 3.109 A.
2.14 575.7 Ω.
2.15 0.025 Ω.
2.16 0.96 MΩ.
2.17 4.2 Ω; 0.3 A; 0.164 A; 0.464 A.

Chapter 3

3.1 (a) 0.3 A; (b) 4.5 W; (c) 22.5 Wh; (d) 0.27 p.
3.2 (a) 8.49 A; (b) 2.12 V; (c) 144.2 Wh; (d) 1.442 p.
3.3 (a) 21 V; (b) 126 W; (c) 252 Wh.
3.4 (a) 3 kW; (b) 12.5 A; (c) 19.2 Ω.
3.5 85.1 per cent
3.6 80 per cent
3.7 21 mW; 10.66 mW.

Chapter 4

4.1 $V_{BA} = 11.92$ V.
4.2 0.516 A (flowing upwards); 3.04 V (top negative with respect to bottom).
4.3 -13 V.
4.4 $R_T = 6.667\ \Omega$; (a) 0.538 A; (b) 0.318 A.
4.5 $E_T = -1$ V; $R_T = 2.5\ \Omega$; (a) -0.222 A; (b) -0.182 A.
4.6 $I_N = 1.25$ A; $R_N = 10\ \Omega$; (a) 0.5 A; (b) 0.417 A.
4.7 $I_N = 0.4$ A; $R_N = 3.75\ \Omega$; (a) 0.171 A; (b) 0.109 A.
4.8 6.667 Ω, 20.42 W; 2.5 Ω, 0.1 W; 10 Ω, 3.9 W; 3.75 Ω, 0.15 W.
4.9 4.615 W.

Chapter 5

5.1 15.5 nF.
5.2 300 cm^2.
5.3 0.2 mm.
5.4 5.02.
5.5 12.5 $\mu C/m^2$.
5.6 75 kV/m.
5.7 6.27.
5.8 63.25 V; 0.632 mC.
5.9 (a) (i) 0.01667 μF, (ii) 0.834 μJ; (b) (i) 0.12 μF, (ii) 6 μJ.
5.10 (a) 0.0429 μF; (b) 0.5 μF.
5.11 (a) 0.429 μC, 2.145 μJ; (b) 5 μC, 25 μJ.
5.12 8.57 nF.
5.13 8.57 V; 28.6 nC.
5.14 0.01 μF; 0.008 μF.
5.15 5.714 μC; 28.57 V, 14.29 V (across the largest value of capacitance).
5.16 0.1 μF; 0.2 μF; 0.4 μF.
5.17 (a) 42.86 V; (b) 7.14 V; 0.429 μC.

Chapter 6

6.1 (a) 0.0075 H; (b) 7500 μH.
6.2 60 mH.
6.3 4 A.
6.4 0.667 mWb.
6.5 1200.
6.6 24.67 μH.
6.7 0.151 H.
6.8 350 V.
6.9 700 V.
6.10 1000.
6.11 1 mWb.
6.12 50 ms.
6.13 5000 V.
6.14 2.5 H.
6.15 50 A.
6.16 0.4 J.
6.17 5 J; 250 W.
6.18 (a) 1 H; (b) 68.97 mH.
6.19 (a) 1 H; (b) 7 H.
6.20 (a) 550 J; (b) 350 J.
6.21 5.32 H.

Chapter 7

7.1 0.8 V.
7.2 (a) 0.693 V; (b) 0.384 V.
7.3 0.332 V.
7.4 100 V.
7.5 5 N.
7.6 0.48 μN m
7.7 200 ampere turns; 1000 ampere turns/m.
7.8 2.65 A.
7.9 0.455×10^6 A/Wb; 325 ampere turns.
7.10 0.035 A.
7.11 (a) 840 ampere turns, 1070 ampere turns/m; (b) 744.
7.12 (a) 2.5 A; (b) 625 ampere turns; (c) 1658 ampere turns/m; (d) 0.75×10^6 A/Wb; (e) 4.17 T.
7.13 1293 ampere turns/m; 2328 ampere turns.
7.14 0.477 m.
7.15 0.4 T; 795.8.
7.16 (a) 1.24 A; (b) 1.51 A.
7.17 1110 ampere turns.
7.18 (a) 2.6 A; (b) 3.8 A.
7.19 1.95 A.
7.20 0.358 J.
7.21 7.2 J.

Chapter 8

8.1 (a) 0.4363 rad; (b) 2.182 rad; (c) −4.29 rad; (d) −15.64 rad; (e) 40.1°; (f) −234.9°, (g) 515.7°.
8.2 (a) 314.2 rad/s, 20 ms; (b) 6283 rad/s, 1 ms; (c) 62.83×10^6 rad/s, 0.1 μs.
8.3 30 Ghz.
8.4 95.5 kHz, 10.47 μs.
8.5 13.17 V (wave *X*); 40.9 V (wave *Y*).
8.6 Wave *X*: 20.43 V, 1.55, 2.13; wave *Y*: 43 V, 1.05, 1.16.
8.7 (a) 628.3 Ω; (b) 6.283 Ω.
8.8 314.2 Ω; 79.6 mA.
8.9 0.127 H.
8.10 159.2 Hz.
8.11 (a) 31.42 Ω; (b) 7.54 Ω.
8.12 398 Ω; 31.8 Ω; 15.9 Ω.
8.13 6.37 μF.
8.14 1.59 kHz.
8.15 (a) 25.1 mA; (b) 1.57 mA.
8.16 (a) 60 Ω; (b) 53.05 μF.
8.17 12 A; 3 A.
8.18 676.64 A at 7.2° relative to the reference direction.
8.19 I_1 : inphase component = 295.44 A, quadrature component = −52.09 A;
I_2 : inphase component = 375.88 A, quadrature component = 136.81 A;
I_T : inphase component = 671.32 A, quadrature component = 84.72 A.
8.20 (a) 18.44 V at −60.2° with respect to the reference direction;
(b) reference component = 9.16V, quadrature component = −16V.
8.21 4.36 mA at 59° relative to the reference direction.
8.22 87.18 V at 173.4° relative to the reference direction.

Chapter 9

9.1 12.5 Ω; 36.9°.
9.2 37.17 Ω; 47.7°.
9.3 10 Ω; 30°.
9.4 (a) 11.81 Ω, 32.1°; (b) 5.17 Ω; (c) 12.25 Ω, 0.039 H; (d) 26.4 Ω, 200 Hz.
9.5 2.69 A; 57.5° (I lagging V_S); $V_R = 26.9$ V, $V_L = 42.2$ V.
9.6 382.5 mH.
9.7 7.81 Ω; 50.2°.
9.8 (a) 87.48 Ω, 46.7°; (b) 142.4 μF; (c) 39.14 Ω, 49.28°; (d) 100 Hz.
9.9 (a) 125 Ω; (b) 75 Ω; (c) 42.44 μF; (d) 36.87° (I leading V_S); (e) $V_R = 200$ V, $V_C = 150$ V.
9.10 (a) 0.24 μF; (b) 774.4 Ω; (c) 12.9 mA; (d) $V_R = 5.17$ V, $V_{C1} = 3.43$ V, $V_{C2} = 5.14$ V.
9.11 120.3 μF.
9.12 (a) $X_L = 31.42$ Ω, $X_C = 39.79$ Ω; (b) 10.3 Ω; (c) 9.71 A (54.4° I leading);
(d) $V_R = 58.26$ V, $V_{COIL} = 310.6$ V, $V_C = 386.4$ V.
9.13 (a) 1.414 kΩ; (b) 1.414 V; (c) 45° (I lagging); (d) $V_R = 1$ V, $V_L = 2$ V, $V_C = 3$ V.
9.14 (a) 15.57 Ω; (b) $V_R = 55$ V, $V_L = 110$ V, $V_C = 54.9$ V; (c) 77.9 V; (d) 45° (I lagging).

9.15 (a) $I_1 = 3.98$ A (90° lagging), $I_2 = 3.03$ A (in phase); (b) 5 A (52.7° lagging); (c) 20 Ω.
9.16 (a) $I_1 = 40$ mA (in phase), $I_2 = 31.4$ mA (90° leading); (b) 50.9 mA (38.1° leading); (c) 196.6 Ω.
9.17 (a) (i) $I_1 = 5.37$ mA (57.5° lagging), $I_2 = 1.58$ mA (90° leading); (ii) 4.13 mA (45.7° lagging); (iii) 2.42 kΩ; (b) (i) $I_1 = 1.57$ mA (81° lagging), $I_2 = 6.28$ mA (90° leading); (ii) 4.74 mA (87° leading); (iii) 2.11 kΩ
9.18 5.01 A at an angle of 35.9°.
9.19 274.3 A leading by 28.3°.
9.20 (a) $I_1 = 7.81$ A (38.7° lagging), $I_2 = 7.81$ A (51.3° leading); (b) 11.05 A (6.3° leading); (c) $V_{R1} = 7.81$ V (lagging V_S by 38.7°), $V_L = 6.25$ V (leading V_S by 51.3°), $V_{R2} = 6.25$ V (leading V_S by 51.3°), $V_C = 7.81$ V (lagging V_S by 38.7°); (d) 9.06 Ω at angle −6.3°.

Chapter 10

10.1 (a) 20 MVA, 17.32 MVAr; (b) 12.1 MVA, −7 MVAr.
10.2 83.9 mW; 130.5 VA.
10.3 80 MW; 60 MVAr.
10.4 11.18 VAr; 0.67.
10.5 223.6 kW; 41.81 deg.
10.6 28 A; 0.928.
10.7 (a) 12 A; (b) 45.6 deg, 0.7 lag; (c) 0.9.
10.8 (a) 3.37 A; (b) 8.42 A; (c) 0.95; (d) 1.48 kVAr.
10.9 1729.4 kVA; 0.95 lagging.
10.10 517.9 kVAr.
10.11 21.63 kVA, 0.832 lagging; (a) 6.08 kVAr; (b) 18.95 kVA; (c) 0.829 leading.
10.12 (a) 35.7 A; (b) 0.59 lagging.
10.13 (a) active component = 2.09 A, quadrature component = 2.13 A; (b) 2.2 A; (c) 166.4 Ω, 1.44 A; (d) 19.13 μF.

Chapter 11

11.1 581.2 Hz; 0.15 A.
11.2 1.83
11.3 0.253 H.
11.4 (a) 46 Hz; (b) 10 A; (c) 1 kW; (d) 5.77.
11.5 10 Ω; 10 mH; 2.53 μF.
11.6 25 μF; 666.7 Hz.
11.7 (a) 5.03 kHz; (b) 10 mA; (c) 10.5 V, 3.16 V; (d) 31.6.
11.8 0.02 μF.
11.9 (a) 5 V; (b) 1.67 V; (c) 1 V.
11.10 3.56 kHz.
11.11 8 μH.
11.12 159 μF.
11.13 1 A; 20 Ω.
11.14 0.02 μF; 4 μA; 0.4 μW.
11.15 (a) 0.255 H; (b) 25.5 μF; (c) 0.3 A; (d) 15 W; (e) 1.33, 166.7 Ω.

Chapter 12

12.1 6.06 A; 83.33 A.
12.2 200 V.
12.3 (a) 4.4 kV; (b) 40 A.
12.4 175.
12.5 2.73 A.
12.6 (a) 9.091 A, 272.7 A; (b) 110 V.
12.7 (a) 330 V; (b) 33 A; (c) 3.3 A.
12.8 (a) 33; (b) 0.61 A.
12.9 11.36 A; 113.6 A; 12.5 kVA.
12.10 (a) 175; (b) 10 V.
12.11 646.5 V; 32.32 V.
12.12 1110 V; 99.9 V.
12.13 (a) 75 cm^2; (b) 1650, 220.
12.14 43.24 Hz.
12.15 (a) 56.3 cm^2; (b) 1267, 160; 7.9 A, 62.5 A.
12.16 97 per cent.
12.17 96.7 per cent.
12.18 (a) 2.57 kW; (b) 3.17 kW.
12.19 (a) 300 V; (b) 33.26 A at 0.7 power factor lagging.

Chapter 13

13.1 19.7 A.
13.2 450 V.
13.3 440 V.
13.4 (a) 230.9 V; (b) 103.9 kVA.
13.5 51.96 A; 254 V.
13.6 (a) 5.56 MVA; (b) 2.43 MVAr; (c) 291.6 A.
13.7 (a) 15.59 kVA, 14.03 kW; (b) 11.55 A.
13.8 76.21 kVA; 49.44 kVAr; 0.761.
13.9 (a) 200 V; (b) 173.2 A; (c) 100 A; (d) 103.92 kVA.
13.10 (a) 18.74 A; (b) 10.82 A.
13.11 (i) (a) 519.6 V; (b) 30 A; (c) 27 kVA; (ii) (a) 300 V; (b) 52 A; (c) 27 kVA.
13.12 (a) 20.6 A; (b) 11.9 A.
13.13 (a) 288.7 V; (b) 22.64 A, 13.1 A; (c) 20.33 kW.
13.14 (a) 20 A; (b) 34.64 A; (c) 24 kW; (d) 0.8.
13.15 (a) 4.15 A; (b) 7.18 A.
13.16 (a) 21.33 Ω, 16 Ω, 26.67 Ω;
(b) 8.31 kW; 45 A; 24.93 kW.
13.17 433 V; 25 A; 9.375 kW.
13.18 $I_R = 25.4$ A in phase with V_{RN}; $I_Y = 16.93$ A lagging V_{YN} by 30°; $I_B = 25.4$ A leading V_{BN} by 30°; $I_N = 12.03$ A leading V_{RN} by 159.4°; 15.763 kW.
13.19 $I_R = 1.0$ A in phase with V_{RN}; $I_Y = 0.96$ A at angle $-136.7°$; $I_B = 0.667$ A at angle 180°; $I_N = 0.751$ A at angle $-119°; 225W$.

Chapter 14

14.2 (a) 103.2 kΩ; (b) 6 V; (c) 5.91 kΩ; (d) 10.11 mA; (e) 600 Ω; (f) 90.91.
14.3 (a) 48.04 kΩ; (b) 5.93 V; (c) 5.91 kΩ; (d) 10.11 mA; (e) 593 Ω; (f) 90.91.
14.4 (a) 60 μA; (b) 1 kΩ; (c) 9.17 kΩ; (d) 1 kΩ; (e) 157.5 kΩ.
14.5 (a) 5.5 mA; (b) 3.54 V; (c) 8.72 kΩ; (d) 645 Ω; (e) 42.8 kΩ.
14.6 (a) $R_{B1} = 48.34$ kΩ, $R_{B2} = 7.02$ kΩ; (b) 8 V; (c) 1.01 V.
14.7 (a) 4 mA; (b) $R_{B1} = 51.18$ kΩ, $R_{B2} = 15.1$ kΩ; (c) $V_E = 1.2$ V.
14.8 7.02 kΩ.
14.9 (a) 6.54 mA; (b) 6.54 V; (c) 42.8 mW; (d) 101.

Chapter 15

15.1 −20.74.
15.2 (a) −22.92; (b) −18.77.
15.3 0.948 V.
15.4 −0.5 V.
15.5 11.
15.6 (a) 12.05; (b) 10.05.
15.7 11.43 kΩ.
15.8 25 kΩ.
15.9 (a) −4.5 V; (b) −7.5 V; 6.667 s.
15.10 470 kΩ.

Chapter 16

16.1 13.82 kHz.
16.2 0.1 μF.
16.3 1.59 kHz.
16.4 1 nF.
16.5 2.6 MHz; 1.0.
16.6 10 nF.
16.7 500 pF.
16.8 132.2 kHz; 0.21.
16.9 50.3 kHz.
16.10 7.1 kHz.
16.11 3.35 ms; 2.35:1.

Chapter 17

17.1 (a) −10.97 dB; (b) 8.45 dB.
17.2 80 mW.
17.3 35 W.
17.4 (a) −38.06 dB; (b) 23.52 dB.

17.5 15.92 dB.
17.6 (a) 6.64 V; (b) 5.27 V.
17.7 0.005.
17.8 0.501; 5.01 mW.
17.9 400 Ω; 3; 9.54.
17.10 178.9 Ω; 2.62; 8.36 dB.
17.11 4; $R_1 = 180\ \Omega$; $R_2 = 160\ \Omega$.
17.12 $P_1 = P_3 = 800\ \Omega$; $P_2 = 533.3\ \Omega$.
17.13 $T_1 = T_2 = 80\ \Omega$; $T_3 = 160\ \Omega$.
17.14 20 kHz.
17.15 (a) 56 mV; (b) 42 mV; (c) 26.8 mV.
17.16 (a) 128.6 mV; (b) 176.8 mV; (c) 218.5 mV.

Chapter 18

18.1 (a) 1 0100 1001 0100; (b) 10111.11; (c) 0.0000 0011 (**note**: the last four bits recur).
18.2 (a) 5798; (b) 1500; (c) 105; (d) 111.
18.3 (a) 0.0001; (b) 0.111; (c) 0.10 recurring; (d) 0.10001.
18.4 (a) 187; (b) 0.4375; (c) 23.25.
18.5 (a) 1735_8; (b) 3220_6; (c) 12001_5; (d) 1021112_3.
18.6 (a) 3124_6; (b) 464_9; (c) 2022_3.
18.7 (a) 1011; (b) 100100; (c) 11110.01; (d) 111.1011; (e) 10010.101.
18.8 (a) 0000 0011; (b) 000 1010; (c) 1111 1111; (d) 1111 1110.
18.9 (a) 10010; (b) 11100.001; (c) 1.0001.
18.10 (a) 10; (b) 11; (c) 0.0011.
18.13 (b) and (e).
18.14 $f = B + A.\bar{C} + \bar{A}.C$.
18.15 (a) $A = 0$, $B = 0$ or $A = 1$, $B = 1$; (b) $A = 1$, $B = 0$ or $A = 0$, $B = 1$.
18.16 $A = 1, B = 0, f = 0$.

Chapter 19

19.2 (a) $250\angle 143.1°$; (b) $63.25\angle(-71.57°)$; (c) $10.77\angle 21.8°$; (d) $7.81\angle(-129.8°)$.
19.3 (a) $1.93 + j2.3$; (b) $-1.93 - j2.3$; (c) $-1.88 - j0.68$; (d) $1 - j1.732$; (e) $2 - j3.46$; (f) $-4.6 + j3.86$; (g) $4.6 + j3.86$.
19.4 (a) $-2 - j2 \equiv 2.828\angle(-135°)$; (b) $-100 - j900 \equiv 905.5\angle(-96.34°)$; (c) $2 + j6 \equiv 6.32\angle 71.57°$.
19.5 (a) $-2 + j3.73 \equiv 4.23\angle 118.2°$; (b) $3 + j2.54 \equiv 3.93\angle 40.2°$; (c) $-5.724 + j15 \equiv 16.05\angle 110.9°$.
19.6 (a) $6 + j2 \equiv 6.33\angle 18.4°$; (b) $7.83 - j1.79 \equiv 8.03\angle(-12.85°)$; (c) $-2.6 - j0.392 \equiv 2.63\angle(-171.4°)$.
19.7 $-7.86 - j9.6 \equiv 12.4\angle(-129.3°)$
19.8 (a) $80\angle 110° \equiv -27.36 + j75.18$; (b) $-576\angle 130° \equiv 370.25 - j441.24$.

19.9 (a) $28.16\angle 106.5 \equiv -8 + j27$; (b) $52.82\angle(-114.7°) \equiv -22 - j48$; (c) $201.15\angle(-123.3°) \equiv -110.4 - j168.1$.

19.10 (a) $1.667\angle 100° \equiv -0.29 + j1.64$; (b) $1.39\angle 146.3° \equiv -1.15 + j0.77$; (c) $1.28\angle 165.6° \equiv -1.24 + j0.32$.

19.11 (a) $2.915\angle(-49.05°) \equiv 1.91 - j2.2$; (b) $0.38\angle(-56.1°) \equiv 0.209 - j0.31$; (c) $4.054\angle 1° \equiv 4.05 + j0.07$.

19.12 $21.05\angle 152° \equiv -18.6 + j9.88$.

19.13 $2 - j4$ V $\equiv 4.472\angle(-63.44°)$ V.

19.14 $6 - j5$ A $\equiv 7.81\angle(-39.81°)$ A.

19.15 $13.31 + j76.96$ V $\equiv 78.1\angle 80.19$ V.

19.16 $20.78\angle 90°$ A $\equiv j20.78$ A.

19.17 (a) (i) $500 + j628.3\ \Omega \equiv 803\angle 51.49°\ \Omega$, (ii) $7.8 - j9.7$ mA $\equiv 12.45\angle(-51.49°)$ mA; (b) (i) $306.1 + j243.6\ \Omega \equiv 391.2\angle 38.5°\ \Omega$, (ii) $20 - j15.9$ mA $\equiv 26.56\angle(-38.5°)$ mA.

19.18 $2\angle 0°$ A $\equiv 2 + j0$ A.

19.19 27 kW; 21 kVAr (lagging); 34.2 kVA.

Chapter 20

20.1

```
Exercise 20.1
Vs      1     0     100V
R1      3     4     8ohms
R2      5     6     1ohm
R3      7     4     10ohm
R4      8     4     20ohms
R5      9     0     70ohms
R6      10    11    40ohms
R7      11    0     20ohms
Va1     1     2     0V
Va2     2     3     0V
Va3     2     5     0V
Va4     6     7     0V
Va5     6     8     0V
Va6     4     9     0V
Va7     2     10    0V
.OPTIONS NOPAGE
.END
```

The currents are as follows: main current, 3.02 A; in 8 Ω, 0.6621 A; in 1 Ω, 0.6908 A; in 10 Ω, 0.4606 A; in 20 Ω, 0.2303 A; in 70 Ω, 1.353 A; in 40 and 20 Ω, 1.667 A;

20.2

```
Exercise 20.2
V1      1    0    10V
V2      4    3    20V
V3      7    0    15V
Va1     1    2    0V
Va2     4    5    0V
Va3     7    6    0v
R1      2    3    10ohm
R2      5    0    15ohm
R3      3    6    20ohm
.OPTIONS NOPAGE
.END
```

The SPICE solutions are: 21.923 V; 10 V; 11.923 V; 0.808 A in the 10 V source; 1.462 A in the 20 V source; 0.654 A in the 15 V source.

20.3

```
Exercise 20.3
Vs      1    0        AC       50V
Vam     1    2        0V
R       2    3        10ohm
L       3    0        0.05H
.AC     LIN  1        50Hz     50Hz
.PRINT  AC   IM(Vam)     IP(Vam)     V(2,3)     V(3,0)
.OPTIONS NOPAGE
.END
```

2.685 A; $-57.52°$; $V_R = 26.85$ V; $V_L = 42.18$ V.

20.4

```
Exercise 20.4
Vs      1    0        AC       20V
Vam1    1    2        0
VamC    2    3        0
R       2    4        10ohm
L       4    0        31.8E-3
C       3    0        159UF
.AC     LIN  1        50       50
.PRINT  AC   IM(Vam1)          IP(Vam1)
.PRINT  AC   IM(VamC)          IP(VamC)
.OPTIONS NOPAGE
.END
```

0.999 A; 19.98 Ω

20.5 1.995 kΩ; 1.091 kΩ; −43.74; 4.374 V.

20.6

```
Exercise 20.6
Vin       3    0    2V
R1        3    4    50Kohm
R2        4    1    50Kohm
R3        1    5    50Kohm
R4        5    0    50Kohm
R5        6    0    50Kohm
R6        6    2    50Kohm
R7        2    0    50Kohm
Eopamp1   1    0    0      4      1000
Rin1      4    0    1MEGohm
Eopamp2   2    0    5      6      1000
Rin2      4    5    1MEGohm
.OPTIONS NOPAGE
.END
```

(a) −3.99 V, −12.8 V; (b) −3.97 V, −12.62 V.

20.7

```
Exercise 20.7
Vin     1     0       AC     1
R1      1     2       10Kohm
C       1     2       0.01UF
R2      2     0       1Kohm
.AC     DEC   5       100Hz  100KHz
.PLOT   AC    VDB(2)      VP(2)
.OPTIONS NOPAGE
.END
```

56.4°; 5.01 Hz

Bibliography

The following books in the Macmillan Masters series provide particularly useful information on the subject of Electronic and Electrical Engineering.

Mastering Electronics by John Watson
Mastering Electrical Engineering by Noel M. Morris
Mastering Mathematics for Electrical and Electronic Engineering by Noel M. Morris

The first two books give vital information about Electronics and Electrical Engineering, respectively. The second and third books contain further information and examples on the use of PSpice software, and the third title provides all the background mathematics not only for this book but also for the first two titles listed.

Index